Mechanical Research of Reinforced Concrete Materials

Mechanical Research of Reinforced Concrete Materials

Editor

Wei Wang

 Basel • Beijing • Wuhan • Barcelona • Belgrade • Novi Sad • Cluj • Manchester

Editor
Wei Wang
Key Laboratory of Impact and
Safety Engineering,
Ministry of Education,
Faculty of Mechanical
Engineering and Mechanics,
Ningbo University
Ningbo, China

Editorial Office
MDPI
St. Alban-Anlage 66
4052 Basel, Switzerland

This is a reprint of articles from the Special Issue published online in the open access journal *Materials* (ISSN 1996-1944) (available at: www.mdpi.com/journal/materials/special_issues/ mechanical_research_of_reinforced_concrete_materials).

For citation purposes, cite each article independently as indicated on the article page online and as indicated below:

Lastname, A.A.; Lastname, B.B. Article Title. *Journal Name* **Year**, *Volume Number*, Page Range.

ISBN 978-3-0365-9405-7 (Hbk)
ISBN 978-3-0365-9404-0 (PDF)
doi.org/10.3390/books978-3-0365-9404-0

© 2023 by the authors. Articles in this book are Open Access and distributed under the Creative Commons Attribution (CC BY) license. The book as a whole is distributed by MDPI under the terms and conditions of the Creative Commons Attribution-NonCommercial-NoDerivs (CC BY-NC-ND) license.

Contents

Preface vii

Wei Wang
Mechanical Research on Reinforced Concrete Materials
Reprinted from: *Materials* 2023, 16, 6892, doi:10.3390/ma16216892 1

Wei Wang, Zhonghao Zhang, Qing Huo, Xiaodong Song, Jianchao Yang, Xiaofeng Wang and et al.
Dynamic Compressive Mechanical Properties of UR50 Ultra-Early-Strength Cement-Based Concrete Material under High Strain Rate on SHPB Test
Reprinted from: *Materials* 2022, 15, 6154, doi:10.3390/ma15176154 5

Wei Zhang, Jize Mao, Xiao Yu, Bukui Zhou and Limei Wang
Study on the Dynamic Mechanical Properties of Ultrahigh-Performance Concrete under Triaxial Constraints
Reprinted from: *Materials* 2023, 16, 6591, doi:10.3390/ma16196591 25

Zhixiang Xiong, Wei Wang, Guocai Yu, Jian Ma, Weiming Zhang and Linzhi Wu
Experimental and Numerical Study of Non-Explosive Simulated Blast Loading on Reinforced Concrete Slabs
Reprinted from: *Materials* 2023, 16, 4410, doi:10.3390/ma16124410 43

Kai Dong, Kun Jiang and Wenjun Ruan
The Strain Rate Effects of Coral Sand at Different Relative Densities and Moisture Contents
Reprinted from: *Materials* 2023, 16, 4217, doi:10.3390/ma16124217 65

Yangyong Wu, Jianhui Wang, Fei Liu, Chaomin Mu, Ming Xia and Shaokang Yang
A Research Investigation into the Impact of Reinforcement Distribution and Blast Distance on the Blast Resilience of Reinforced Concrete Slabs
Reprinted from: *Materials* 2023, 16, 4068, doi:10.3390/ma16114068 81

Zailin Yang, Chenxi Sun, Guanxixi Jiang, Yunqiu Song, Xinzhu Li and Yong Yang
Analytical Study of SH Wave Scattering by a Circular Pipeline in an Inhomogeneous Concrete with Density Variation
Reprinted from: *Materials* 2023, 16, 3693, doi:10.3390/ma16103693 97

Kai Dong, Kun Jiang, Chunlei Jiang, Hao Wang and Ling Tao
Study on Mass Erosion and Surface Temperature during High-Speed Penetration of Concrete by Projectile Considering Heat Conduction and Thermal Softening
Reprinted from: *Materials* 2023, 16, 3604, doi:10.3390/ma16093604 115

Xuhui Zhang, Xun Wu and Yang Wang
Corrosion-Effected Bond Behavior between PVA-Fiber-Reinforced Concrete and Steel Rebar under Chloride Environment
Reprinted from: *Materials* 2023, 16, 2666, doi:10.3390/ma16072666 135

Xinxin Ma, Jianheng Sun, Fengshuang Zhang, Jing Yuan, Mingjing Yang, Zhiliang Meng and et al.
Mechanical Behaviour Evaluation of Full Iron Tailings Concrete Columns under Large Eccentric Short-Term Loading
Reprinted from: *Materials* 2023, 16, 2466, doi:10.3390/ma16062466 155

Fei Wang, Jianmin Hua, Xuanyi Xue, Neng Wang, Feidong Yan and Dou Feng
Effect of Superfine Cement Modification on Properties of Coral Aggregate Concrete
Reprinted from: *Materials* **2023**, *16*, 1103, doi:10.3390/ma16031103 . . . . **177**

Huawei Lai, Zhanjiang Wang, Liming Yang, Lili Wang and Fenghua Zhou
Determining Dynamic Mechanical Properties for Elastic Concrete Material Based on the Inversion of Spherical Wave
Reprinted from: *Materials* **2022**, *15*, 8181, doi:10.3390/ma15228181 . . . . **195**

Lijun Wang, Shuai Cheng, Zhen Liao, Wenjun Yin, Kai Liu, Long Ma and et al.
Blast Resistance of Reinforced Concrete Slabs Based on Residual Load-Bearing Capacity
Reprinted from: *Materials* **2022**, *15*, 6449, doi:10.3390/ma15186449 . . . . **207**

Preface

This reprint describes the mechanical properties of reinforced concrete materials, including theoretical, experimental and numerical studies, to evaluate the general deformation response, damage evolution and failure patterns of ordinary and high-performance reinforced concrete materials under various loading conditions (e.g., quasi-static, dynamic, fatigue, and impact). Its target audience is civil engineering and related professionals in higher education institutions, as well as relevant engineering and technical personnel, such as material scientists, structural protection personnel, and restoration personnel. The published articles are results of collective works of research teams; however, the list of first authors (in alphabetical order) includes K. Dong, Y. Wu, X. Ma, X. Zhang, F. Wang, L. Wang, H. Lai, W. Zhang, Z. Z. Xiong and W. Wang. The editors are grateful for the collaboration of the authors and the excellent help of the MDPI publishers.

Wei Wang
Editor

Editorial

Mechanical Research on Reinforced Concrete Materials

Wei Wang

Key Laboratory of Impact and Safety Engineering, Ningbo University, Ministry of Education, Ningbo 315211, China; wangwei7@nbu.edu.cn

Citation: Wang, W. Mechanical Research on Reinforced Concrete Materials. *Materials* **2023**, *16*, 6892. https://doi.org/10.3390/ma16216892

Received: 23 October 2023
Revised: 24 October 2023
Accepted: 26 October 2023
Published: 27 October 2023

Copyright: © 2023 by the author. Licensee MDPI, Basel, Switzerland. This article is an open access article distributed under the terms and conditions of the Creative Commons Attribution (CC BY) license (https://creativecommons.org/licenses/by/4.0/).

Reinforced concrete (RC) is a commonly used construction material in civilian and military buildings due to its superior material characteristics compared to steel and timber (e.g., higher durability, corrosion resistance, and fire resistance). These inherent properties of reinforced concrete make it suitable for the construction of most civil engineering structures, for example, bridges, dams, nuclear containment structures, protective/defense structures, and residential/embassy buildings. Concrete is a frequently used material that is subjected to intense dynamic loading in civil and defense engineering, such as blast and impact loadings, which can induce high pressure, high strain rates, and a large amount of strain in concrete structures. The response of such structures is very complex due to the effects of high inertia, high strain rate, high temperature, and shock waves traveling through the reinforced concrete. Although the mechanical behavior of reinforced concrete has been investigated by many researchers using experimental and theoretical approaches for 200 years, establishing an accurate and comprehensive description of the actual mechanical behavior exhibited by reinforced concrete under service and ultimate conditions remains a challenge.

The purpose of this Special Issue is to investigate theoretical, experimental, and numerical studies on the mechanical properties of reinforced concrete materials, to evaluate the general deformation response, damage evolution, and failure patterns of ordinary and high-performance reinforced concrete materials under various loading conditions (e.g., quasi-static, dynamic, fatigue, and impact).

Dong et al. [1] studied the influence of relative density and water content on the dynamic characteristics of coral sand using an SHPB device, and obtained a stress–strain curve of the material under uniaxial strain compression with different relative densities and water contents. The results showed that the strain rate became less sensitive to the stiffness of coral sand with an increase in relative density.

Wu et al. [2]. studied the effects of different reinforcement distributions and blasting distances on the antiknock performance of RC plates. The results showed that the degree of damage on a single-layer steel plate was more serious than that on a double-layer steel plate under contact explosion and non-contact explosion. When the scale distance was the same, the degree of damage on single-layer and double-layer steel bars first increased, and then, decreased with an increase in distance.

Ma et al. [3] studied the structural behavior of full-iron tailing-reinforced concrete (FITRC) columns under high eccentric loading. Six FITRC column and two CRC column specimens were examined to investigate the effects of different raw materials, section dimensions, and eccentricities on the mechanical behavior of RC columns under high eccentric compression. They found that under high eccentric short-term loads, the failure modes of the FITRC and CRC columns were similar, and their failure manifested as yielding of the tensile and compressive rebars and concrete crushing in the compression zone. Additionally, the moment augmentation factor of the FITRC columns should be 1.15 for safety reasons.

Zhang et al. [4] investigated the bond behavior between PVA-fiber-reinforced concrete and steel rebar corroded in a chloride environment. The effects of PVA fibers and corrosion loss on bond behavior were clarified. They found that with the increase in the fibers, the corrosive cracking became more obvious. Moreover, the PVA fiber generally showed a negative effect on bond behavior but a positive effect on the descending branches in the case of splitting failure.

In their innovative study, Wang et al. [5] used superfine cement to modify coral aggregates. The effects of the water–cement ratio and curing time on the water absorption and strength of modified coral aggregates were investigated. The experimental results showed that when the water–cement ratio exceeded 1.25, the slurry did not form a shell on the surface of the coral aggregates and the water absorption of the coral aggregates increased significantly. The strength of the modified coral aggregates cured for a short duration was slightly lower than that of unmodified coral aggregates, while the strength of the coral aggregates cured for 28 days was approximately 20% higher than that of the unmodified coral aggregates.

Wang et al. [6] analyzed the anti-explosion performance of RC slabs by studying the residual bearing capacity of RCs slab under close-in blast loading, and it was found that the load distribution of the RC slabs was extremely uneven under these conditions. The range and degree of damage of the low-reinforcement-ratio slab were significantly higher than those of the high-reinforcement-ratio slab. Increasing the reinforcement ratio can inhibit the crack extension and reduce the residual displacement of a component, and reduce the decrease in bearing capacity after damage.

Lai et al. [7] presented a new method to study the dynamic mechanical properties of concrete under low pressure and a high strain rate via the inversion of spherical wave propagation. The dynamic parameters of the rate-dependent constitutive relation of elastic concrete were determined by measuring the velocity histories of spherical waves. The numerical constitutive relation was expressed in the form of distortion, and it was found that the distortion law had an obvious rate effect. The research results showed that the strain rate of concrete had a considerable effect when the strain rate of concrete was in the range of 10^2 s^{-1}.

Zhang et al. [8] used a true triaxial split–Hopkinson pressure bar system to conduct dynamic compression tests on ultra-high-performance concrete with different steel fiber contents (0.5%, 1% and 2%) under triaxial constraint. The results showed that the dynamic peak axial stress–strain and dynamic peak lateral stress–strain of UHPC are very sensitive to the strain rate, and the dynamic strength failure criterion of UHPC under triaxial constraints was established.

Xiong et al. [9] proposed a non-explosive method to simulate the explosion load of reinforced concrete slabs, and conducted experiments and numerical simulations to evaluate the effectiveness of this method. The experimental results showed that there was almost no difference between the pressure wave and the pressure peak produced by the simulated explosion load and the actual explosion. In addition, cone rubber was more suitable than plane rubber as an impact buffer material when simulating the explosion load.

Wu et al. [2] carried out field chemical explosion experiments on RC slabs with the same reinforcement ratio but different reinforcement distributions, and with the same blast distance but different scale distances. The effects of different reinforcement distributions and explosion distances on the antiknock performance and damage forms of RC plates were studied. It was found that double-layer and multi-layer reinforcement improved the blasting resistance of RC plates.

Yang et al. [10] studied shear horizontal (SH) wave scattering using a circular pipeline in an inhomogeneous concrete with density variation. A model of inhomogeneous concrete with density variation was established in the form of a polynomial–exponential coupling function. The results showed that the inhomogeneous density parameters, the wave number of the incident wave, and the angle of the incident wave in concrete were important factors affecting the distribution of dynamic stress around the circular pipe in concrete with inhomogeneous density.

Dong et al. [11] established theoretical models to calculate the mass erosion and heat conduction of projectile noses, including models of cutting, melting, the heat conduction of flash, and the conversion of plastic work into heat. The coupling numerical calculation of the erosion and heat conduction of the projectile nose showed that melting erosion was the main factor affecting mass loss at high-speed penetration, and the mass erosion ratio of melting and cutting was related to the initial velocity.

Wang et al. [12] used SHPB to study the dynamic mechanical properties of UR50 ultra-early-strength cement-based self-compacting high-strength material. The experimental results showed that the dynamic compressive strength increased with an increase in strain rate, which had an obvious strain-rate-strengthening effect. The damage variables at different strain rates were fitted, and it was found that an increase in strain rate had an obstructive effect on the increase in the damage variable and the increase rate.

The research papers included in this Special Issue demonstrate the usefulness of global institutional collaborations in helping to share knowledge and deepen our understanding of various topics in the field of reinforced concrete materials.

Finally, the editors would like to express their sincere thanks to the authors who actively participated in the peer review process, and to the editorial team of *Materials* for their cooperation and support.

Funding: This research was funded by National Natural Science Foundation of China (grant number. 11302261 and 11972201).

Acknowledgments: The editors would like to thank the authors who submitted their papers to this Special Issue, "Mechanical Research on Reinforced Concrete Materials", as well as the reviewers and editors for their contributions to improving these submissions.

Conflicts of Interest: The author declares no conflict of interest.

References

1. Dong, K.; Jiang, K.; Ruan, W. The Strain Rate Effects of Coral Sand at Different Relative Densities and Moisture Contents. *Materials* **2023**, *16*, 4217. [CrossRef] [PubMed]
2. Wu, Y.; Wang, J.; Liu, F.; Mu, C.; Xia, M.; Yang, S. A Research Investigation into the Impact of Reinforcement Distribution and Blast Distance on the Blast Resilience of Reinforced Concrete Slabs. *Materials* **2023**, *16*, 4068. [CrossRef] [PubMed]
3. Ma, X.; Sun, J.; Zhang, F.; Yuan, J.; Yang, M.; Meng, Z.; Bai, Y.; Liu, Y. Mechanical Behaviour Evaluation of Full Iron Tailings Concrete Columns under Large Eccentric Short-Term Loading. *Materials* **2023**, *16*, 2466. [CrossRef] [PubMed]
4. Zhang, X.; Wu, X.; Wang, Y. Corrosion-Effected Bond Behavior between PVA-Fiber-Reinforced Concrete and Steel Rebar under Chloride Environment. *Materials* **2023**, *16*, 2666. [CrossRef] [PubMed]
5. Wang, F.; Hua, J.; Xue, X.; Wang, N.; Yan, F.; Feng, D. Effect of Superfine Cement Modification on Properties of Coral Aggregate Concrete. *Materials* **2023**, *16*, 1103. [CrossRef] [PubMed]
6. Wang, L.; Cheng, S.; Liao, Z.; Yin, W.; Liu, K.; Ma, L.; Wang, T.; Zhang, D. Blast Resistance of Reinforced Concrete Slabs Based on Residual Load-Bearing Capacity. *Materials* **2022**, *15*, 6449. [CrossRef] [PubMed]
7. Lai, H.; Wang, Z.; Yang, L.; Wang, L.; Zhou, F. Determining Dynamic Mechanical Properties for Elastic Concrete Material Based on the Inversion of Spherical Wave. *Materials* **2022**, *15*, 8181. [CrossRef] [PubMed]
8. Zhang, W.; Mao, J.; Yu, X.; Zhou, B.; Wang, L. Study on the Dynamic Mechanical Properties of Ultrahigh-Performance Concrete under Triaxial Constraints. *Materials* **2023**, *16*, 6591. [CrossRef] [PubMed]
9. Xiong, Z.; Wang, W.; Yu, G.; Ma, J.; Zhang, W.; Wu, L. Experimental and Numerical Study of Non-Explosive Simulated Blast Loading on Reinforced Concrete Slabs. *Materials* **2023**, *16*, 4410. [CrossRef] [PubMed]
10. Yang, Z.; Sun, C.; Jiang, G.; Song, Y.; Li, X.; Yang, Y. Analytical Study of SH Wave Scattering by a Circular Pipeline in an Inhomogeneous Concrete with Density Variation. *Materials* **2023**, *16*, 3693. [CrossRef] [PubMed]
11. Dong, K.; Jiang, K.; Jiang, C.; Wang, H.; Tao, L. Study on Mass Erosion and Surface Temperature during High-Speed Penetration of Concrete by Projectile Considering Heat Conduction and Thermal Softening. *Materials* **2023**, *16*, 3604. [CrossRef] [PubMed]
12. Wang, W.; Zhang, Z.; Huo, Q.; Song, X.; Yang, J.; Wang, X.; Wang, J.; Wang, X. Dynamic Compressive Mechanical Properties of UR50 Ultra-Early-Strength Cement-Based Concrete Material under High Strain Rate on SHPB Test. *Materials* **2022**, *15*, 6154. [CrossRef] [PubMed]

Disclaimer/Publisher's Note: The statements, opinions and data contained in all publications are solely those of the individual author(s) and contributor(s) and not of MDPI and/or the editor(s). MDPI and/or the editor(s) disclaim responsibility for any injury to people or property resulting from any ideas, methods, instructions or products referred to in the content.

Article

Dynamic Compressive Mechanical Properties of UR50 Ultra-Early-Strength Cement-Based Concrete Material under High Strain Rate on SHPB Test

Wei Wang [1,2], Zhonghao Zhang [1], Qing Huo [1], Xiaodong Song [1], Jianchao Yang [3,*], Xiaofeng Wang [3], Jianhui Wang [3] and Xing Wang [3]

1. Key Laboratory of Impact and Safety Engineering, Ningbo University, Ministry of Education, Ningbo 315211, China
2. Institute of Advance Energy Storage Technology and Equipment, Ningbo University, Ningbo 315211, China
3. Institute of Defence Engineering AMS, PLA, Luoyang 471023, China
* Correspondence: jiebao9630@163.com

Abstract: UR50 ultra-early-strength cement-based self-compacting high-strength material is a special cement-based material. Compared with traditional high-strength concrete, its ultra-high strength, ultra-high toughness, ultra-impact resistance, and ultra-high durability have received great attention in the field of protection engineering, but the dynamic mechanical properties of impact compression at high strain rates are not well known, and the dynamic compressive properties of materials are the basis for related numerical simulation studies. In order to study its dynamic compressive mechanical properties, three sets of specimens with a size of $\Phi 100 \times 50$ mm were designed and produced, and a large-diameter split Hopkinson pressure bar (SHPB) with a diameter of 100 mm was used to carry out impact tests at different speeds. The specimens were mainly brittle failures. With the increase in impact speed, the failure mode of the specimens gradually transits from larger fragments to small fragments and a large amount of powder. The experimental results show that the ultra-early-strength cement-based material has a greater impact compression brittleness, and overall rupture occurs at low strain rates. Its dynamic compressive strength increases with the increase of strain rates and has an obvious strain rate strengthening effect. According to the test results, the relationship curve between the dynamic enhancement factor and the strain rate is fitted. As the impact speed increases, the peak stress rises, the energy absorption density increases, and its growth rate accelerates. Afterward, based on the stress–strain curve, the damage variables under different strain rates were fitted, and the results show that the increase of strain rate has a hindering effect on the increase of damage variables and the increase rate.

Keywords: ultra-early-strength concrete; split Hopkinson pressure bar (SHPB); impact mechanics; dynamic response; strain rate effect

1. Introduction

With the rapid development of precision-guided weapons, facilities in the field of protection engineering are facing a serious threat of weapons, "accurately and aggressively". Therefore, the design and construction of military protection engineering must consider the dynamic response characteristics and parameters of concrete. Ultra-early-strength cement-based materials, because of their excellent performance, have attracted great attention in airport engineering repairs and protection engineering. Ultra-early-strength cement-based materials are usually subjected to strong dynamic load impact compressions during service, like ammunition penetration and explosion. Under blast loads, concrete structures can crater on the surface, peel off the concrete on the back, or even crack [1–3]. The dynamic compression performance of the material plays an important role in studying the dynamic mechanical response process of materials under penetration and explosion.

Therefore, investigating the performance of ultra-early-strength cement base under dynamic load and establishing a dynamic stress–strain curve has important military value and engineering significance.

At present, although domestic and foreign researchers have carried out many experimental studies on concrete materials, there are few studies on the dynamic mechanical properties of UR50 ultra-early-strength cement materials under high dynamic loads. Pang et al. [4] studied activated fly ash concrete under high strain rates with an SHPB device. Its dynamic strength is affected by both temperature and the water–binder ratio. At room temperature, the dynamic strength is directly proportional to the water–binder ratio, but there is a negative correlation at high temperatures. Hu et al. [5] applied an SHPB device to study the spalling strength and strain rate effects of concrete materials finding that spalling strength and strain rate are positively correlated. Zhu et al. [6] conducted a numerical simulation study on the SHPB test of cement materials. The strain value and strain rate of cement materials will be strongly affected by the amplitude of the incident wave. Bragov et al. [7] investigated the mechanical properties of fine-grained concrete at high strain rates with an SHPB device. Levi-Hevroni et al. [8] used the SHPB test to explore the dynamic reinforcement factor and tensile strength of concrete first, then applied the test data to calibrate the three concrete material model parameters in LS-DYNA. Zhang et al. [9] studied the influence of the specimen shape on the dynamic increase factor under a high strain rate based on the SHPB test device. The results showed that the dynamic enhancement factor of the tubular specimen was lower than that of the cube. Wang et al. [10] studied the physical mechanism of the static–dynamic composite multiaxial strength of concrete under the premise of considering the cohesive and frictional strength. Gu et al. [11] introduced the theory of non-local circumferential dynamics to the analysis of dispersion and impact failure of elastic waves in SHPB tests and verified the feasibility of the analysis by experiments and numerical simulations. Hassan and Wille [12] studied the dynamic mechanical properties of ultra-high-performance concrete (UHPC) at high strain rates based on the SHPB test device. Erzar and Forquin [13] studied the effects of aggregate and free water on the mechanical properties of concrete materials under high strain rates through experiments and numerical simulations.

With the development of research, some scholars have begun to explore the dynamic mechanical properties of new composite concrete materials. Wang et al. [14] studied the failure mode and energy absorption mechanism of autoclaved aerated concrete under low-velocity impact. Kang [15] explored the mechanical behavior of foam-insulated concrete sandwich panels under uniform loads through experiments and numerical simulations. Wang et al. [16] studied the mechanical properties of lightweight aggregate foam concrete at different compression rates and found that the compressive strength is directly proportional to the density of the foam concrete. Shafigh et al. [17] used oil palm shells to prepare lightweight concrete and tested the compressive strength under different curing times. Cao et al. [18] discussed the influence of specimen size on the dynamic compression performance of fiber-reinforced reactive powder concrete at high strain rates. Xiong et al. [19] studied the dynamic mechanical properties of Carbon Fiber Reinforced Polymer (CFRP) confined concrete at a high strain rate based on an SHPB test device with a diameter of 155 mm. The results show that CFRP-confined concrete is not sensitive to the strain rate effect. Liu et al. [20] used a separate SHPB device with a diameter of 100 mm to study the influence of the content of the redispersible polymer emulsion powder on the dynamic mechanical properties of Carbon Fiber Reinforced Polymer Concrete (CFRPC). The dynamic compressive strength of carbon fiber composites increases firstly and then decreases with the increase of polymer content. Wei et al. [21] used SHPB to study the dynamic response of a ceramic shell for titanium investment casting under high strain rates. Ceramic shells are highly sensitive to the strain rate effect, and the path of crack propagation is different under quasi-static and high strain rate loads. Sun et al. [22] used a 75 mm diameter SHPB to study the dynamic mechanical properties of steel fiber-reinforced concrete at different strain rates and steel fiber content. As the strain rate or steel fiber content increases, the ductility, strength, and toughness will increase. Scott et al. [23] established a constitutive model of the dynamic response characteristics of concrete materials based on a large amount of experimental data. Georgin and Reynouard [24] established a viscoelastic model of the strain rate effect and applied it in a numerical simulation. Xiuli et al. [25] used SHPB to conduct a uniaxial dynamic compression test of concrete materials, and based on this, established a non-linear uniaxial dynamic strength criterion for concrete materials. The temperature has a certain influence on the dynamic properties of materials. Aiming at the dynamic mechanical properties of concrete specimens under the coupled action of high temperature and impact, Huo et al. [26] used SHPB to carry out the impact resistance test of concrete-filled steel tube specimens at a high temperature of 400 °C, and the results showed that the restraint of steel tubes improved the impact resistance of concrete specimens.

With the development of material technology, convenient, efficient, and excellent-performance concrete materials have begun to attract people's attention. Ultra-early-strength cement-based self-compacting high-strength material has the advantages of good fluidity, ultra-fast hardening, and high strength. It is easy to use and can be mechanically stirred by adding water on site. It has attracted the attention of the field of protection engineering. However, there is little research on ultra-early-strength cement-based self-compacting high-strength materials. In contrast to previous studies, the dynamic compressive properties of the material were not investigated. In order to study the dynamic mechanical response process of materials under the action of penetrating explosions and lay the foundation for numerical simulation research, this paper conducts related research. In this paper, a separate SHPB test device with a diameter of 100 mm and an ultra-early-strength cement-based self-compacting high-strength material with a product code of "UR50" were used to conduct impact tests at different loading speeds to explore the dynamic mechanical properties of this new type of concrete material and provide data support for studying the impact resistance and numerical simulation of UR50 ultra-early-strength cement-based materials.

2. UR50 Ultra-Early-Strength Concrete Material

UR50 ultra-early-strength cement-based self-compacting high-strength material is a special cement-based material. It is a pre-dry mixed powder composed of aggregate, cement, functional mineral powder, nano filler, specially modified additives, and special steel fiber. The maximum particle size of the aggregate is less than 5 mm. The product is processed and mixed in the factory and packed in bags, and the shelf life is about 6 months. After adding water and mixing on site, it has good fluidity, ultra-high strength, ultra-high toughness, ultra-impact resistance, and ultra-high durability. Wang et al. [27] found that the microstructure was greatly improved compared to conventional high-strength concrete, the pores were eliminated, and the nano-microstructure was strengthened. The design of dry mixing and pre-dispersed low-proportion components greatly improves the strength and durability of concrete, and this method makes the microstructure of concrete denser.

UR50 ultra-early-strength cement-based self-compacting high-strength material is convenient to use, and after adding water and mechanical mixing, concrete materials with excellent fluidity, super-fast hardening, and high strength can be obtained immediately. The amount of water added is $9.3 \pm 0.5\%$ of the weight of the dry powder. The slump extension of the mixed concrete can reach 770–830 mm, and the pouring can be self-compacting without vibration.

The early strength of UR50 ultra-early-strength cement-based materials has developed rapidly, and the compressive strength can reach 50 MPa in 2 h, 70 MPa in 24 h, and even the later compressive strength exceeds 80 MPa. The compressive strength changes with time, as shown in Table 1.

Table 1. Mechanical properties of UR50.

Times	Compressive Strength (MPa)	Flexural Strength (MPa)
2 h	54.0	7.2
24 h	71.0	9.7
7 d	80.0	9.7
28 d	81.2	10.1

3. Dynamic Test with SHPB

3.1. Experimental Specimen

The size of the experimental specimen was designed as $\Phi 100 \times 50$ mm in order to avoid the side wall effect caused by the specimen mold during the curing process and the discreteness of the specimen itself caused by the inconsistent curing and vibrating conditions. It was ensured that the test specimen met the relevant international specifications [28]. Adopting the method of pouring ultra-early-strength cement-based material panels and after curing for 28 days, the core was taken from the panel by using the coring machine to take it out of the poured ultra-early-strength cement-based material panel. The thickness of the ultra-early-strength cement-based material plate was about 60 mm, and the plate was vibrated evenly with a vibrating table and then placed in the pool for curing for a scheduled time and then taken out. The inner diameter of the core bit was 100 mm. After the core is taken, the upper and below faces were ground to a thickness of 50 mm with a grinder and polished to make a test specimen with a size of $\Phi 100 \times 50$ mm. There were three sets of test specimens and each group had three specimens. The finished test specimens are shown in Figure 1.

Figure 1. Physical image of processed specimens.

3.2. Experimental Device

The impact compression test equipment was a $\Phi 100$ mm SHPB of the Engineering Protection Research Institute. The test device is mainly composed of an operating console, launching device, impact bar (bullet), speed measuring device, incident bar, transmission bar, support, absorption bar, buffer device, measuring device, etc. The pressure bar is made of high-strength spring steel, as shown in Figure 2a. In order to effectively eliminate the influence of friction on the support, rolling bearings were installed at the support of the device. The length of the incident bar was 4500 mm, and the aspect ratio of the transmission rod was 25, which meets the relevant requirements [29]. The length of the transmission bar was 2500 mm, and the length of the striker bar was 500 mm and 800 mm, respectively. The compressed gas pressure was controlled by the operating console to control the impact speed of the bullet. The impact speed was measured by the speed measuring device. Strain gauges were attached to the incident bar and transmission bar to measure the incident wave, reflected wave, and transmitted wave. The strain gauge was connected to the super dynamic strain gauge through the Huygens bridge, and after being amplified by the strain gauge, it was saved as a transient record. The original waveform was analyzed and processed by a self-compiled data processing program to obtain the stress–strain rate relationship, as shown in Figure 2b.

Figure 2. An SHPB device separated by 100 mm. (**a**) Physical map of the SHPB device. (**b**) Schematic diagram of the SHPB device.

3.3. Experimental Design

In the experiment, the impact velocity of the bullet was altered by changing the driving air pressure so as to obtain the stress–strain curve of the material under different strain rates. A total of three kinds of loading speeds (impact speeds of 5 m/s, 10 m/s, and 15 m/s) for the SHPB impact compression experiments were carried out. Three repeated experiments were carried out for each speed state, and the average curve of these three experiments was taken as the stress–strain curve of the material under the strain rate. The experiment process of ultra-early-strength cement-based material is shown in Figure 3. The experiment loading speed and test specimen number are shown in Table 2.

Figure 3. Physical photos of the test process.

Table 2. Experimental program.

Speeds	Specimen Number	Measured Size of Test Specimen
5 m/s	S_5-1	Φ101.50 × 48.84 mm
	S_5-2	Φ100.76 × 49.44 mm
	S_5-3	Φ101.08 × 51.44 mm
10 m/s	S_10-1	Φ101.40 × 49.24 mm
	S_10-2	Φ101.05 × 50.40 mm
	S_10-3	Φ101.40 × 50.54 mm
15 m/s	S_15-1	Φ101.32 × 50.12 mm
	S_15-2	Φ101.38 × 50.32 mm
	S_15-3	Φ101.42 × 51.10 mm

3.4. Calibration of SHPB

The basic principle of SHPB is to decouple the wave propagation effect and the strain rate effect of the material and then separate the strain rate effect of the material. Figure 4a shows the electrical signal curve collected by the data acquisition system, which represents the curve of voltage change over time. The recorded data can be restored to stress and strain curves by data processing software. Figure 4b is the three-wave diagram of the incident wave, reflected wave, and transmitted wave obtained from the experiment. The comparison wave is the incident wave + the reflected wave. It can be seen from the figure that the transmitted wave is in good agreement with the comparison wave, which meets the criterion of $\varepsilon_t(t) = \varepsilon_r(t) + \varepsilon_i(t)$, proving that the test data are valid and can be used for analysis.

In order to make the stress pulse have enough time to reflect back and forth before the failure of the ultra-early-strength cement-based material specimen to obtain a uniform distribution of stress in the specimen, wave shapers were installed on the impacted end of the incident bar, which can eliminate the overshoot and wave oscillation of the stress wave caused by the dispersion effect of the large-size SHPB device, and it is helpful to obtain the true response characteristics of the material. In the experiment, under different loading conditions, different shapers were selected. Among them, the shaper used under low strain rate loading conditions (impact velocity 5 m/s) was a Φ16 × 0.5 mm copper sheet. Under the condition of medium strain rate loading (impact velocity 10 m/s), a Φ30 × 2 mm copper sheet shaper was used. Under high strain rate loading conditions (impact velocity 15 m/s), the shaper is a Φ30 × 4 mm copper sheet. By choosing a proper shaper, the dispersion effect can be eliminated effectively, the change in the waveform when the wave propagates in the waveguide bar is reduced, and the accuracy of the experiment is improved.

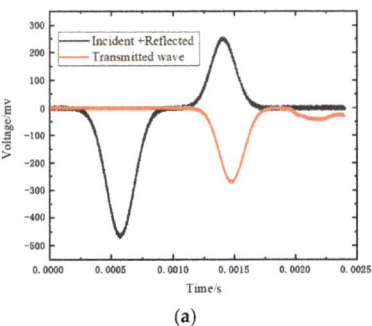 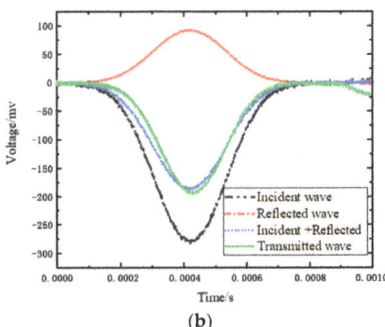

(a)   (b)

Figure 4. Original stress wave. (**a**) SHPB test collection wave. (**b**) The three waves graph.

Assuming that the strain rate is constant during the loading process, the average strain rate during loading can be determined by Formula (1):

$$\dot{\varepsilon} = \varepsilon/t \tag{1}$$

If $t = 100$ µs and the failure strain $\varepsilon = 8500$ µε, it can be determined that the highest strain rate that can satisfy the stress uniformity is $85\ \text{s}^{-1}$. It is particularly pointed out that the strain rate of the specimen is not constant during the experiment. During the data processing, the average strain rate of the loading stage before the failure of the specimen is taken as the average strain rate.

4. Test Results

4.1. Average Strain Rate of 7.5 s^{-1}

Table 3 shows the macro morphology of UR50 ultra-early-strength cement-based specimens after impact compression at different stages when the impact velocity is 5 m/s (the average strain rate is 7.5 s^{-1}) in three repeated experiments. It can be seen from the table that at an impact velocity of 5 m/s, the specimens were damaged under the impact compression stress wave. The specimen ruptured into a number of larger fragments, indicating that the ultra-early-strength cement-based material has a greater impact compression brittleness. Under the low strain rate, the overall fracture will occur, but under the same strain rate conditions, ordinary concrete will generally not fail. The stress–strain curve and the average stress–strain curve obtained by repeating the experiment three times at an impact velocity of 5 m/s are shown in Figure 5.

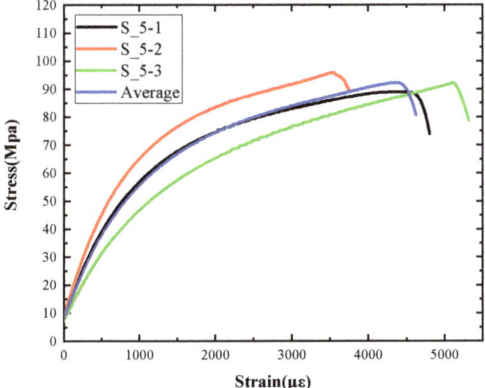

Figure 5. Stress–strain curve of 5 m/s.

Table 3. Test results at an impact velocity of 5 m/s.

Test No.	Before the Test	After the Test	Recycled
S_5-1			
S_5-2			
S_5-3			

4.2. Average Strain Rate of 15.3 s^{-1}

The macro morphology of the ultra-early-strength cement-based material specimen after three repeated experiments, when the impact velocity is 10 m/s (the average strain rate is 15.3 s^{-1}), can be seen in Table 4. It can be seen from the table that, at an impact velocity of 10 m/s, the specimen was severely damaged under the impact of the compression stress wave. Because of the short time of the load, cracks appeared throughout the whole specimen at the moment of impact and rapidly expanded until it broke into many smaller pieces. The stress–strain curve and the average stress–strain curve obtained by three repeated experiments are shown in Figure 6. The consistency of the repeated experiments is great.

Table 4. Test results at an impact velocity of 10 m/s.

Test No.	Before the Test	After the Test	Recycled
S_10-1			
S_10-2			
S_10-3			

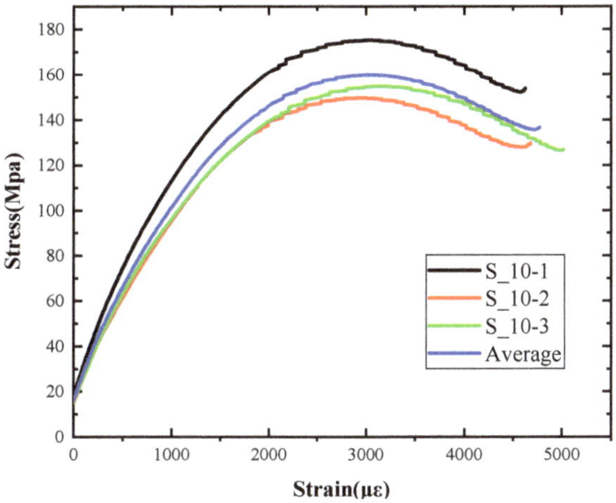

Figure 6. Stress–strain curve of 10 m/s.

4.3. Average Strain Rate of 23.5 s^{-1}

The macro morphology of the ultra-early-strength cement-based material specimens in each stage of the three repeated experiments, when the impact velocity is 15 m/s (the average strain rate is 23.5 s^{-1}), can be seen in Table 5. At an impact velocity of 15 m/s, the specimen was impacted within a very short time after the bullet was launched, accompanied by a loud noise. Afterward, it can be observed that the concrete specimens were a comminuted failure under the impact compression stress wave, and the specimens were broken into a large amount of powder and small pieces. Figure 7 shows the stress–strain curve and the average stress–strain curve obtained from three repeated experiments at an impact velocity of 15 m/s. It can be seen from the figure that the curve rises sharply at the beginning of the impact, and subsequently, the compressive strength of the specimen quickly reached the peak point due to its own strain rate sensitivity, and the second half of the curve began to gradually decrease after the end of the loading.

Table 5. Test results at an impact velocity of 15 m/s.

Test No.	Before the Test	After the Test	Recycled
S_15-1			
S_15-2			
S_15-3			

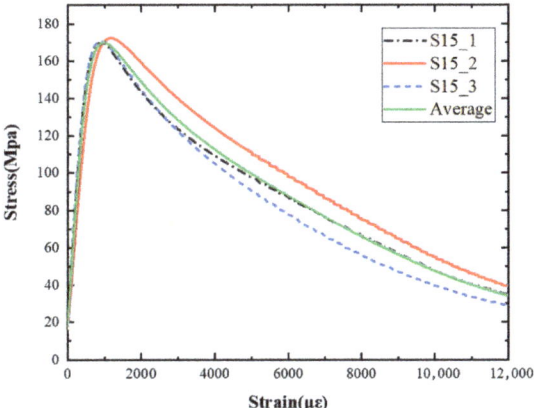

Figure 7. Stress–strain curve of 15 m/s.

5. Discussion

5.1. Strain Rate Effect and Analysis of Compressive Strength

Figure 8 is a summary of the stress–strain curves of ultra-early-strength cement-based materials at different strain rates. The impact velocities are 5 m/s, 10 m/s, and 15 m/s, and the corresponding average strain rates are 7.5 s^{-1}, 15.3 s^{-1}, and 23.5 s^{-1}, respectively. Experimental results show that ultra-early-strength cement-based materials are strain-rate sensitive materials, and the stress–strain curves of the materials at different strain rates are significantly different. When the strain rate is 7.5 s^{-1} (impact velocity 5 m/s), the stress–strain curve shows a yield platform, and it shows that under this loading condition, the ultra-early-strength cement-based material specimen enters an obvious yielding stage from the elastic stage.

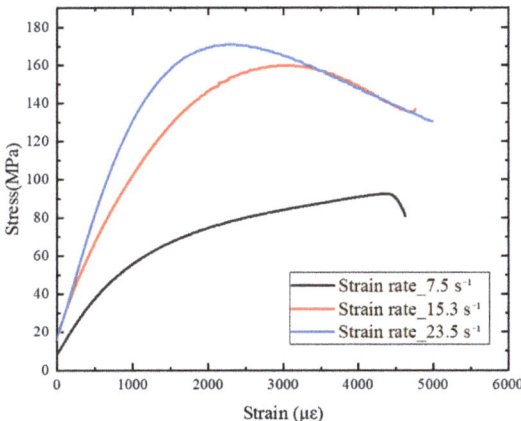

Figure 8. Dynamic stress–strain curve at different strain rates.

After reaching the peak point, the specimen was damaged. The stress–strain curve cannot be unloaded to the zero point, indicating that during the loading process, damage evolution occurred inside the concrete, and the specimen produced plastic deformation.

When loading at 10 m/s and above, the stress strain curve of the ultra-early-strength cement-based material does not show an obvious yield platform, and it transitions directly from the elastic stage to the yield stage. An obvious strain softening phenomenon appears after the peak stress (the stress decreases with increasing strain), indicating that the specimen still has the load-bearing capacity. The stress–strain curve at the strain software stage is no longer the mechanical response of the initial complete material, but the specimen still has a certain residual strength at this time. At this stage, the stress–strain curve still has engineering significance for analyzing the damage and destruction of concrete structures under the explosion and impact load, so the curve of the strain-softening stage is still retained in data processing.

The phenomenon that the dynamic compressive strength of concrete materials increases with the increase in strain rate has been confirmed by extensive experiments, but there is no unified conclusion on the mechanism of the strain rate effect of strength. The increase in the strain rate causes an increase in the strength of concrete materials, which is generally caused by the free water viscosity effect and the crack propagation inertia effect. The inertial effect of crack propagation is generally caused by the concrete matrix and aggregates. Considering that the ultra-early-strength cement-based materials do not contain aggregates, which is different from ordinary concrete, the main reason that the dynamic compressive strength of ultra-early-strength cement-based materials increases with the increase of the strain rate may be the effect of free water viscosity inside the material.

The dynamic increase factor, DIF, the ratio of the dynamic strength to the static strength of concrete, is used to characterize the dynamic characteristics of brittle materials frequently. Table 6 shows the dynamic compressive strength of ultra-early-strength cement-based materials obtained by SHPB impact compression loading experiments at three different strain rates.

Table 6. Dynamic compressive strength under different strain rates.

Strain Rates	Dynamic Compressive Strength (Mpa)		
7.5 (s^{-1})	93.954	95.982	91.836
15.3 (s^{-1})	146.194	149.262	144.668
23.5 (s^{-1})	173.013	175.845	170.697

In this paper, the DIF model Formula (2) of the concrete under the one-dimensional stress state approved by the Euro-International Committee for Concrete (the CEB) is used to fit the experimental data in Table 6.

$$DIF = \frac{f_c}{f_{co}} = \begin{cases} \left(\frac{\dot{\varepsilon}}{\dot{\varepsilon}_s}\right)^{\alpha_c} & \dot{\varepsilon} < k \\ \beta_c \left(\frac{\dot{\varepsilon}}{\dot{\varepsilon}_s}\right)^{\gamma_c} & \dot{\varepsilon} \geq k \end{cases} \quad (2)$$

In the formula, f_c is the corresponding compressive strength (MPa) when the strain rate is $\dot{\varepsilon}$, f_{c0} is the static compressive strength (MPa), $\dot{\varepsilon}$ is the strain rate, $\dot{\varepsilon}_s (\dot{\varepsilon}_s = 3 \times 10^{-5} \text{ s}^{-1})$ is the quasi-static strain rate, α_c, β_c and γ_c are fitting parameters, and k is the critical strain rate. The fitting results are as follows:

$$DIF = \frac{f_c}{f_{co}} = 0.0018 \left(\frac{\dot{\varepsilon}}{\dot{\varepsilon}_s}\right)^{0.5211} \quad 7.5 \text{ s}^{-1} \leq \dot{\varepsilon} \quad (3)$$

Figure 9 shows the relationship between the dynamic enhancement factor of ultra-early-strength cement-based materials and the strain rate. It can be seen from the figure that the dynamic compressive strength of the material has a significant strain rate effect, and as the strain rate increases, the dynamic compressive strength increases significantly.

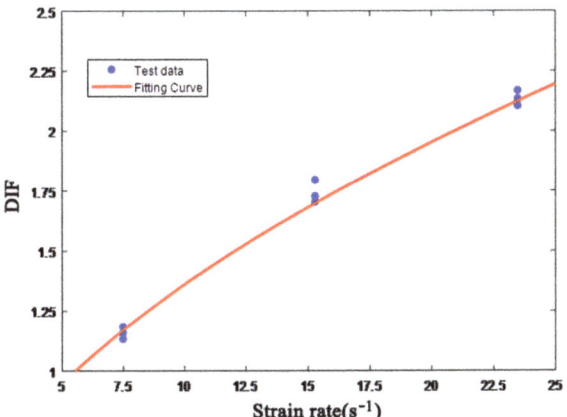

Figure 9. Relationship between DIF and strain rate.

5.2. Energy Absorption Density Analysis

Toughness is the ability of a material or structure to absorb energy under a load until it fails. It not only depends on the bearing capacity but also on the deformation capacity [30]. The methods to determine the toughness index include the energy method, intensity method, energy ratio method, characteristic point method, etc. In this paper, the energy method was used, and the area enclosed under the stress–strain curve is used to represent the characterization method of absorbed energy. The energy absorption density can be calculated by Formula (4).

$$\omega = \int \sigma * (\varepsilon) d\varepsilon \qquad (4)$$

Figure 10 shows the energy absorption density curve at different loading speeds obtained by Formula (4). It can be seen from the figure that under different impact speeds (strain rates), the energy absorption density increases with the increase of strain, and as the impact velocity increases, the growth rate of the energy absorption density accelerates. Figure 10a shows the relationship between energy absorption density and strain when the impact velocity is 5 m/s. The average energy absorption density at the stress peak point is about 1.5×10^5 J/m^3, and the corresponding peak strain is 3100 με. When the impact velocity is 10 m/s, as shown in Figure 10b, the corresponding curve of each test piece is relatively concentrated, and the corresponding energy absorption density at the peak stress point is about 3×10^5 J/m^3.

Relative to the impact velocity of 5 m/s, the energy density value doubles but the peak strain hardly changes. When the impact velocity increases to 15 m/s, the corresponding curves of each test piece are scattered slightly. As shown in Figure 10c, the curve becomes steep, compared with the low-speed impact, the corresponding energy absorption density at the peak stress point is about 2.3×10^5 J/m^3, and the corresponding peak strain is about 1500 με. According to the energy density absorption value, peak stress, and corresponding peak strain of the specimen under different impact speeds, it can be concluded that, as the impact velocity increases, the peak stress rises, the energy absorption density increases and its growth rate accelerates, as shown in Figure 10d. The peak strain at an impact velocity of 15 m/s is lower than that of low-speed impact (5 m/s and 10 m/s).

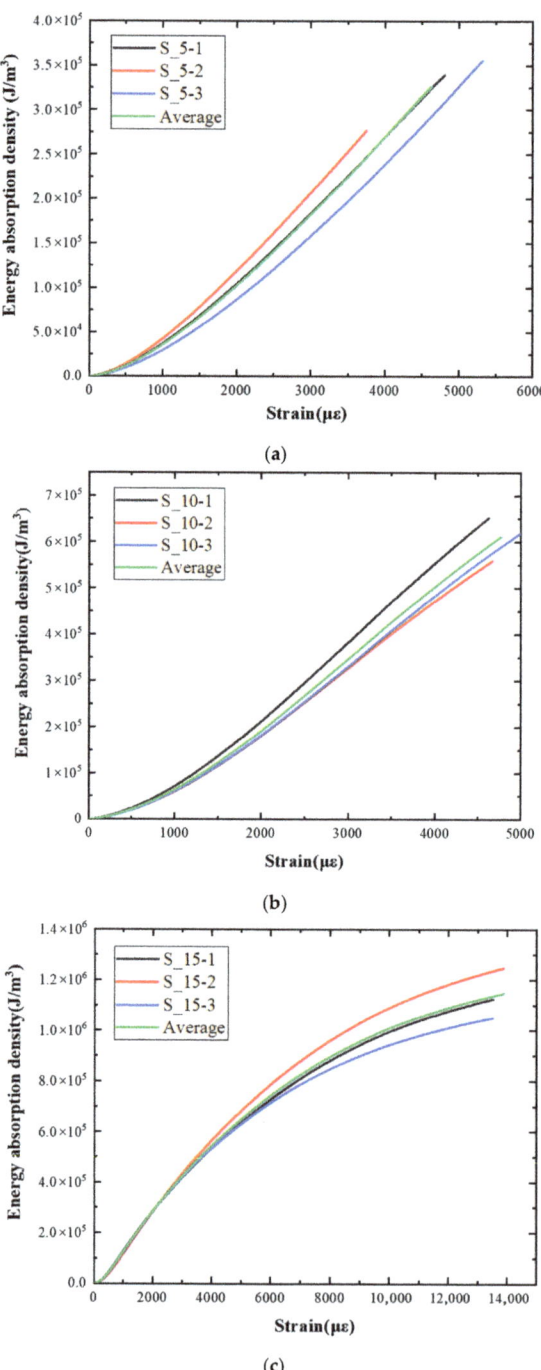

Figure 10. Cont.

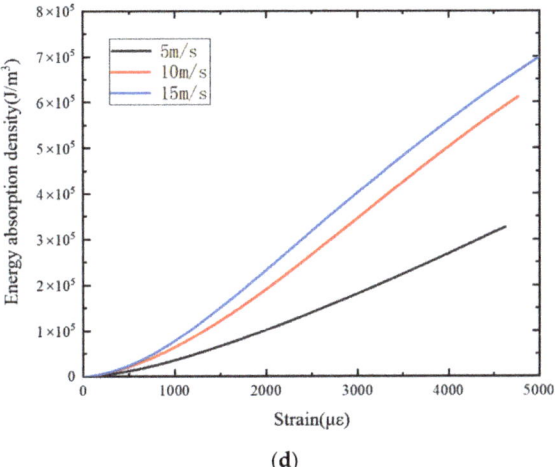

(d)

Figure 10. Energy absorption density–strain curve. (**a**) Impact velocity of 5 m/s. (**b**) Impact velocity of 10 m/s. (**c**) Impact velocity of 15 m/s. (**d**) Average value.

5.3. Damage Evolution Process Analysis

Under the impact load, cracks will appear inside the specimen, and the cracks will gradually propagate from the inside to the outside. When the cracks propagate to the boundary of the specimen, they will cause penetration and breakage. In this section, from the perspective of continuum mechanics, the region containing many scattered micro-cracks is regarded as a local uniform field, the overall effect of the crack in the field is considered, and the damage state of the uniform field is described by defining an irreversible related field variable, which is the damage variable D. Under uniaxial compression, the constitutive relation of concrete material after damage can be expressed as [31]

$$\sigma = E\varepsilon(1-D) \tag{5}$$

where σ is stress, ε is strain, E is elastic modulus, and D is the damage variable. The damage variable is obtained by the transformation of Equation (5):

$$D = 1 - \frac{\sigma}{E\varepsilon} \tag{6}$$

The strain ε can be expressed as

$$\varepsilon = \varepsilon_e + \varepsilon_p \tag{7}$$

where ε_e is the elastic strain and ε_p is the plastic strain. The elastic strain ε_e can be derived from stress σ.

$$\varepsilon_e = \frac{\sigma}{E} \tag{8}$$

The plastic strain ε_p is expressed as

$$\varepsilon_p = \varepsilon - \frac{\sigma}{E} \tag{9}$$

The relationship between the damage variable and the plastic strain under different loading speeds (strain rates) is shown in Figure 11. The damage variable D is calculated by calculating the elastic modulus E according to the linear elastic segment of the stress–strain curve and then substituting the stress and strain into Equation (6).

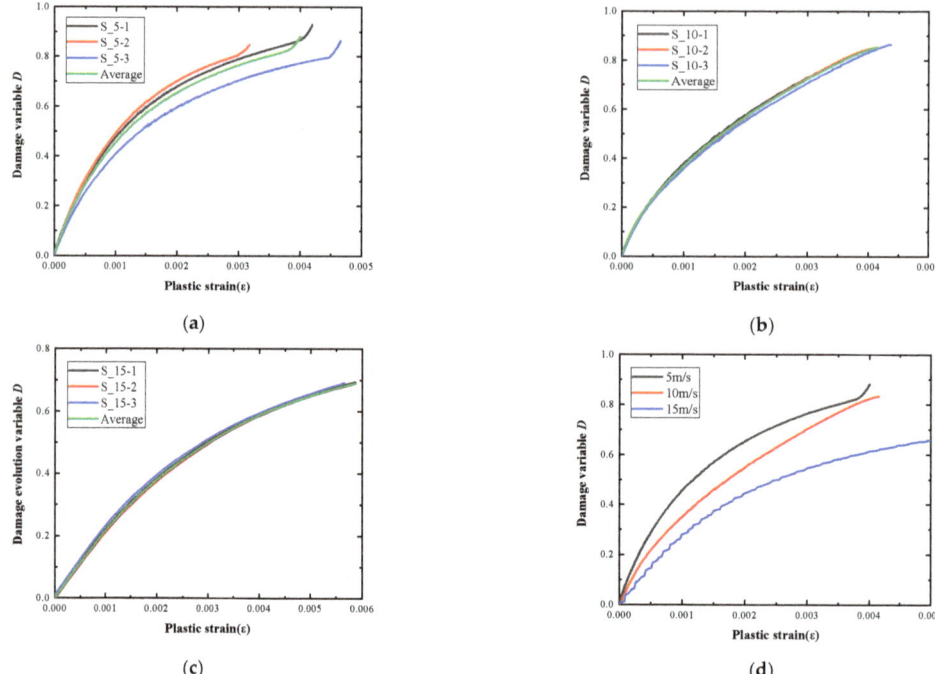

Figure 11. Damage evolution variable. (**a**) Impact velocity of 5 m/s. (**b**) Impact velocity of 10 m/s. (**c**) Impact velocity of 15 m/s. (**d**) Average value.

With the increase of the impact velocity, the crack propagation inside the specimen is hindered, and the evolution speed of the damage variable decreases, and under the same plastic strain, the damage variable corresponding to the high strain rate is lowered. When the stress of the concrete material is 70% of its peak value, it will enter the plastic stage [31]. According to the test results, when the ultra-early-strength cement-based material loses its bearing capacity, the corresponding damage variable is between 0.8 and 0.9, approximately.

According to the result of the analysis, the damage variable is related to strain rate $\dot{\varepsilon}$ and plastic strain ε_p. On this basis, the average value of the three damage variables under different strain rates was fitted. The comparison result of the fitting is shown in Figure 12. It can be seen from the figure that both are in a high degree of agreement. The fitting formula is as follows:

$$D = 0.363 \times \ln(1 + A\varepsilon_p)$$
$$A = 2437 * \sin(0.054 * \dot{\varepsilon} + 1.39) \quad 7.5 \text{ s}^{-1} \leq \dot{\varepsilon} \leq 23.5 \text{ s}^{-1} \tag{10}$$

From the average value of the stress–strain curve of three repeated tests at different strain rates, the elastic modulus E (GPa) of the ultra-early-strength material at different strain rates is obtained. The relationship between elastic modulus E and strain rate is shown in Figure 13, and R^2 is 0.98. By fitting the elastic modulus E at different strain rates, it is found that the relationship between the strain rate and the elastic modulus can be expressed as:

$$E = 3.297\dot{\varepsilon} + 33.53 \quad 7.5 \text{ s}^{-1} \leq \dot{\varepsilon} \leq 23.5 \text{ s}^{-1} \tag{11}$$

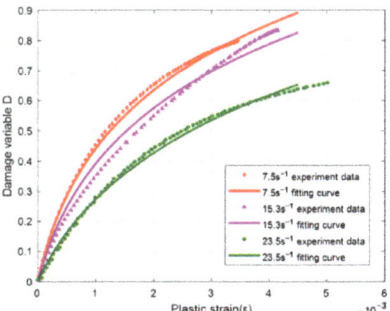

Figure 12. Comparison of damage variable fitting results.

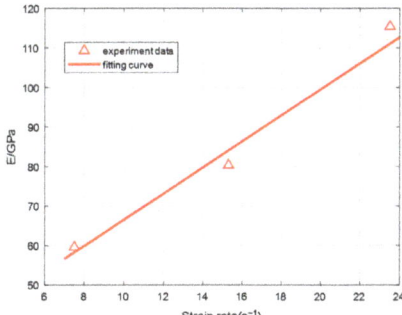

Figure 13. The relationship between elastic modulus and strain rate.

By substituting the evolution variable expressions (8) and (10) under different strain rates into Equation (9), the stress–strain expression of the ultra-early-strength cement-based material can be obtained. The comparison between the results calculated by Equation (9) and the measured stress–strain results under different strain rates is shown in Figure 14. It can be found from the figure that the stress–strain curve calculated from the damage variables is in good agreement with the test results. Therefore, the internal damage evolution process of ultra-early-strength cement-based materials can be characterized by the damage variables.

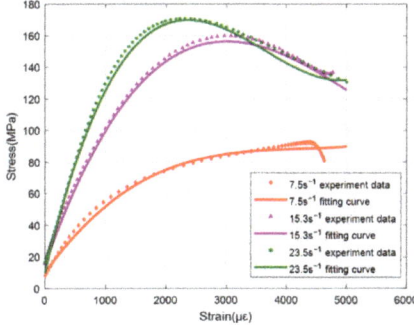

Figure 14. Comparison of fitting results with test results.

6. Conclusions

In order to explore the dynamic mechanical properties of UR50 ultra-early-strength cement materials, an experimental study on the dynamic mechanical properties of ultra-early-strength cement-based materials at high strain rates was carried out using a large-diameter SHPB. The macroscopic failure morphology, dynamic stress–strain curve, and the relationship between dynamic compressive strength and strain rate of specimens under different strain rates (impact velocity) were obtained. According to the experimental results, the main conclusions are as follows:

(1) Under different loading conditions, different types of copper sheets were selected as the shaper to eliminate the dispersion effect. Under the same impact velocity, the stress–strain curves of the three specimens are in good agreement, which ensures the validity and reliability of the experimental results;

(2) The compression brittleness of ultra-early-strength cement-based specimens is relatively large, and the failure mode is a mainly brittle fracture, and with the increase of loading speed, the failure mode of the specimens gradually transited from larger fragments to small fragments, with an eventual large amount of powder;

(3) The dynamic compressive strength of ultra-early-strength cement-based materials increases with the increase of the strain rate, which has an obvious strain rate strengthening effect. Fitting determines the relationship curve between the dynamic increase factor, DIF, and the strain rate. It has a linear relationship with the logarithm of the strain rate. The higher the strain rate, the larger the DIF, indicating that it has the advantage of impact-resistant mechanical properties;

(4) The concept of absorption density is introduced to facilitate a better understanding of the toughness of ultra-early-strength cement-based materials. As the impact velocity increases, the peak stress rises, the energy absorption density value increases, and its growth rate accelerates. The peak strain at an impact velocity of 15 m/s is lower than that of low-speed impact (5 m/s and 10 m/s);

(5) Based on the theory and method of continuum mechanics, the evolution process of the damage variables of ultra-early-strength cement-based materials was analyzed from a macro perspective. The damage variable equations at different strain rates were fitted according to the test results and based on the stress–strain curve, and the rationality of the damage evolution process was proved. With the increase in strain rate, the internal crack propagation of the specimen is hindered, and the increase rate of the damage variable decreases. Under the same plastic strain, the damage variable of the specimen under the high strain rate is relatively small.

In general, UR50 ultra-early-strength cement-based materials are more brittle in shock compression and will undergo an overall fracture at low strain rates. The dynamic compressive strength increases with the increase of the strain rate and has an obvious strain rate strengthening effect. However, further research on UR50 ultra-high early-strength concrete (UHESC) under SHPB-impact tests should be conducted. Additional research is also needed for the UR50 concrete base with high-speed impact to explore impact performance. The results of this research will further the development of dynamic material simulation methods and material models.

Author Contributions: Formal analysis, Q.H. and X.S.; Investigation, X.W. (Xing Wang); Methodology, X.S.; Resources, J.Y.; Supervision and project administration, J.W.; Validation, X.W. (Xiaofeng Wang); Writing—original draft, Q.H. and J.Y.; Writing—review & editing, W.W. and Z.Z. All authors have read and agreed to the published version of the manuscript.

Funding: This research was funded by the National Natural Science Foundation of China (Grant Nos. 11302261 and 11972201). This paper is also funded by the Key Laboratory of Impact and Safety Engineering (Ningbo University) project, Ministry of Education. The project number is CJ202011.

Institutional Review Board Statement: Not applicable.

Informed Consent Statement: Not applicable.

Data Availability Statement: Not applicable.

Conflicts of Interest: The authors declare no conflict of interest.

References

1. Hong, J.; Fang, Q.; Chen, L.; Kong, X. Numerical predictions of concrete slabs under contact explosion by modified K&C material model. *Constr. Build. Mater.* **2017**, *155*, 1013–1024.
2. Tai, Y.; Chu, T.; Hu, H.; Wu, J. Dynamic response of a reinforced concrete slab subjected to air blast load. *Theor. Appl. Fract. Mech.* **2011**, *56*, 140–147. [CrossRef]
3. Zhou, X.; Kuznetsov, V.; Hao, H.; Waschl, J. Numerical prediction of concrete slab response to blast loading. *Int. J. Impact Eng.* **2008**, *35*, 1186–1200. [CrossRef]
4. Pang, B.J.; Wang, L.W.; Yang, Z.Q. The effect on dynamic properties of reactive powder concrete under high temperature burnt and its micro-structure analysis. *Key Eng. Mater.* **2010**, *452*, 109–112. [CrossRef]
5. Hu, S.S.; Zhang, L.; Wu, H.J.; Wu, X.T. Experimental Study on Spalling Strength of Concrete Materials. *Eng. Mech.* **2004**, *21*, 128–132.
6. Zhu, J.; Hu, S.; Wang, L. An analysis of stress uniformity for concrete-like specimens during SHPB tests. *Int. J. Impact Eng.* **2009**, *36*, 61–72. [CrossRef]
7. Bragov, A.M.; Konstantinov, A.Y.; Lamzin, D.A.; Lomunov, A.K.; Gonov, M.E. Determination of the mechanical properties of concrete using the split Hopkinson pressure bar method. *Procedia Struct. Integr.* **2020**, *28*, 2174–2180. [CrossRef]
8. Levi-Hevroni, D.; Kochavi, E.; Kofman, B.; Gruntman, S.; Sadot, O. Experimental and numerical investigation on the dynamic increase factor of tensile strength in concrete. *Int. J. Impact Eng.* **2018**, *114*, 93–104. [CrossRef]
9. Zhang, M.; Wu, H.J.; Li, Q.M.; Huang, F.L. Further investigation on the dynamic compressive strength enhancement of concrete-like materials based on split Hopkinson pressure bar tests. Part I: Experiments. *Int. J. Impact Eng.* **2009**, *36*, 1327–1334. [CrossRef]
10. Wang, G.; Lu, D.; Li, M.; Zhou, X.; Wang, J.; Du, X. Static–Dynamic Combined Multiaxial Strength Criterion for Concrete. *J. Eng. Mech.* **2021**, *147*, 04021017. [CrossRef]
11. Gu, X.; Zhang, Q.; Huang, D.; Yv, Y. Wave dispersion analysis and simulation method for concrete SHPB test in peridynamics. *Eng. Fract. Mech.* **2016**, *160*, 124–137. [CrossRef]
12. Hassan, M.; Wille, K. Experimental impact analysis on ultra-high performance concrete (UHPC) for achieving stress equilibrium (SE) and constant strain rate (CSR) in Split Hopkinson pressure bar (SHPB) using pulse shaping technique. *Constr. Build. Mater.* **2017**, *144*, 747–757. [CrossRef]
13. Erzar, B.; Forquin, P. Experiments and mesoscopic modelling of dynamic testing of concrete. *Mech. Mater.* **2011**, *43*, 505–527. [CrossRef]
14. Wang, B.; Chen, Y.; Fan, H.; Jin, F. Investigation of low-velocity impact behaviors of foamed concrete material. *Compos. Part B Eng.* **2019**, *162*, 491–499. [CrossRef]
15. Kang, J. Composite and non-composite behaviors of foam-insulated concrete sandwich panels. *Compos. Part B Eng.* **2015**, *68*, 153–161. [CrossRef]
16. Wang, X.; Liu, L.; Zhou, H.; Song, T.; Qiao, Q.; Zhang, H. Improving the compressive performance of foam concrete with ceramsite: Experimental and meso-scale numerical investigation. *Mater. Des.* **2021**, *208*, 109938. [CrossRef]
17. Shafigh, P.; Jumaat, M.Z.; Mahmud, H.B.; Alengaram, U.J. A new method of producing high strength oil palm shell lightweight concrete. *Mater. Des.* **2011**, *32*, 4839–4843. [CrossRef]
18. Cao, S.; Hou, X.; Rong, Q.; Zheng, W.; Abid, M.; Li, G. Effect of specimen size on dynamic compressive properties of fiber-reinforced reactive powder concrete at high strain rates. *Constr. Build. Mater.* **2019**, *194*, 71–82. [CrossRef]
19. Xiong, B.; Demartino, C.; Xiao, Y. High-strain rate compressive behavior of CFRP confined concrete: Large diameter SHPB tests. *Constr. Build. Mater.* **2019**, *201*, 484–501. [CrossRef]
20. Liu, G.-J.; Bai, E.-L.; Xu, J.-Y.; Yang, N.; Wang, T.-J. Dynamic compressive mechanical properties of carbon fiber-reinforced polymer concrete with different polymer-cement ratios at high strain rates. *Constr. Build. Mater.* **2020**, *261*, 119995. [CrossRef]
21. Wei, Y.; Lu, Z.; Hu, K.; Li, X.; Li, P. Dynamic response of ceramic shell for titanium investment casting under high strain-rate SHPB compression load. *Ceram. Int.* **2018**, *44*, 11702–11710. [CrossRef]
22. Sun, X.; Zhao, K.; Li, Y.; Huang, R.; Ye, Z.; Zhang, Y.; Ma, J. A study of strain-rate effect and fiber reinforcement effect on dynamic behavior of steel fiber-reinforced concrete. *Constr. Build. Mater.* **2018**, *158*, 657–669. [CrossRef]
23. Scott, B.D.; Park, R.; Priestley, M.J.N. Stress-strain behavior of concrete confined by overlapping hoops at low and high strain rates. *ACI Struct. J.* **1982**, *79*, 13–27.
24. Georgin, J.F.; Reynouard, J.M. Modeling of structures subjected to impact: Concrete behaviour under high strain rate. *Cem. Concr. Compos.* **2003**, *25*, 131–143. [CrossRef]
25. Xiuli, D.; Yang, W.; Dechun, L. Non-linear uniaxial dynamic strength criterion for concrete. *J. Hydraul. Eng.* **2010**, *41*, 300–309.
26. Huo, J.S.; Zheng, Q.; Chen, B.S. Tests on impact behaviour of micro-concrete-filled steel tubes at elevated temperatures up to 400 °C. *Mater. Struct.* **2009**, *42*, 1325–1334. [CrossRef]
27. Wang, W.; Huo, Q.; Yang, J.C.; Wang, J.H.; Wang, X. Experimental investigation of ultra-early-strength cement-based self-compacting high strength concrete slabs (URCS) under contact explosions. *Def. Technol.* **2022**. [CrossRef]

28. ASTM International. *ASTM C39/C39M-18 Standard Test Method for Compressive Strength of Cylindrical Concrete Specimens*; ASTM International: West Conshohocken, PA, USA, 2018.
29. Davies, R.M. A critical study of the Hopkinson pressure bar. *Philos. Trans. R Soc. A* **1948**, *240*, 375–457.
30. Huang, B.; Xiao, Y. Compressive impact tests of lightweight concrete with 155-mm-diameter spilt hopkinson pressure bar. *Cem. Concr. Compos.* **2020**, *114*, 103816. [CrossRef]
31. Wang, C.L.; Xu, B.G.; Li, S.L. Study on damage constitutive model of steel fiber reinforced concrete under uniaxial compression. *Geotech. Mech.* **2006**, *27*, 151–154.

Article

Study on the Dynamic Mechanical Properties of Ultrahigh-Performance Concrete under Triaxial Constraints

Wei Zhang [1], Jize Mao [1], Xiao Yu [2], Bukui Zhou [2,*] and Limei Wang [2,*]

[1] College of Aerospace and Civil Engineering, Harbin Engineering University, Harbin 150001, China
[2] Institute of Defense Engineering, Academy of Military Sciences, Beijing 100850, China
* Correspondence: zbk751225@sina.com (B.Z.); wanglmengineer@163.com (L.W.)

Abstract: To confirm the effect of confining pressure on the dynamic mechanical behavior of ultrahigh-performance concrete (UHPC), this study used a true triaxial split Hopkinson pressure bar test system to perform dynamic compression tests on UHPC under triaxial constraints. The confining pressure range considered was 5~10 MPa, the strain rate range was 35~80 s^{-1}, and the steel fiber contents were 0.5%, 1% and 2%. The three-dimensional dynamic engineering stress-strain relationship and equivalent stress-strain relationship of UHPC under different confining pressures and different strain rates were obtained and analyzed in detail. The results show that under the confinement condition, the dynamic peak axial stress–strain and dynamic peak lateral stress–strain of UHPC have strong sensitivity to the strain rate. In addition, the dynamic peak lateral stress–strain is more sensitive to the confining pressure than the dynamic axial stress. An empirical strength enhancement factor (DIFc) that considers the strain rate effect and confining pressure was derived, and the impact of the coupling between the enhancement caused by the confining pressure and the strain rate effect on the dynamic strength of the UHPC under triaxial confinement was discussed. A dynamic strength failure criterion for UHPC under triaxial constraint conditions was established.

Keywords: ultrahigh-performance concrete (UHPC); dynamic impact; static triaxial constraint; strain rate effect; failure criterion

1. Introduction

In current construction projects, concrete materials (normal reinforced concrete (NRC), mortar and ultrahigh-performance concrete (UHPC)) are still widely used. With the development of society and the increase in the global population, conflicts in some parts of the world are increasing. To protect the safety and property of the public, the dynamic behavior of concrete materials under impact loads caused by projectile impacts and explosions has attracted much attention. In such cases, the concrete material undergoes multiaxial compression with high confinement [1,2]. Therefore, the study of the mechanical response of concrete materials under triaxial compression not only is very important for the structural design of military and civil protection engineering projects, but also is conducive to the development and verification of constitutive models.

Researchers have performed many studies on the static response of concrete materials under triaxial compression conditions and have obtained many important conclusions. These studies are helpful for understanding the effect of constraint conditions on concrete strength, deformation capacity and failure mechanisms. Wang et al. [3] conducted a series of triaxial compression tests on a 100 mm cube test specimen with an unconfined compressive strength of approximately 10 MPa and obtained an approximately linear relationship between the normalized triaxial compressive strength and the constraint ratio. Noori et al. [4] prepared two kinds of steel fiber-reinforced mortar and ordinary high-performance concrete cylindrical specimens with 1% and 2% steel fiber volume fractions and conducted triaxial compression tests. The results show that the increase in confining pressure increases the peak value of axial stress-strain, making the test specimen exhibit stronger ductility. The addition of steel fibers is beneficial to improve the energy absorption capacity of these materials. Ren et al. [5] and Lu et al. [6] obtained a triaxial stress–strain relationship through triaxial compression tests on UHPC, and discussed the failure criterion and toughness of UHPC under triaxial compression. Chi et al. [7] developed a failure envelope based on the five-parameter failure criterion through a true triaxial compression test of hybrid steel–polypropylene fiber-reinforced concrete (HFRC). The proposed envelope curve can provide accurate approximations of the ultimate strength for plain concrete and fiber-reinforced concrete under static loading. Sirijaroonchai et al. [8] studied various fiber types (high-strength hooked steel fibers and ultrahigh-molecular-weight polyethylene fibers) and fiber volume fractions (1~2%) through passive triaxial testing of high-performance fiber-reinforced cement-based composites under different confining pressures (41 MPa and 52 MPa). The results showed that for a higher confinement ratio, the confinement effect introduced by various types and volume fractions of fibers decreases. Jiang et al. [9] conducted triaxial compression tests under passive confinement on cube specimens of normal reinforced concrete (NRC) by means of a newly developed triaxial test system. Compared to the traditional triaxial compression calibrated model, the peak strength prediction error was within ±10%. These results provided new ideas for studying the failure mechanism of concrete under passive confinement.

In addition, Imran et al. [10], Vu et al. [11] and Piotrowska et al. [12] systematically studied the influence of water content, fiber, aggregate and cement paste on the triaxial compressive properties of concrete.

However, to date, the dynamic response of concrete under triaxial confinement has seldom been discussed [13], mainly because it is still challenging to perform tests by coupling confining pressure and dynamic loading. At present, based on traditional split Hopkinson pressure bar (SHPB) test technology, there are two methods to achieve triaxial compression: namely, the active confining pressure method and the passive confining pressure method. The active confining pressure loading system applies radial confining pressure to the specimen before dynamic loading. Gran et al. [14] studied the dynamic behavior of concrete when considering confinement using an active confining pressure device. The results showed that, compared with the quasistatic condition, the strain rate effect was not observed when the strain rate was $0.5~\text{s}^{-1}$. However, when the strain rate reached $1.3~5~\text{s}^{-1}$, the shear failure envelope established by the triaxial compression data was 30~40% higher than that under static loading. Zeng et al. [15] used an MTS device to obtain a series of complete stress-strain curves for specimens subjected to different strain rates and confining pressure combinations. The results confirmed that there was a significant coupling effect between the strength enhancements controlled by the strain rate and confining pressure. Based on the 1D SHPB test system, Xue and Hu [16] precharged the cement mortar test specimen with a constant liquid confining pressure through hydraulic cylinders and found that the strain rate effect was significant for cement mortar under confining pressure. Similarly, Marvern et al. [17] and Gary et al. [18] applied confining pressures of 3~10 MPa to concrete specimens. The experiments found that the concrete specimen strength had strain rate sensitivity under this range of confining pressures. Under a confining pressure of 5 MPa, the strength of concrete increased by 30% when the strain rate increased from $250~\text{s}^{-1}$ to $600~\text{s}^{-1}$. Yan et al. [2] and Fujikake et al. [19] implemented a series of triaxial dynamic compression tests of concrete under low-strain-rate loading conditions ($<2~\text{s}^{-1}$) and found that with an increasing confining pressure, the strain rate effect was weakened. When the confining pressure exceeded the uniaxial compressive strength of concrete, the strain rate sensitivity of the concrete basically disappeared. Fu et al. [20] studied the dynamic compression characteristics of hybrid fiber-reinforced concrete under confining pressure and established an empirical dynamic strength criterion by using the active confining pressure applied via the SHPB test. The passive confining pressure method is a relatively new technology that has been widely used in recent years. In this method, a lateral ring is added around the sample to limit its expansion, thus generating a constraining force, which is the lateral confining pressure [11]. Liu et al. [21] used an SHPB dynamic loading device to investigate the dynamic mechanical properties of NRC under constraint conditions. The confining pressure was obtained by imposing a transverse metal sleeve on the axially loaded specimen. The test results show that the dynamic peak axial stress, dynamic peak lateral stress and peak axial strain of concrete are very sensitive to the strain rate under constraint conditions. Li et al. [22] simulated the SHPB test of concrete materials using the finite element method and the Drucker–Prager (DP) model. They found that due to the limitation of the lateral inertia effect, the apparent strength increase in the concrete material was due to the influence of the hydrostatic pressure rather than the strain rate sensitivity of the material.

Notably, in the active confining pressure method, the pressure ($\sigma_2 = \sigma_3$) is considered to be a constant value, but it is difficult to simultaneously measure the changes during the impact loading process, and the hydraulic oil is prone to leak under a high confining pressure [15,18]. Under passive constraints, the confining pressure depends on the gap between the specimen and the sleeve and the material properties and thickness of the sleeve. It is relatively difficult to precisely control the gap between the specimen and the sleeve, and the lateral stress varies during the dynamic loading process [23].

Researchers have made enormous efforts to obtain the true triaxial stress state of the test specimen and the stress response state in different directions during the dynamic impact process. Cui et al. [24] used a newly developed three-dimensional SHPB (3D-SHPB) device to simultaneously compress a test specimen with the same amplitude in three mutually perpendicular directions to obtain the volume characteristics of concrete under dynamic hydrostatic pressure. Liu et al. [25] conducted a series of experiments on cement and concrete specimens using the 3D-SHPB loading system to investigate the mechanical properties and fracture behavior of concrete-based materials under coupled static-dynamic loading conditions. The test results showed that the value of the dynamic uniaxial compressive strength of concrete was much higher than the static uniaxial compressive strength, but that it decreased with axial prestress. The dynamic compressive strength under a triaxial constraint is approximately 3 times higher than the quasistatic uniaxial strength. Xu et al. [26] used their own developed dynamic test system under static triaxial constraints to investigate the dynamic compressive performance of NRC specimens under different static triaxial constraint conditions. The three-dimensional dynamic engineering stress–engineering strain relationship, dynamic volumetric strain–hydrostatic pressure relationship and equivalent stress–equivalent strain relationship were obtained and analyzed in detail. The findings indicated that the load path dependence and the strain rate dependence were significant. Liu et al. [27] carried out a series of dynamic and static compression tests on NRC with the 3D-SHPB system. It was concluded that there was an obvious strain rate effect on NRC when it was under triaxial confining pressure. The confining pressure has an effect on the dynamic stress of concrete, but with the increase in strain rate, the effect gradually decreases.

In summary, the mechanical response of concrete materials under triaxial confining pressure that has been mostly focused on is the quasistatic mechanical response. Due to the limitations of the test conditions, only a few dynamic tests under triaxial confining pressure have been carried out at present. The research target has mostly been NRC. However, the application of UHPC in the field of modern protective engineering is increasing. Research on the dynamic mechanical response of UHPC under triaxial compression is still lacking. In this work, dynamic mechanical tests of UHPC under triaxial confining pressure were carried out by using a true triaxial split Hopkinson pressure bar system. Considering the influence of the confining pressure stress and strain rate, the engineering stress-strain and equivalent stress-strain relationships of UHPC were analyzed. The dynamic strengthening mechanism of UHPC under triaxial constraints was discussed, and the dynamic failure criterion was fitted based on the test results.

2. True Triaxial Test

2.1. Sample Materials

The UHPC used in this paper was composed of Chinese standard grade 52.5 P.II Portland cement, silica fume, fine aggregate and a water reducer, and the mix ratios are shown in Table 1. The steel fibers were copper-plated straight steel fibers, and the volume additions were 0.5%, 1% and 2%. The fiber length, diameter and tensile strength were 12 mm, 0.2 mm and 2850 MPa, respectively.

Table 1. Mix proportions of concrete (unit: kg/m^3).

	Cement	Silica Fume	Fine Aggregate	Superplasticizer	Water	Steel Fibers
U0.5	850	200	1000	40	150	39
U1	850	200	1000	40	150	79
U2	850	200	1000	40	150	158

The size of the SHPB test specimen used in this experiment was 50 mm × 50 mm × 50 mm, as shown in Figure 1. To ensure the smoothness of the test specimen and reduce the stress concentration during the experiment, the two end faces of the test specimen were polished after curing so that the parallelism deviation of the three opposite faces was less than 0.25 mm, and the dimensional accuracy of the test specimen was kept within ±0.5%. The end-face flatness deviation was less than 0.05 mm, the adjacent surfaces of the cubic test specimen had good verticality and the maximum deviation was less than 0.3°. To reduce the discreteness of the test results, 2 specimens were tested for each confining pressure and each strain rate, for a total of 54 specimens.

Figure 1. Trial mold and test specimen.

In addition, a total of nine 100 mm × 100 mm × 100 mm cube test specimens were prepared. The average strength of three specimens of each type was taken as the quasistatic compressive strength of each type of UHPC material. The measured strength and average strength are shown in Table 2.

Table 2. UHPC's 28 d compressive strength (unit: MPa).

	Measured Strength			Average Strength
U0.5	110.4	110.0	115.6	112.0
U1	126.4	130.5	131.0	129.3
U2	146.4	150.2	149.8	148.8

2.2. Triaxial Test

2.2.1. Test Equipment

This work adopted a true triaxial split Hopkinson pressure bar test system, and a diagram of the test setup is shown in Figure 2.

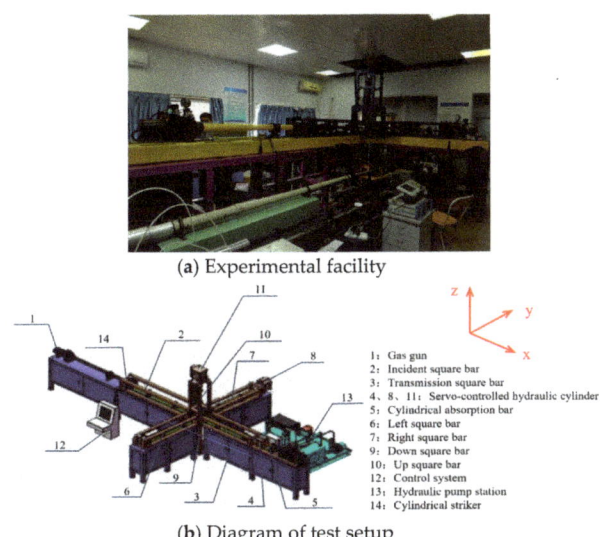

(a) Experimental facility

(b) Diagram of test setup

Figure 2. The true triaxial split Hopkinson pressure bar loading system.

The experimental system included two parts: ① A true triaxial static load application system, which was composed of hydraulic cylinders ((4), (8) and (11) in Figure 2) that were oriented in three orthogonal directions, and their corresponding reaction supports, which could test the cube compressive stress in three directions. ② A bullet launching and signal test system, which was mainly composed of a high-pressure gas cannon (1) in the impact direction (x-direction); an incident square rod (2) and a supporting square rod (3) horizontal to the impact direction; a supporting square bar ((6) and (7)) in the y-direction; and a supporting square bar ((9) and (10)) in the z-direction, where the y- and z-directions are perpendicular to the impact direction. The experiment console (12) controls the hydraulic system and the launch system. The hydraulic station (13) provides hydraulic oil for the three hydraulic cylinders in the servo control process.

As shown in Figure 3, to reduce the loss of equipment, a gasket made of the same material as the steel rod was added to the surface of the specimen, and its surface was evenly coated with Vaseline to reduce the interface friction [28]. During the test, the specimen is placed at the intersection of the six bar axes. The incident rod, specimen, transmission rod and servo-controlled hydraulic cylinder are constrained by the x-axis steel frame, and the static constraint in the x-axis direction ($\sigma_{x\text{-static}}$) can be realized by pumping the hydraulic cylinder. A cylindrical impactor impacts the incident square rod through the predesigned hole in the steel frame. The cylindrical absorbing rod passes through the hole in the hollow servo-controlled hydraulic cylinder, where it makes contact with the transmission rod and transfers a part of the transmitted waves to the momentum trap. The static constraints ($\sigma_{y\text{-static}}$ and $\sigma_{z\text{-static}}$) in the y-axis and z-axis directions are loaded in the same way as that in the x-axis direction, both by pumping hydraulic cylinders. The diameter of the cylindrical impacting rod was 42 mm, and the length was 500 mm. The lengths of the incident rod and the transmission rod were 2500 mm and 2000 mm, respectively. Their cross-sections were squares of 50 mm × 50 mm. The diameter of the cylindrical absorption rod was 42 mm, and the length was 800 mm. In this study, the lengths of the left and right rods along the y-axis were both 2000 mm, and their cross-sections were squares of 50 mm × 50 mm; the lengths of the upper rod and the lower rods along the z-axis are both 1600 mm, and their cross-sections were also squares of 50 mm × 50 mm [26].

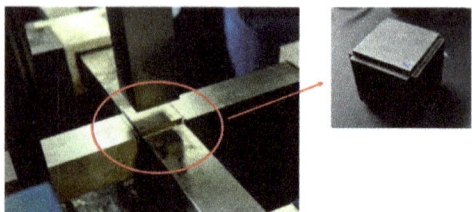

Figure 3. Standard cubic specimen placed between the bars of the loading element.

2.2.2. Testing Techniques

In this work, the difference between the traditional SHPB test and 3D-SHPB test was that the 3D-SHPB test was mainly divided into two stages. In the first stage, the static load was preloaded, the six surfaces of the cube were in contact with six square steel rods, and then the load was applied to the test specimen through the hydraulic device and six steel rods. The static loads were $\sigma_{x\text{-static}}$, $\sigma_{y\text{-static}}$ and $\sigma_{z\text{-static}}$. In the second stage, a dynamic load was applied, and the impacting rod was launched in the x-direction. The impact of the impacting rod on the incident rod generated elastic compression waves, named incident waves (ε_i), which propagate toward the specimen. When the incident wave reached the incident rod/specimen interface, due to the impedance mismatch between the incident rod/transmission rod and the test specimen, the incident wave was divided into the reflected wave ε_r and the transmitted wave ε_t, and the reflected wave was propagated in the opposite direction of the incident rod. The transmitted waves passed through the specimen and entered the transmission rod, which was also the cause of the plastic deformation of the test specimen. At the same time, due to the dynamic load applied to the x-axis, the test specimen experienced lateral expansion (i.e., the Poisson effect). Therefore, a wave was measured with the four square bars in the y-axis and z-axis directions: $\varepsilon_{y\text{-}left}$, $\varepsilon_{y\text{-}right}$, $\varepsilon_{z\text{-}up}$ and $\varepsilon_{z\text{-}down}$.

At the same time, the basic principle of the true triaxial split Hopkinson pressure bar system is the same as that of the traditional SHPB test technique, and the theory of one-dimensional wave propagation in an elastic bar also needs to be followed. Therefore, the stress along the x-axis σ_{x-dyn}, the corresponding strain ε_{x-dyn} and the corresponding strain rate $\dot{\varepsilon}_{x-dyn}$ are still calculated using the three-wave method [24]:

$$\sigma_{x-dyn}(t) = \frac{P_{s-x}}{A_{s-x}} = \frac{EA_0}{2A_{s-x}}[\varepsilon_i(t) + \varepsilon_r(t) + \varepsilon_t(t)] \tag{1}$$

$$\varepsilon_{x-dyn}(t) = \frac{C_0}{l_{s-x}}\int_0^t [\varepsilon_i(t) - \varepsilon_r(t) - \varepsilon_t(t)]dt \tag{2}$$

$$\dot{\varepsilon}_{x-dyn}(t) = \frac{d\varepsilon_{x-dyn}(t)}{dt} = \frac{C_0}{l_{s-x}}[\varepsilon_i(t) - \varepsilon_r(t) - \varepsilon_t(t)] \tag{3}$$

where C_0, E and A_0 are the wave velocity, elastic modulus and cross-sectional area of the pressure bar, respectively. P_{s-x}, l_{s-x} and A_{s-x} are the pressure in the x-axis direction and the length and area of the sample along the x-axis, respectively. ε_i, ε_r and ε_t are the incident wave, reflected wave and transmitted wave, respectively.

For the y-axis and z-axis directions, the particle velocity and stress state at both ends of the test specimen can be directly obtained from the strain gauge signals on the square bar. Therefore, the stress, strain and strain rate along the y-axis and z-axis are calculated as follows [29]:

$$\sigma_k(t) = \frac{A_0}{2A_k}E_0|\varepsilon_{k,1}(t) + \varepsilon_{k,2}(t)| \tag{4}$$

$$\varepsilon_k(t) = -\frac{C_0}{L_k}\int_0^t |\varepsilon_{k,1}(t) + \varepsilon_{k,2}(t)|dt \tag{5}$$

$$\dot{\varepsilon}_k(t) = \frac{|\Delta V|}{L_k} = \frac{C_0|\varepsilon_{k,1}(t) + \varepsilon_{k,2}(t)|}{L_k} \tag{6}$$

where k represents the y-axis or z-axis and L_k and A_k are the length and area of the sample in the k-direction (y-axis or z-axis), respectively. Δv is the velocity difference between opposing surfaces in the k-direction. When k represents the y-axis, $\varepsilon_{k,1}$ and $\varepsilon_{k,2}$ are $\varepsilon_{y,left}$ and $\varepsilon_{y,right}$, and when k represents the z-axis, they are $\varepsilon_{z,up}$ and $\varepsilon_{z,down}$. The other parameters are the same as in Equations (1)~(3).

3. Test Results and Analysis

Notably, the application condition of the above equations is that the stress in the x-direction in the specimen should reach the equilibrium state, and the two waves in the y-direction and the two waves in the z-direction both have good consistency, which will be discussed in Section 3.1.

3.1. Test Waveform Analysis

Before the dynamic test, a stress equilibrium check must be performed [30]. Formulas (7)~(9) can be used to check the stress equilibrium state of the test specimen.

$$\sigma_{x-t} = \sigma_{x-i} + \sigma_{x-r} \tag{7}$$

$$\sigma_{y-left} = \sigma_{y-right} \tag{8}$$

$$\sigma_{z-up} = \sigma_{z-down} \tag{9}$$

where σ_{x-t}, σ_{x-i} and σ_{x-r} represent the transmitted stress, incident stress and reflected stress of the test specimen in the x-axis direction, respectively; σ_{y-left} and $\sigma_{y-right}$ represent the dynamic stress on the left and right sides of the test specimen in the y-axis direction, respectively; and σ_{z-up} and σ_{z-down} represent the dynamic stress on the upper and lower sides of the test specimen in the z-axis direction, respectively.

Figure 4 shows a typical stress equilibrium examination performed by U0.5 when the triaxial static pressure is 5 MPa and impact velocity V = 13 m/s. The test specimen reached the stress equilibrium, which confirmed the validity of the dynamic mechanical property test results under true triaxial static loading in the present study.

3.2. Dynamic Stress-Strain Relationship

Table 3 lists the dynamic mechanical properties of UHPC under triaxial compression. The strain rates in the table are the strain rates in the x-axis direction. It can be seen from the table that the dynamic compressive behavior of UHPC with three fiber dosages was very sensitive to the strain rate. Under the same fiber dosage, the peak stress of UHPC in the x-axis direction increased significantly with an increasing strain rate. The increase in confining pressure has less of an effect on the strength in the x-axis direction but has a greater impact on the strength in the y-axis and z-axis directions. The dynamic response of the U0.5 test specimen under triaxial confining pressure was taken as an example for detailed analysis.

Figure 5a shows the typical experimental results of the specimen under triaxial confining pressure. It can be seen from the figure that the dynamic response of the specimen under confining pressure goes through five stages. In the first stage, the stress–strain curve is slightly concave, indicating that the internal voids of the concrete are compacted under loading. However, it should be noted that the dynamic response of this stage will be different at high strain rates, as there is no obvious compression stage at high strain rates [31]. In the second stage, the stress increases linearly with strain. In the third stage, the growth trend of stress slows with increasing strain, probably because the existence of the confining pressure limits the free development of internal cracks in the specimen and increases the energy dissipation rate of concrete. In the fourth stage, the stress reaches the peak value, and with the increase in strain, the stress does not increase, showing an obvious plastic stage. In the fifth stage, with the further increase in strain, the stress decreases suddenly. Compared with the traditional 1D SHPB test, the strain softening behavior in this study was not obvious. Figure 5b shows the corresponding typical strain time-history curve. It can be seen that the strain increased rapidly with the loading time until the peak strain was reached. With continuous loading, the peak strain increased slowly until the specimen failed.

X-direction

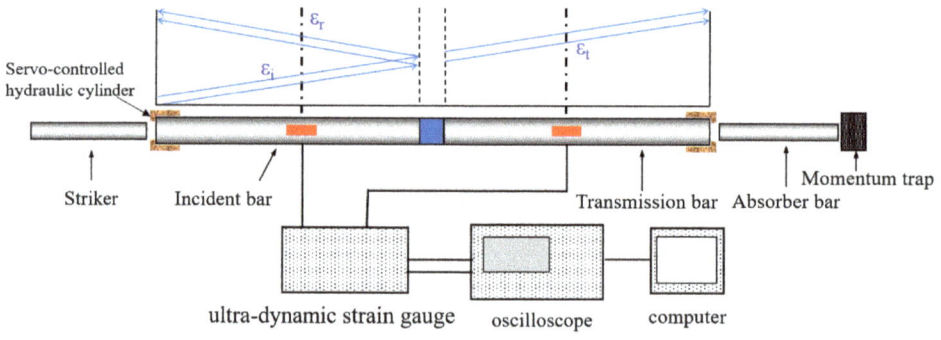

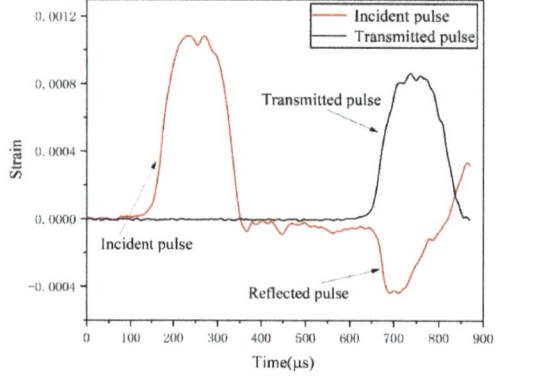
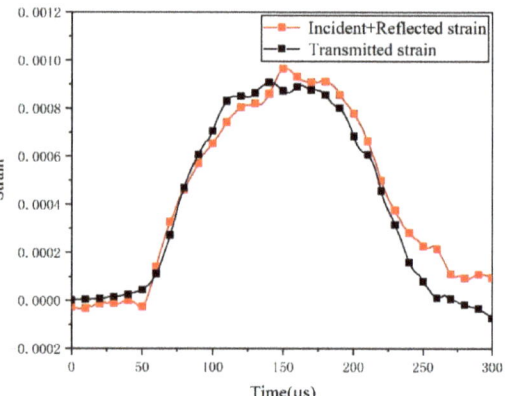

x-direction

Y-direction

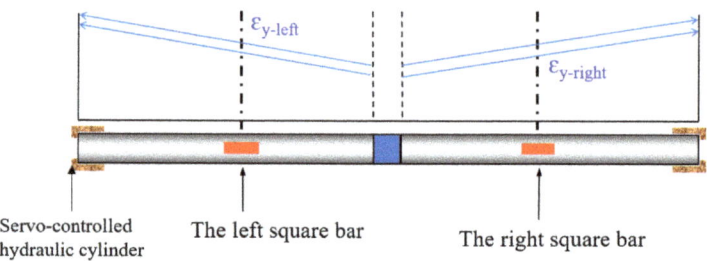

Figure 4. *Cont.*

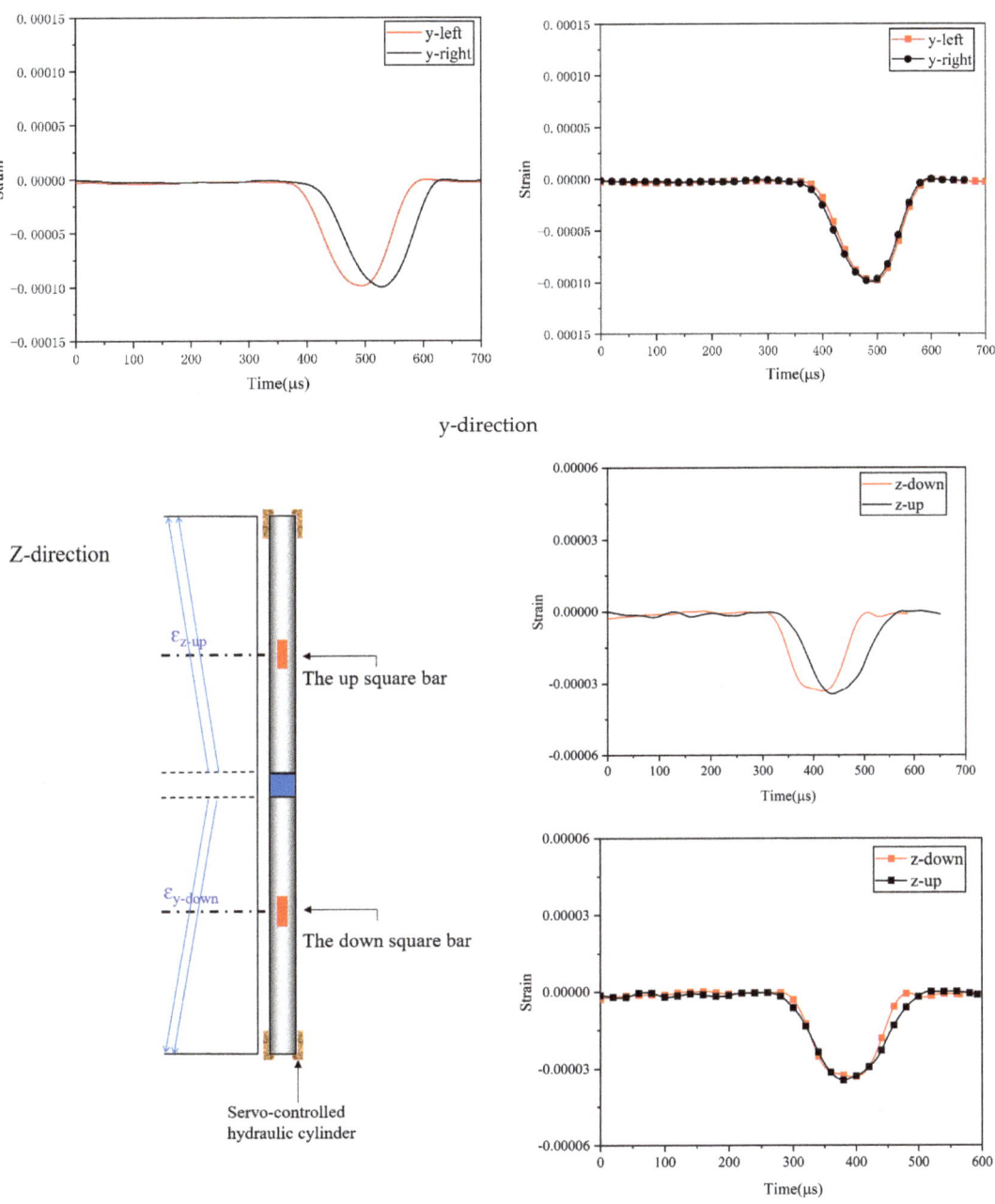

Figure 4. Stress equilibrium.

Table 3. Mechanical properties of UHPC under true triaxial dynamic compression.

Sample No.	Strain Rate (s^{-1})	Confining Pressure (MPa)			Dynamic Pressure (MPa)		
		x	y	z	x	y	z
U0.5	~35	5.0	5.0	5.0	124.8	7.2	5.4
		7.5	7.5	7.5	118.8	9.3	10.0
		10.0	10.0	10.0	119.1	11.0	8.1
	~60	5.0	5.0	5.0	183.6	8.9	8.3
		7.5	7.5	7.5	181.7	11.2	9.4
		10.0	10.0	10.0	183.1	13.9	10.8
	~80	5.0	5.0	5.0	257.2	14.7	15.9
		7.5	7.5	7.5	263.6	17.4	18.1
		10.0	10.0	10.0	264.4	22.5	17.3
U1	~35	5.0	5.0	5.0	138.0	7.8	7.8
		7.5	7.5	7.5	143.2	7.8	9.86
		10.0	10.0	10.0	144.9	8.5	11.8
	~60	5.0	5.0	5.0	187.7	10.2	8.8
		7.5	7.5	7.5	188.4	10.9	10.8
		10.0	10.0	10.0	190.7	11.7	14.0
	~80	5.0	5.0	5.0	273.7	15.8	15.7
		7.5	7.5	7.5	272.3	14.9	17.2
		10.0	10.0	10.0	283.1	16.2	16.1
U2	~35	5.0	5.0	5.0	153.1	7.3	11.1
		7.5	7.5	7.5	151.4	9.8	8.2
		10.0	10.0	10.0	154.1	14.5	8.9
	~60	5.0	5.0	5.0	188.5	11.4	10.6
		7.5	7.5	7.5	192.3	11.8	11.0
		10.0	10.0	10.0	185.6	12.1	8.6
	~80	5.0	5.0	5.0	288.1	16.8	18.5
		7.5	7.5	7.5	288.8	19.4	20.2
		10.0	10.0	10.0	290.0	18.6	18.0

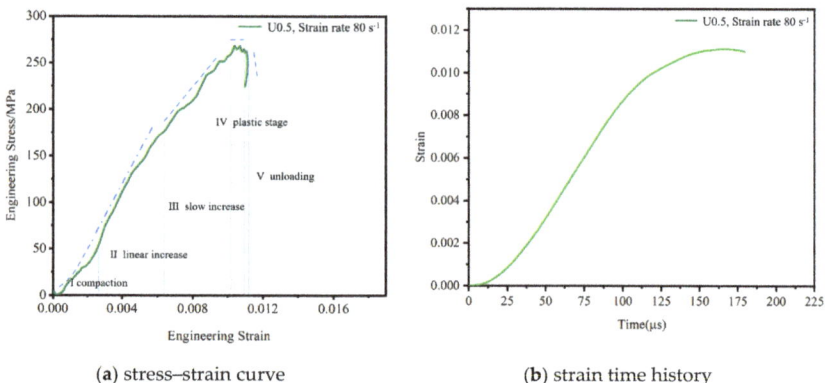

(a) stress–strain curve (b) strain time history

Figure 5. Typical curve in x-direction.

Figure 6 shows the dynamic stress-strain relationship along the x-axis direction under the action of triaxial confining pressure. During the test, the strain rate range was 35~80 s^{-1}, and the static load was in the range of 5~10 MPa. The test results show that under the same static load, the dynamic stress and dynamic strain of the test specimen in the x-axis direction increased with an increasing strain rate; for example, when [$\sigma_{x\text{-static}}$, $\sigma_{y\text{-static}}$, $\sigma_{z\text{-static}}$] = [5, 5, 5], as the strain rate increased from 35 s^{-1} to 80 s^{-1}, the dynamic stress increased from 138 MPa to 264.4 MPa, and the dynamic strain increases from 0.005 to 0.01, showing a significant strain rate effect, which was also consistent with the results of the traditional 1D SHPB test. Under the same strain rate, as the lateral confining pressure increased, the dynamic stress-strain in the x-axis direction increased slightly, but the effect was insignificant.

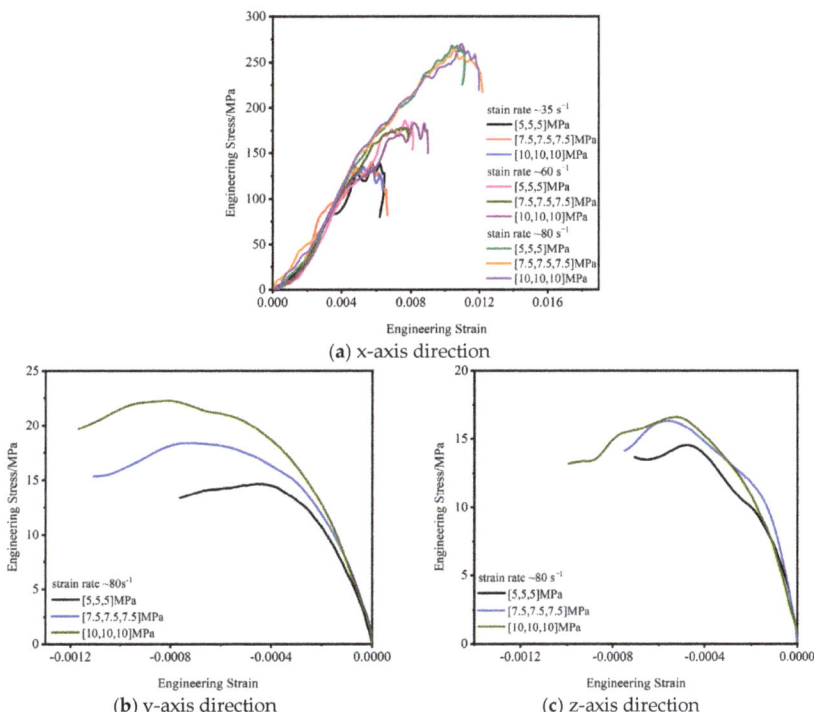

Figure 6. Stress-strain curve.

Figure 6a shows that under the same confining pressure, with the increase in the strain rate the elastic modulus increased slightly, which was consistent with the conclusion of the 1D SHPB test. At the same strain rate, the increase in confining pressure had a small effect on the elastic modulus, which may be due to the limitations of the experimental conditions. The strain rate range and the preset confining pressure range in the study were narrow, resulting in similar test results. Reference [21] suggested that the confining pressure limits the deformation of concrete to improve the energy dissipation rate of concrete under compression, improve the deformation ability of concrete and reduce the deformation modulus. A follow-up study will further analyze and discuss the relationship between the elastic modulus and the size of the preloaded pressure through simulation research.

The impact load was applied to the test specimen along the x-axis. Due to the dynamic Poisson effect, deformation of the test specimen along the y-axis and z-axis was inevitable. Figure 6b,c show the dynamic response of the y-axis and z-axis, respectively, when the impact velocity was 30 m/s; that is, the strain rate was about 80 s^{-1}. With the increase in the static load and the confining pressure, the lateral stress-strain of the test specimen increased. Compared to the dynamic response on the x-axis, the dynamic stress and strain of the y-axis and z-axis are more sensitive to the static load of the confining pressure. For example, when the static load of the confining pressure [$\sigma_{x\text{-static}}$, $\sigma_{y\text{-static}}$, $\sigma_{z\text{-static}}$] = [5, 5, 5] increased to [$\sigma_{x\text{-static}}$, $\sigma_{y\text{-static}}$, $\sigma_{z\text{-static}}$] = [10, 10, 10], the peak stress along the y-axis increased from 15 MPa to 22.5 MPa, an increase of approximately 50%, and the peak strain increased from 0.0004 to 0.0008, an increase of approximately 100%. Since the static load was the same $\sigma_{y\text{-static}} = \sigma_{z\text{-static}}$, the stress-strain relationship of the z-axis was similar to that of the y-axis.

3.3. Relationship between Dynamic Equivalent Stress and Equivalent Strain

The equivalent stress and equivalent strain are [32]

$$\bar{\sigma} = \sqrt{\frac{1}{2}\left[(\sigma_x - \sigma_y)^2 + (\sigma_y - \sigma_z)^2 + (\sigma_z - \sigma_x)^2\right]} \qquad (10)$$

where $\sigma_x = \sigma_{x-dyn} + \sigma_{x-static}$, $\sigma_y = \sigma_{y-dyn} + \sigma_{y-static}$ and $\sigma_z = \sigma_{z-dyn} + \sigma_{z-static}$, and

$$\bar{\varepsilon} = \sqrt{\frac{2}{9}\left[(\varepsilon_x - \varepsilon_y)^2 + (\varepsilon_y - \varepsilon_z)^2 + (\varepsilon_z - \varepsilon_x)^2\right]} \qquad (11)$$

where $\varepsilon_x = \varepsilon_{x-dyn} + \varepsilon_{x-static}$, $\varepsilon_y = \varepsilon_{y-dyn} + \varepsilon_{y-static}$ and $\varepsilon_z = \varepsilon_{z-dyn} + \varepsilon_{z-static}$.

Figure 7 shows the relationship curve of the equivalent stress and effective strain of the U1 test specimen under different strain rates when the hydrostatic pressure was [10, 10, 10]. Similar to the engineering stress-strain relationship, the equivalent stress-strain was roughly divided into five stages, namely, the compaction stage, linear increase stage, slow increase stage, plastic stage and stress unloading stage. As the impact velocity increased from 13 m/s to 30 m/s and the strain rate increased from 30 s^{-1} to 80 s^{-1}, the slope before the peak was almost unchanged, while the equivalent peak strength increased from 134.7 MPa to 266.95 MPa, an increase of 98.2%; additionally, the equivalent peak strain increased from 0.0031 to 0.0085, an increase of 174.2%. This exhibited a clear strain rate effect.

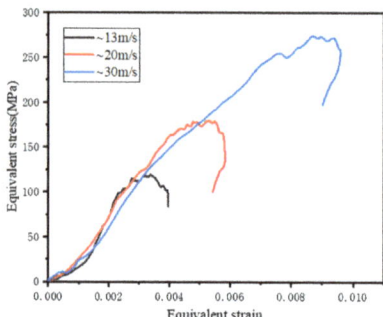

Figure 7. Equivalent stress-strain curves of U1 under different strain rates.

Figure 8 shows the relationship between the equivalent stress and the effective strain of the U1 specimen when the strain rate was 60 s^{-1} and the confining pressure increased from 5 MPa to 10 MPa. The equivalent peak strength increased from 177.5 MPa to 178.2 MPa, an increase of 0.3%; the equivalent peak strain increased from 0.0046 to 0.0048, an increase of 4.3%. This shows that when the strain rate was 30~100 s^{-1}, the effect of the confining pressure on the equivalent stress-strain was insignificant. This was consistent with the study by Xu et al. [26] on the dynamic strength response of normal strength concrete under triaxial confining pressure.

3.4. Dynamic Strength Enhancement Factor

Based on the traditional 1D SHPB test, the dynamic increase factor (DIF) of strength is defined as the ratio of the dynamic strength to the quasistatic strength of the test specimen under uniaxial compression, and it is widely used to analyze the mechanical response of material under dynamic impact loading [22,33]. Notably, the dynamic mechanical properties of traditional concrete only use the dynamic strength in the impact direction, and it is difficult to reflect the influence of static loading. Therefore, the test data in this paper were analyzed by using the equivalent strength enhancement factor (DIFc), which can be expressed as:

$$DIF_C = \bar{\sigma}/\sigma_c \qquad (12)$$

where $\bar{\sigma}$ is the dynamic equivalent strength and σ_c is the unconfined quasistatic compressive strength of concrete. The dynamic equivalent strength factor DIFc of UHPC measured in this experiment under the action of true triaxial confining pressures of 5~10 MPa is plotted in Figure 9. As shown in Figure 9, as the strain rate increased, the DIFc increased rapidly. At the same rate, the DIFc was significantly higher than the DIF value under conventional SHPB compression.

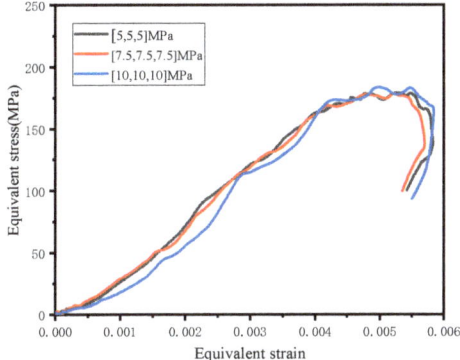

Figure 8. Equivalent stress-strain curve of U1 when the strain rate was 60 s^{-1}.

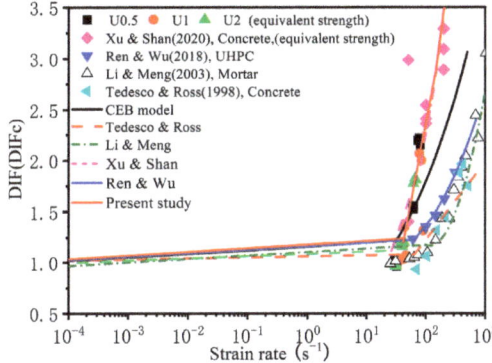

Figure 9. DIF (DIF$_c$) for the compressive strength of UHPC [22,26,33–35].

This indicates that the enhancement of concrete strength under the coupling action of confining pressure and strain rate is greater than the enhancement effect of a single strain rate effect.

For concrete-like materials, several empirical formulas have been proposed based on conventional SHPB tests to estimate the influence of the strain rate effect on compressive strength, and the results are summarized in Table 4. Based on the data studied in this paper, we fitted the DIFc. Due to the lack of test data for UHPC under dynamic loading and confining pressure, it is difficult to accurately determine the transition strain rate in this study, so we refitted the DIFc based on the CEB model. The empirical formula of the DIFc is shown in Table 4, and the CEB model specification is used for the strain rates lower than the transition strain rate.

Table 4. DIF(DIF$_C$) empirical formula.

Refs	DIF(DIF$_c$) Relations
CEB mode [34]	$DIF = \begin{cases} (\dot{\varepsilon}/\dot{\varepsilon}_0)^{0.014} & for\ \dot{\varepsilon} \leq 30\ s^{-1} \\ 0.012(\dot{\varepsilon}/\dot{\varepsilon}_0)^{1/3} & for\ \dot{\varepsilon} \leq 30\ s^{-1} \end{cases}$
Tedesco & Ross [35]	$DIF = \begin{cases} 0.00965 \log \dot{\varepsilon} + 1.058 \geq 1.0 & for\ \dot{\varepsilon} \leq 63.1\ s^{-1} \\ 0.758 \log \dot{\varepsilon} - 0.289 \leq 2.5 & for\ \dot{\varepsilon} > 63.1\ s^{-1} \end{cases}$
Li & Meng [22]	$DIF = \begin{cases} 1 + 0.03438(\log \dot{\varepsilon} + 3) & for\ \dot{\varepsilon} \leq 10^2\ s^{-1} \\ 1.729(\log \dot{\varepsilon})^2 - 7.1372 \log \dot{\varepsilon} + 8.5303 & for\ \dot{\varepsilon} \geq 10^2\ s^{-1} \end{cases}$
Xu & Shan [26]	$DIF_c = \begin{cases} 1 + 0.02192(\log \dot{\varepsilon} + 3.771) & for\ \dot{\varepsilon} \leq 25.7\ s^{-1} \\ 2.147(\log \dot{\varepsilon})^2 - 5.408 \log \dot{\varepsilon} + 4.466 & for\ \dot{\varepsilon} > 25.7\ s^{-1} \end{cases}$
Ren & Wu [33]	$DIF = \begin{cases} (\dot{\varepsilon}/\dot{\varepsilon}_0)^{0.014} & for\ \dot{\varepsilon} \leq 30\ s^{-1} \\ 0.5835 \log (\dot{\varepsilon})^2 - 1.5905 \log \dot{\varepsilon} + 2.1988 & for\ \dot{\varepsilon} > 30\ s^{-1} \end{cases}$
Present study	$DIF_c = \begin{cases} (\dot{\varepsilon}/\dot{\varepsilon}_0)^{0.014} & for\ \dot{\varepsilon} \leq 30\ s^{-1} \\ 1.7883 \log (\dot{\varepsilon})^2 - 3.7413 \log \dot{\varepsilon} + 2.5824 & for\ \dot{\varepsilon} > 30\ s^{-1} \end{cases}$

4. Discussion

4.1. Analysis of the Dynamic Enhancement Mechanism under Confining Pressure

Many experimental and theoretical studies have been conducted on the sensitivity of concrete materials to the loading rate [36,37]. The physical mechanisms of the strain rate effect can be classified into three aspects [21]: the thermal activation mechanism, viscosity mechanism (Stefan effect) and inertia mechanism. Additionally, studies have shown that there is a coupling effect between the enhancement caused by the confining pressure and the strain rate effect. However, there is still no consensus on the mechanism of the concrete strain rate effect when considering the confining pressure.

The thermal activation mechanism causes a thermal vibration effect on an atomic scale. Thermal vibration will break the atomic bonds, resulting in microcracks. The stronger the dynamic load is, the more microcracks that form. When a test specimen is in the low strain rate range, the failure of the concrete is mainly caused by the development of a small number of cracks, and the cracks will develop along the weakest path. With an increasing strain rate, the stress wave cannot be transmitted out of the specimen in a short time, and many microcracks will be generated in the test specimen to dissipate this energy. Therefore, macroscopically, the increase in dynamic strength is small under a low strain rate, and the increase in dynamic strength is large at a high strain rate. When confining pressure exists in the test specimen, the magnitude of the strength increase caused by the strain rate effect is reduced because the confining pressure limits the development of microcracks, which is similar to the conclusion of Liu et al. [21]; when the strain rate increases, the magnitude of the strength increase caused by the strain rate effect is reduced. Due to the change in the energy consumption mode, this weakening effect gradually decreases.

The viscosity mechanism is related to the water content of concrete and can be simply summarized as follows: there is a thin layer of viscous film (free water) between the concrete matrix units on both sides of the microcrack, with a certain distance h between them. When two matrix units are separated by velocity v, an opposing force is generated [38]. The greater the speed, the greater the opposing force. In the compression test performed in this work, the separation and slip of the surface parallel to the inclined crack were considered, as shown in Figure 10. It can be seen from the figure that in the test specimen with added confining pressure, the existence of the confining pressure reduced the slip rate of the concrete matrix elements on both sides of the microcrack, thus weakening the dynamic enhancement of compressive strength.

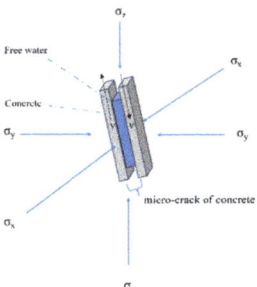

Figure 10. Weakening of the Stefan effect induced by confinement.

Researchers generally believe that the inertia effect mainly controls the dynamic strength of concrete under a high strain rate. When concrete is in the high strain rate range, the macroscopic bearing capacity of a concrete material increases with the increase in strain rate, while the true dynamic strength has a limit value at a very high strain rate. In the high strain rate range, the concrete can still bear a load, but ultimately the concrete will fail after unloading. Therefore, the inertia effect was not considered within the scope of the present study.

In summary, when the strain rate is low, the strength increase caused by the strain rate effect is small, while the strength increase caused by the confining pressure is the main reason for the improvement in concrete strength. With an increasing strain rate, the increase in strength caused by the strain rate effect becomes the main reason for the improvement in concrete strength, as shown in Figure 11. However, the contribution of the confining pressure and strain rate effects to the increase in strength at different stages needs to be further study. In addition, in the dynamic impact process, the confining pressure will have a negative effect on the strength increase caused by the strain rate effect, but with the increase in strain rate, this negative effect will gradually weaken.

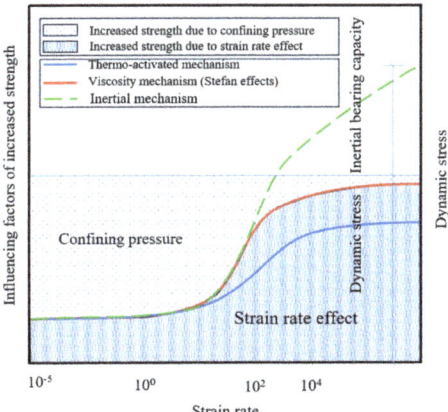

Figure 11. Mechanism of confining pressure and strain rate effects of concrete.

4.2. Failure Criterion

According to the three-parameter failure criterion proposed by Bresler–Pister in an octahedral stress space [39], the failure envelope surface of this failure criterion tends to be a quadratic surface, which is smooth and convex in shape. The model expression is as follows:

$$\tau_0 = a - b\sigma_0 + c\sigma_0^2 \tag{13}$$

where a, b and c are material parameters, which together determine the shape and size of the failure envelope. To be suitable for general situations, normalization is performed by the corresponding uniaxial compressive strength fc to determine the failure surface parameter. By fitting the normalized compression test data, the UHPC failure surface parameters related to the uniaxial compressive strength are obtained. And, τ_0 and σ_0 are calculated as follows:

$$\tau_0 = \frac{\tau_{oct}}{f_c};\ \sigma_0 = \frac{\sigma_{oct}}{f_c} \quad (14)$$

where

$$\tau_{oct} = \frac{\sqrt{(\sigma_x - \sigma_y)^2 + (\sigma_y - \sigma_z)^2 + (\sigma_z - \sigma_x)^2}}{3};\ \sigma_{oct} = \frac{\sigma_x + \sigma_y + \sigma_z}{3} \quad (15)$$

To consider the effect of the strain rate on the failure criterion of concrete under true triaxial dynamic loading, Equation (13) is rewritten as:

$$\tau_0 = a_{\dot{\varepsilon}} - b_{\dot{\varepsilon}} \sigma_0 + c_{\dot{\varepsilon}} \sigma_0^2 \quad (16)$$

where $a_{\dot{\varepsilon}}$, $b_{\dot{\varepsilon}}$ and $c_{\dot{\varepsilon}}$ are the rate effect parameters and τ_0 and σ_0 are calculated according to Equations (14) and (15). Equation (16) was used to perform regression analysis on the test data, and the obtained rate effect parameter values are shown in Table 5. The relationship of the corresponding values $\tau_0 \sim \sigma_0$ is shown in Figure 12.

Table 5. Values of fitting parameters under different strain rates.

Strain Rate/s^{-1}	a	b	c	R^2
35	−0.69	−4.31	−3.82	0.84
60	−0.32	−2.28	−1.12	0.91
80	−1.21	−4.02	−1.76	0.91

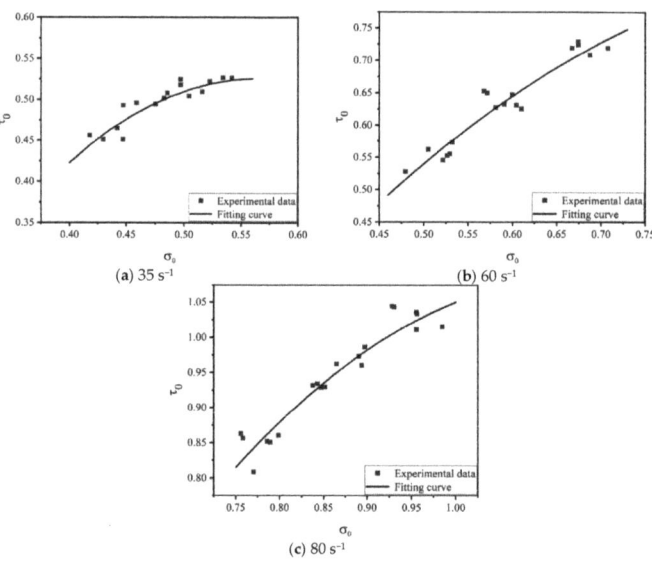

Figure 12. True triaxial dynamic failure criterion in octahedral stress space.

5. Conclusions

The dynamic compression test of UHPC was conducted using a true triaxial SHPB test system setup. The strain rates were in the range of 35~80 s^{-1}, and the confining pressure was 5~10 MPa. Based on the results of the dynamic compression test, the dynamic strength, equivalent stress-strain, DIFc, enhancement mechanism and failure criterion of UHPC were investigated. The main conclusions are as follows:

1. The peak stress, peak strain, equivalent peak stress and equivalent peak strain of UHPC increase obviously with an increasing strain rate in the x-axis loading direction. The confining pressure has little influence on the dynamic response in the x-axis direction, but has a greater influence on the stress and strain in the y-axis and z-axis directions.
2. The equivalent strength enhancement factor DIF_c of UHPC under confining pressure is established and fitted. Under the same strain rate, the equivalent strength, DIF_c, is larger than the DIF obtained from the 1D SHPB test. Based on this, the empirical formula of the DIF_c of UHPC under confining pressure is fitted.
3. There is a coupling effect between the enhancement caused by the confining pressure and the strain rate effect. When the strain rate is low, the extent of the dynamic strength increase caused by the strain rate effect is small, and the strength increase caused by the confining pressure is the main reason for the increase in concrete strength. As the strain rate increases, the weakening effect of the confining pressure gradually weakens, and the strength increase caused by the strain rate effect becomes the main reason for the increase in concrete strength.
4. An improved three-parameter dynamic failure criterion is established and calibrated for this failure criterion.

Author Contributions: Conceptualization, W.Z., B.Z. and L.W.; methodology, W.Z. and X.Y.; validation, J.M., X.Y., B.Z. and L.W.; formal analysis, W.Z., J.M. and L.W.; investigation, B.Z. and L.W.; data curation, J.M., X.Y. and L.W.; writing—original draft preparation, W.Z.; writing—review and editing, W.Z., B.Z. and L.W.; supervision, B.Z. and L.W; funding acquisition, X.Y. and B.Z. All authors have read and agreed to the published version of the manuscript.

Funding: This research was funded by the National Natural Science Foundation of China, grant number (12102476) and the National Key Research and Development Program of China, grant number (2021YFC3100802).

Institutional Review Board Statement: Not applicable.

Informed Consent Statement: Not applicable.

Data Availability Statement: Not applicable.

Conflicts of Interest: The authors declare no conflict of interest.

References

1. Poinard, C.; Malecot, Y.; Daudeville, L. Damage of concrete in a very high stress state: Experimental investigation. *Mater. Struct.* **2010**, *43*, 15–29. [CrossRef]
2. Yan, D.; Lin, G.; Chen, G. Dynamic Properties of Plain Concrete in Triaxial Stress State. *ACI Mater. J.* **2009**, *106*, 89–94.
3. Wang, C.Z.; Guo, Z.H.; Zhang, X.Q. Experimental investigation of biaxial and triaxial compressive concrete strength. *ACI Mater. J.* **1987**, *84*, 92–100.
4. Noori, A.; Shekarchi, M.; Moradian, M.; Moosavi, M. Behavior of steel fiber-reinforced cementitious mortar and high-performance concrete in triaxial loading. *ACI Mater. J.* **2015**, *112*, 95. [CrossRef]
5. Ren, G.M.; Wu, H.; Fang, Q.; Liu, J.Z.; Gong, Z.M. Triaxial compressive behavior of UHPCC and applications in the projectile impact analyses. *Constr. Build. Mater.* **2016**, *113*, 1–14. [CrossRef]
6. Lu, X.; Hsu, C.T.T. Behavior of high strength concrete with and without steel fiber reinforcement in triaxial compression. *Cem. Concr. Res.* **2006**, *36*, 1679–1685. [CrossRef]
7. Chi, Y.; Xu, L.; Mei, G.; Hu, N.; Su, J. A unified failure envelope for hybrid fibre reinforced concrete subjected to true triaxial compression. *Compos. Struct.* **2014**, *109*, 31–40. [CrossRef]
8. Sirijaroonchai, K.; El-Tawil, S. Parra-Montesinos, Behavior of high performance fiber reinforced cement composites under multi-axial compressive loading, Cem. *Concr. Compos.* **2010**, *32*, 62–72. [CrossRef]
9. Jiang, J.; Xiao, P.; Li, B. True-triaxial compressive behaviour of concrete under passive confinement. *Constr. Build. Mater.* **2017**, *156*, 584–598. [CrossRef]
10. Imran, I.; Pantazopoulou, S.J. Experimental study of plain concrete under triaxial stress. *ACI Mater. J.* **1996**, *93*, 589–601.

11. Vu, X.H.; Daudeville, L.; Malecot, Y. Effect of coarse aggregate size and cement paste volume on concrete behavior under high triaxial compression loading. *Constr. Build. Mater.* **2011**, *25*, 3941–3949. [CrossRef]
12. Piotrowska, E.; Malecot, Y.; Ke, Y. Experimental investigation of the effect of coarse aggregate shape and composition on concrete triaxial behavior. *Mech. Mater.* **2014**, *79*, 45–57. [CrossRef]
13. Frew, D.J.; Akers, S.A.; Chen, W.; Green, M.L. Development of a dynamic triaxial Kolsky bar. *Meas. Sci. Technol.* **2010**, *21*, 105704. [CrossRef]
14. Gran, J.K.; Florence, A.L.; Colton, J.D. Dynamic triaxial tests of high-strength concrete. *J. Eng. Mech.* **1989**, *115*, 891–904. [CrossRef]
15. Zeng, S.; Ren, X.; Li, J. Triaxial behavior of concrete subjected to dynamic compression. *J. Struct. Eng.* **2013**, *139*, 1582–1592. [CrossRef]
16. Xue, Z.; Hu, S. Dynamic behavior of cement mortar under active confinement. *Explos. Shock. Waves* **2008**, *6*, 561–564. (In Chinese)
17. Malvern, L.E.; Jenkins, D. *Dynamic Testing of Laterally Confined Concrete*; Deptartment of Information Sciences, California Insti-tute of Technology: Pasadena, CA, USA, 1990.
18. Gary, G.; Bailly, P. Behaviour of quasi-brittle material at high strain rate. Experiment and modelling. *Eur. J. Mech. A Solids* **1998**, *17*, 403–420. [CrossRef]
19. Fujikake, K.; Mori, K.; Uebayashi, K.; Ohno, T.; Mizuncr, J. Dynamic properties of concrete materials with high rates of triaxial compressive loads. *WIT Trans. Built Environ.* **2000**, *48*. [CrossRef]
20. Fu, Q.; Xu, W.; Li, D.; Li, N.; Niu, D.; Zhang, L.; Guo, B.; Zhang, Y. Dynamic compressive behaviour of hybrid basalt-polypropylene fibre-reinforced concrete under confining pressure: Experimental characterisation and strength criterion. *Cem. Concr. Compos.* **2021**, *118*, 103954.
21. Liu, P.; Zhou, X.; Qian, Q. Experimental investigation of rigid confinement effects of radial strain on dynamic mechanical properties and failure modes of concrete. *Int. J. Min. Sci. Technol.* **2021**, *31*, 939–951. [CrossRef]
22. Li, Q.M.; Meng, H. About the dynamic strength enhancement of concrete-like materials in a split Hopkinson pressure bar test. *Int. J. Solids Struct.* **2003**, *40*, 343–360. [CrossRef]
23. Zhang, Q.B.; Zhao, J. A review of dynamic experimental techniques and mechanical behaviour of rock materials. *Rock Mech. Rock Eng.* **2014**, *47*, 1411–1478.
24. Cui, J.; Hao, H.; Shi, Y.; Zhang, X.; Huan, S. Volumetric properties of concrete under true triaxial dynamic compressive loadings. *J. Mater. Civ. Eng.* **2019**, *31*, 04019126. [CrossRef]
25. Liu, P.; Liu, K.; Zhang, Q.B. Experimental characterisation of mechanical behaviour of concrete-like materials under multiaxial confinement and high strain rate. *Constr. Build. Mater.* **2020**, *258*, 119638.
26. Xu, S.; Shan, J.; Zhang, L.; Zhou, L.; Gao, G.; Hu, S.; Wang, P. Dynamic compression behaviors of concrete under true triaxial confinement: An experimental technique. *Mech. Mater.* **2020**, *140*, 103220.
27. Liu, P.; Liu, J.; Bi, J. Experimental and theoretical study of dynamic mechanical behavior of concrete subjected to triaxial confining and impact loads. *J. Build. Eng.* **2023**, *64*, 105715. [CrossRef]
28. Lu, F.; Lin, Y.; Wang, X.; Lu, L.; Chen, R. A theoretical analysis about the influence of interfacial friction in SHPB tests. *Int. J. Impact Eng.* **2015**, *79*, 95–101. [CrossRef]
29. Cadoni, E.; Albertini, C. Modified Hopkinson bar technologies applied to the high strain rate rock tests. In *Advances in Rock Dynamics and Applications*; CRC Press: Boca Raton, FL, USA, 2011; pp. 79–104.
30. Lu, Y.B.; Li, Q.M. Appraisal of pulse-shaping technique in split Hopkinson pressure bar tests for brittle materials. *Int. J. Prot. Struct.* **2010**, *1*, 363–390. [CrossRef]
31. Li, S.H.; Zhu, W.C.; Niu, L.L.; Yu, M.; Chen, C.F. Dynamic characteristics of green sandstone subjected to repetitive impact loading: Phenomena and mechanisms. *Rock Mech. Rock Eng.* **2018**, *51*, 1921–1936.
32. Wang, R.; Zhu, H.X. *Theory of Plasticity*; Science China Press Ltd.: Beijing, China, 1998; pp. 93–101.
33. Ren, G.M.; Wu, H.; Fang, Q.; Liu, J.Z. Effects of steel fiber content and type on dynamic compressive mechanical properties of UHPCC. *Constr. Build. Mater.* **2018**, *164*, 29–43. [CrossRef]
34. Comite Euro-International du Beton. *Fib Model Code for Concrete Structures 2010*; Redwood Books: Trowbridge, UK, 2013.
35. Tedesco, J.W.; Ross, C.A. Strain-rate-dependent constitutive equations for concrete. *ASME J. Press. Vessel Technol.* **1998**, *120*, 398–405. [CrossRef]
36. Qi, C.Z.; Wang, M.Y.; Qian, Q.H. Strain-rate effects on the strength and fragmentation size of rocks. *Int. J. Impact Eng.* **2009**, *36*, 1355–1364. [CrossRef]
37. Qi, C.Z.; Qian, Q.H. Physical mechanism of dependence of material strength on strain rate for rock-like material. *Chin. J. Rock Mech. Eng.* **2003**, *22*, 177–181.
38. Rossi, P. A physical phenomenon which can explain the mechanical behaviour of concrete under high strain rates. *Mater. Struct.* **1991**, *24*, 422–424. [CrossRef]
39. Bresler, B.; Pister, K.S. Strength of concrete under combined stresses. *ACI Struct. J.* **1958**, *55*, 321–345.

Disclaimer/Publisher's Note: The statements, opinions and data contained in all publications are solely those of the individual author(s) and contributor(s) and not of MDPI and/or the editor(s). MDPI and/or the editor(s) disclaim responsibility for any injury to people or property resulting from any ideas, methods, instructions or products referred to in the content.

Article

Experimental and Numerical Study of Non-Explosive Simulated Blast Loading on Reinforced Concrete Slabs

Zhixiang Xiong [1,2], Wei Wang [3], Guocai Yu [1,2,*], Jian Ma [1,2], Weiming Zhang [4] and Linzhi Wu [1,2,*]

1. Key Laboratory of Advanced Ship Materials and Mechanics, Harbin Engineering University, Harbin 150001, China; xiongzx@hrbeu.edu.cn (Z.X.)
2. Department of Engineering Mechanics, College of Aerospace and Civil Engineering, Harbin Engineering University, Harbin 150001, China
3. Key Laboratory of Impact and Safety Engineering, Ningbo University, Ministry of Education, Ningbo 315211, China
4. Center for Composite Materials, Harbin Institute of Technology, Harbin 150080, China
* Correspondence: yuguocai@hrbeu.edu.cn (G.Y.); wulinzhi@hrbeu.edu.cn (L.W.)

Abstract: This study presents a non-explosive method for simulating blast loading on reinforced concrete (RC) slabs. The method involves using a newly developed blast simulator to apply a speedy impact load on the slab, which generates a pressure wave similar to that of an actual blast. Both experimental and numerical simulations were carried out to evaluate the effectiveness of the method. The experimental results showed that the non-explosive method can produce a pressure wave with a peak pressure and duration analogous to those of an actual blast. The numerical simulations also showed good agreement with the experimental results. Additionally, parameter studies were conducted to evaluate the effects of the rubber shape, the impact velocity, the bottom thickness, and the upper thickness on the impact loading. The results indicate that pyramidal rubber is more suitable as an impact cushion for simulating blast loading than planar rubber. The impact velocity has the widest range of regulation for peak pressure and impulse. As the velocity increases from 12.76 to 23.41 m/s, the corresponding range of values for peak pressure is 6.457 to 17.108 MPa, and for impulse, it is 8.573 to 14.151 MPa·ms. The variation in the upper thickness of the pyramidal rubber has a more positive effect on the impact load than the bottom thickness. With the upper thickness increasing from 30 mm to 130 mm, the peak pressure decreased by 59.01%, and the impulse increased by 16.64%. Meanwhile, when the bottom part's thickness increased from 30 mm to 130 mm, the peak pressure decreased by 44.59%, and the impulse increased by 11.01%. The proposed method provides a safe and cost-effective alternative to traditional explosive methods for simulating blast loading on RC slabs.

Keywords: blast simulator; impact loading; reinforced concrete slabs; parameter studies

Citation: Xiong, Z.; Wang, W.; Yu, G.; Ma, J.; Zhang, W.; Wu, L. Experimental and Numerical Study of Non-Explosive Simulated Blast Loading on Reinforced Concrete Slabs. *Materials* **2023**, *16*, 4410. https://doi.org/10.3390/ma16124410

Academic Editor: Miguel Ángel Sanjuán

Received: 31 May 2023
Revised: 13 June 2023
Accepted: 14 June 2023
Published: 15 June 2023

Copyright: © 2023 by the authors. Licensee MDPI, Basel, Switzerland. This article is an open access article distributed under the terms and conditions of the Creative Commons Attribution (CC BY) license (https://creativecommons.org/licenses/by/4.0/).

1. Introduction

Critical infrastructure in sectors, such as energy, communications and government, are highly vulnerable to the threat of improvised explosives, as demonstrated by the ongoing terrorist attacks around the world. To ensure the security and stability of societies, governments are faced with a serious challenge and there is an urgent need for a cost-effective testing tool to support the study of blast effects in the near zone of structures, so that new blast-resistant structures can be developed, validated and deployed more quickly.

Blast testing has long been the primary means of studying close-in blast loading [1–3]. The explosive is capable of exerting a pressure of several tens to hundreds of megapascals on the specimen within a short period of time after detonation, and such test conditions are difficult to replicate by other means. Schenker [4] conducted full-scale blast tests on concrete slabs to obtain dynamic response test data for concrete elements and verified these via numerical calculations. Dharmasena et al. [5] used stainless steel honeycomb sandwich panels and investigated the extent of damage to specimens at 100 mm blast distance and trinitrotoluene (TNT) doses of 1.0, 2.0 and 3.0 kg respectively. However, this research tool is characterized by the difficulty of measuring dynamic mechanical parameters, large scatter, low repeatability and poor visibility of component damage. Therefore, non-explosive methods that are safe, controllable and can be studied in the laboratory, such as the gas gun, the shock tube and blast simulator, has emerged. Since the blast shock wave causes damage to the human brain at pressures lasting a few milliseconds in the approximate range of 68.95 to 690.48 Pa [6], the gas gun is well suited for shock wave testing in this low-pressure range. Bartyczak [7] improved the existing gas gun method and investigated the impact resistance of helmet materials. Schleyer [8] summarized a series of tests with shock tubes on large structures, which proved that the tubes were suitable for simulating the loads generated by far ranges explosions. In order to investigate the effect of concrete strength on the dynamic response of concrete slabs under blast loading, Thiagarajan [9] conducted impact tests on four types of concrete slabs using a shock tube to simulate air blast loading.

To solve the problem of not being able to load full-scale components with gas guns, and to compensate the limitations of shock tubes for near-zero blast studies, the first blast simulation device [10] has been developed (2006) at the University of California, San Diego (UCSD) for the investigation of the blast resistance of different walls, columns and composite structures. Oesterle [11] investigated the impact resistance of concrete masonry walls containing different reinforcements at impulse of 1000–20,000 Pa· s. Gram [12] carried out impact and blast tests on reinforced concrete columns and compared the results of the two tests, which showed good similarity in deformation and failure modes of the specimens. Rodriguezníkl [10] first captured the formation of shear cracks and the spalling of the concrete protective layer when studying reinforced concrete columns under 6800–15,700 Pa· s impulse. This impulse corresponds to the blast load of 560 kg of high explosive on a vehicle at 0.9 m from the ground and 3.5–6.1 m blast distance. Freidenberg [13] carried out impact tests on high-strength prototype walls, all of which simulated blast distances of several meters and charge sizes of several hundred kilograms equivalents, and produced planar shock waves similar to those produced by car bombs. Huson [14] demonstrated the use of water bags to apply loads to composite sandwich members and member joints. Wolfson [15] briefly discuss the dynamic response and damage patterns of honeycomb structures under the action of a proximity blast by firing different types of impact modules. The European Commission Joint Research Centre (JRC) has proposed the development of explosion simulation capabilities similar to those of the United States. The European Laboratory for Structural Assessment (ELSA) has constructed a new testing facility called the electronic blast simulator (e-BLAST) [16–18], which uses servo-controlled drive technology to simulate the loading of structures by air shock waves without the use of explosives. The facility uses three synchronized electric linear motors to accelerate the impact module and load the structure under study. Elastic materials are placed between the impact mass and the structure to maintain consistency with relevant air shock wave pressures. This method will not only describe the effects of explosions on structural systems but also evaluate reinforcement and retrofitting techniques for buildings and bridges to resist terrorist bomb attacks. It will also help investigate the problem of progressive collapse, where local damage spreads disproportionately and leads to overall destruction, as seen in the Oklahoma City bombing.

Currently, there is a lack of research on RC slabs using blast simulators, and the range of impact intensities for simulated blast loading is not sufficient to cover a wide range of blast environments. To address this issue, this paper conducted nine impact tests on reinforced concrete slabs using the newly developed VMLH Blast Simulator. The study also included numerical calculations to investigate the effects of rubber shape, impact velocity, and the bottom and top thicknesses of the pyramidal rubber on the impact loading. Additionally, two equivalence criteria are proposed and used to compare the test results with ideal blast loading. The VMLH Blast Simulator has the ability to safely and precisely apply blast-like loading to components within a controlled laboratory setting, addressing the limitations of traditional explosion testing in terms of data collection, testing duration, and expenses.

2. Experiment Program

2.1. Non-Explosive Blast Simulator

A new impact-based facility has been developed that vertically fires multi-mass impact modules to load large components at precise high velocities (VMLH). The VMLH Blast Simulator is powered by a high-speed gas-hydraulic actuator, as shown in Figure 1. Compared to traditional chemical explosion tests, this equipment has the advantage of easy data acquisition, high reproducibility and reliability, especially without the fireball generated by the explosion, and the ability to use high-speed cameras to capture the deformation–destruction evolution of the component.

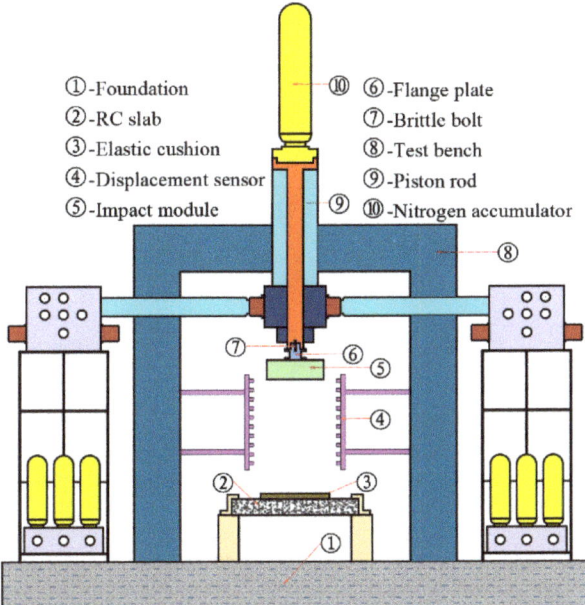

Figure 1. Diagram of the VMLH Blast Simulator.

Before the test, the nitrogen cylinder is inflated and stored. During the test, the high pressure nitrogen is released instantaneously, and the piston rod and impact module connected by the brittle bolt are accelerated together. Then, during the acceleration process, a number of high-speed switching valves in the hydraulic system are opened simultaneously, and the oil in the impact cylinder flows back rapidly; when the impact module reaches the set speed value, the oil pressure rises, producing a buffering effect on the piston rod to slow down. At this point, the brittle bolt breaks, and the impact module, which is separated from the piston rod, starts a short free fall movement until it hits the specimen vertically. The test process is shown in Table 1.

The VMLH Blast Simulator is designed to simulate the effects of an explosion on a component or structure. Similar to other blast simulators [12,17], it achieves this by creating non-elastic collisions between the impact module and the specimen. This is carried out by adding a shock-absorbing layer between them, which transmits the load to a reinforced concrete slab in the form of waves. In this article, silicone rubber is arranged between the impact module and the specimen to adjust the shape of the force during the collision.

When the impact module falls on the rubber, the rubber undergoes elastic deformation, producing lateral waves or shear waves. These waves propagate along the rubber until they reach the edge of the reinforced concrete slab and are reflected back. The rubber also produces longitudinal waves or compression waves, which propagate through the reinforced concrete slab, reach the lower surface of the slab, and are reflected back. These waves cause vibration and deformation of the slab. When the explosion load is applied to the component, a pressure wave called the explosion wave will be generated. This pressure wave will propagate inside the component and the surrounding medium, causing damage to the component and the surrounding environment. Under the action of explosion load, the stress and deformation inside the component will undergo drastic changes, which may lead to the fracture, deformation or collapse of the component.

Table 1. Test procedure of the VMLH Blast Simulator.

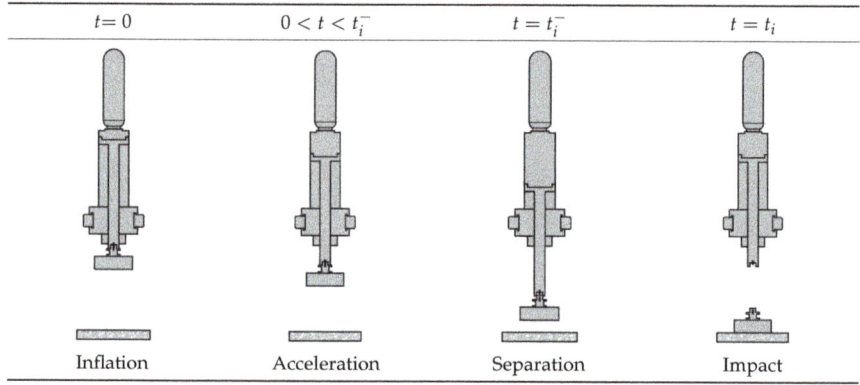

$t=0$	$0 < t < t_i^-$	$t = t_i^-$	$t = t_i$
Inflation	Acceleration	Separation	Impact

2.2. Test Specimen

The dimension of these slabs are 1200 mm length, 900 mm width and 180 mm depth. They are reinforced in both directions with a 10 mm steel bar reinforcement mesh with 100 mm spacing and 25 mm of concrete cover. All the RC slabs were cast from the same batch of commercial concrete with a strength class of C40. Five 150 mm cubes were cast and tested. The average strength of the concrete cubes after 28 days of curing is 45.6 MPa. The type of reinforcement is HRB400E, which has a yield strength of 435 Mpa and a Young's modulus of 209 Gpa. The dimensions of the slab, in addition to the reinforcement detailing and supporting conditions, are shown in Figure 2.

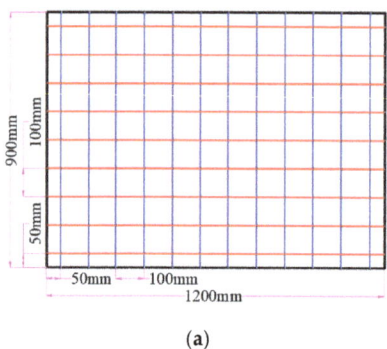
(a)

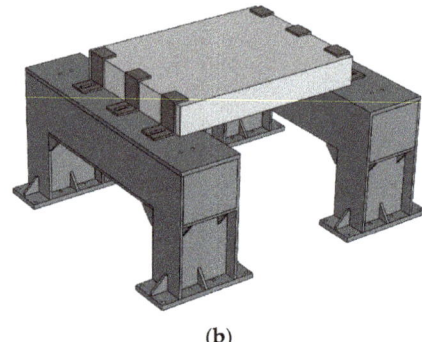
(b)

Figure 2. Preparation of RC slabs. (**a**) Dimensions and reinforcement; and (**b**) supporting structure.

2.3. Impact Cushion

Silicone rubber (referred to as rubber for short) is well-established, inexpensive to prepare, and can be reused in trials. The silicone rubber is placed on the surface of the specimen. On one hand, it prevents direct contact between the impact module and the specimen, which may cause damage to the metal material. On the other hand, the existence of the rubber achieves flexible contact with the specimen, avoiding small angle deviations that may occur during the falling process of the impact module, which may cause uneven loading of the specimen. More importantly, the viscoelastic material properties of the rubber determine the waveform of the load transmitted by the impact. Therefore, rubber was chosen as the impact cushion, and two shapes were designed, as shown in Figure 3. The planar rubber size is 500 mm × 500 mm with a thickness range of 20~100 mm. The pyramidal rubber size is 100 mm × 100 mm, and the height is 50 mm, where the values of bottom thickness h_1 and upper thickness h_2 are 20 mm and 30 mm, respectively.

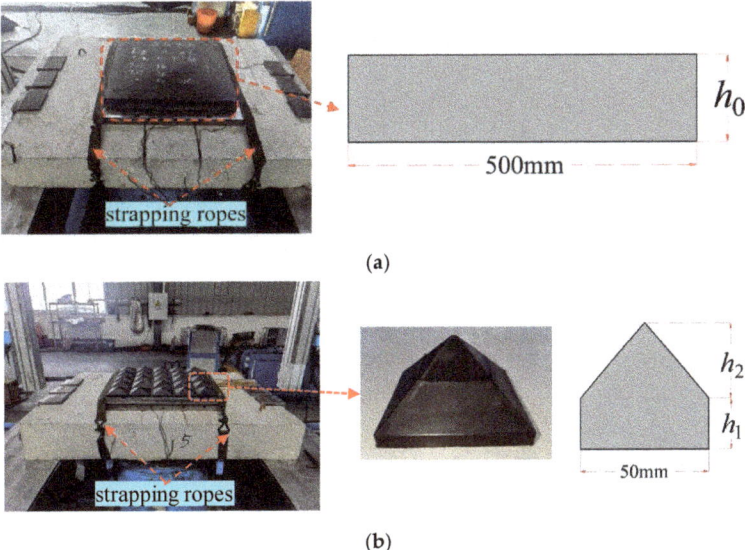

Figure 3. Impact cushion and installation. (**a**) Planar rubber; and (**b**) pyramidal rubber.

2.4. Experimental Setup

A set of displacement sensors was used to measure the impact module speed, as shown in Figure 1. Two accelerometers (SD1407) were attached to the impact module for measuring acceleration, with a range of 5000 g and a sensitivity of 2.2~2.64 pc/g, as shown in Figure 4a. Four load cells (KD3050) were used to measure the impact forces, with a range of 5000 g and a sensitivity of 19.6~19.88 mv/kN, as shown in Figure 4b. These were placed in the holes of the mounting platform, followed by the cover plate and pre-tightened with a screw. To prevent the mounting platform from oscillating significantly during the test, it was secured using two strapping ropes, as shown in Figure 3.

The test data were acquired using the super dynamic signal test and analysis system (DH5960), and the PCO high-speed camera was used to record the impact module from the acceleration to impact with the specimen mass, as shown in Figure 4c. The data acquisition system was set up with six channels, two of which were accelerometers and four were load cells, and the sampling rate was set to 500 kHz. The high-speed camera was connected to a computer with the Camware4 commercial software installed to manage the recording process, as shown in Figure 4d. The high-speed camera used in this test has a frame rate of up to 4500 fps and a resolution of 680 × 1200 pixels.

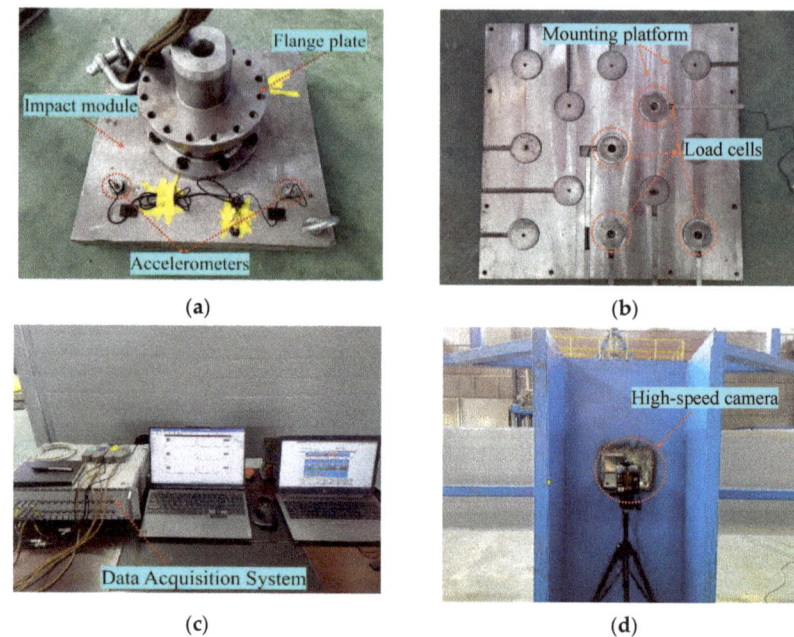

Figure 4. Composition of the measurement system. (**a**) Accelerometer sensors; (**b**) Impact force sensors; (**c**) Data acquisition system; and (**d**) High-speed cameras.

3. Experimental Results and Discussion

The impact tests were carried out using a 100 kg impact module with an adjustable impact speed and impact cushion to apply a range of impact loading to the specimens. A total of nine tests were carried out on four specimens. For tests 1 to 5, planar rubber thicknesses ranging from 20 to 100 mm, and impact velocities of approximately 15 m/s were selected. Tests 6 to 9 used pyramidal rubber in a 5 × 5 arrangement, with impact velocities of 10 to 25 m/s. Tests 10 and 11 were repeat control tests of Test 3. Table 2 summarizes the impact velocity, contact force and loading time for each test and provides the results for the pressure and shock volumes. The calculation of the parameters and the analysis of the test results are presented below.

Table 2. Experimental results (force, time, peak pressure, and impulse).

Tests No.	Specimen No.	h (mm)	v (m/s)	F (kN)	T (ms)	P (MPa)	I (MPa·ms)
1	1	20	15.58	2526.507	6.034	10.106	8.482
2	1	40	15.56	2121.595	7.801	8.486	8.720
3	2	50	15.56	1991.483	8.164	7.966	9.119
4	2	70	15.61	1690.673	9.166	6.763	9.755
5	2	100	15.63	1108.908	12.326	4.436	10.452
6	3	50	12.76	1611.070	6.848	6.457	8.573
7	3	50	15.61	2072.210	5.594	8.766	10.402
8	3	50	19.49	2894.249	5.481	11.577	11.915
9	4	50	23.41	4277.063	5.322	17.108	14.151
10	5	50	15.53	2103.832	7.862	8.415	9.779
11	5	50	15.65	2064.585	8.258	8.258	9.224

Notes: h—Planar rubber thickness; v—Velocity; F—Force; T—Time; P—Peak pressure; I—Impulse.

Based on the assumption of uniform loading, we do not want the specimen to experience significant deformation and damage, as this would not be conducive to the repeatability of our measurement results. As you can see, this is reflected in the test speed, which is controlled below 25 m/s, far below the equipment's speed limit, making the specimen less prone to damage. Figure 5 shows the damage to the specimen before and after a single impact, with several cracks appearing but with no significant deformation occurring. Due to the distance of the specimen's bottom from the buffer bar being only about 100 mm, it was not possible to effectively place the displacement sensors to obtain the mid-span deflection value of the specimen.

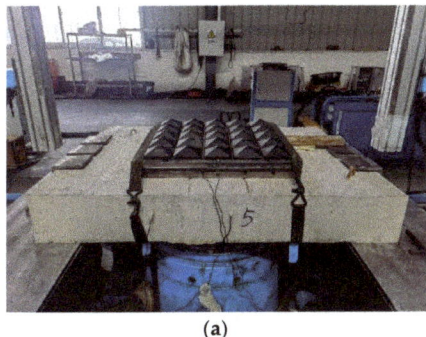
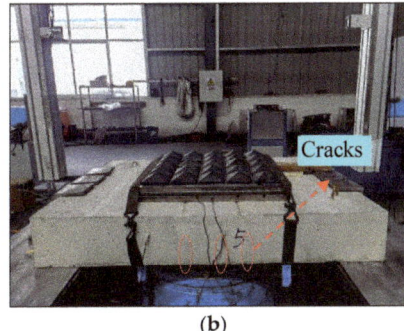

(a) (b)

Figure 5. Damage condition of the specimen after a single impact. (**a**) Before the test; and (**b**) After the test.

3.1. Methodologies for Calculated Parameters

3.1.1. Load

The accelerometers and load cells were strategically placed to capture test data, which was then used to calculate pressure and impulse. Assuming that the VMLH Blast Simulator applies a uniform load to the surface of the specimen, the accelerometer data was converted to force by multiplying it with the weight of the impact module, and to pressure by dividing it by the impact area. Similarly, load cell data was converted directly to pressure by dividing it by the impact area. For instance, the pressure and impulse data obtained from Test 8 are illustrated in Figure 6. The impact loads obtained from both methods were found to be very similar, which is why the load cell data was primarily used for subsequent analysis.

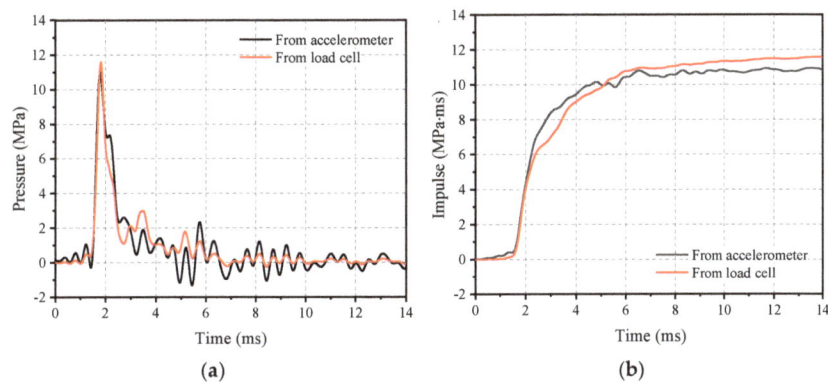

(a) (b)

Figure 6. Impact loading measured using accelerometers and load cells. (**a**) Pressure profile; and (**b**) Impulse profile.

3.1.2. Velocity

A high-speed camera was used to record the impact test procedure. Figure 7 illustrates the four distinct stages of compression, acceleration, separation, and impact. Following the separation from the brittle bolt, the impacting module is still 0.9 m away from the specimen surface and begins to fall freely. As a result, the velocity of the impact module at the moment of impact with the specimen can be determined by calculating the final velocity of the displacement sensor.

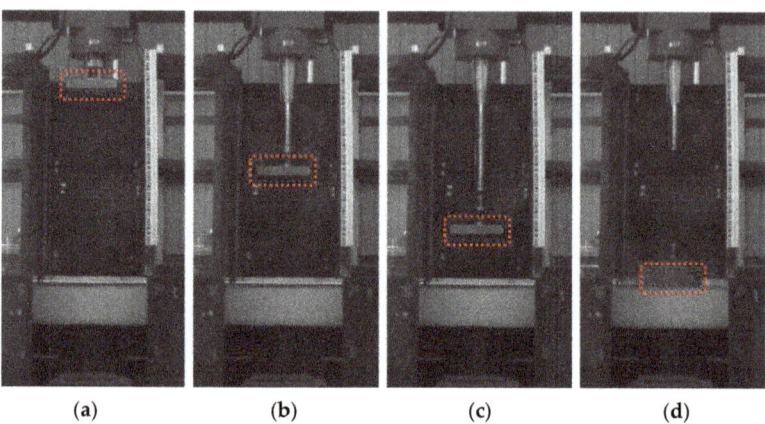

Figure 7. High-speed video of impact test. (**a**) Inflation; (**b**) Acceleration; (**c**) Separation; and (**d**) Impact.

3.2. Comparison of Impact Loading and Blast Loading

A typical blast scenario is shown in Figure 8, which includes a spherical charge of TNT weight, W, at a standoff distance, R, away from a structure [19,20]. The detonation of the explosive creates a shock wave that forms a reflected wave when it reaches the surface of the structure. Under these conditions, an example of a typical reflected pressure profile at a point on the structure is also shown in Figure 8, where P_r is the peak reflected overpressure, and T_p is the positive phase duration. The area under the pressure–time history is the specific impulse (hereafter simply referred to as impulse). As the value of the negative pressure is much smaller than the positive pressure [21], this study will only focus on the positive phase of the impulse, I_r. Various methods have been used to evaluate the true values of P_r, T_p and I_r [22].

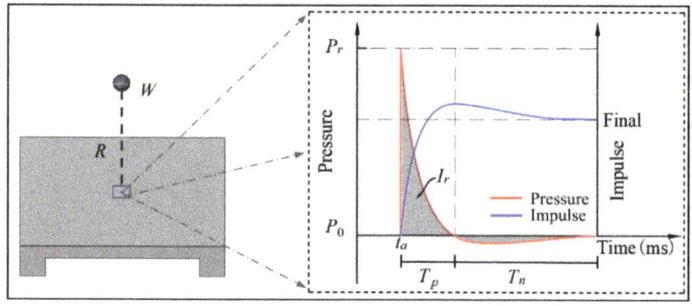

Figure 8. Blast scenario with representative pressure profile.

Close-in charges, such as roadside car bombs, last between 2 and 4 ms and have an impulse maximum of about 11 Mpa·ms to 15 Mpa·ms [19,23,24]. As long as the characteristic response time of the specimen is greater than four times the duration of the impulse, the impulse will dominate the response of the specimen, regardless of the exact shape of the pressure–time history [25]. Civil structures, including individual elements, such as beams and slabs, meet this condition. Therefore, this paper discusses two equivalence criteria for simulating blast loading, one that considers only the impulse force without regard to the exact shape of the pressure curve, and the other that considers both the impulse and the pressure–time curves. An example of equivalent conversions for Test 4 is given in the Table 3, where the parameters of blast environment have been obtained using graphical methods in TM5-1300 [24]. The data in the TM5-1300 manual is based on real test data and empirical formulas, and has been verified and applied multiple times, and widely cited in a series of studies [26–29]. In Figure 9, the pressure–time history of Test 6 is compared to the corresponding ideal blast profile, which is calculated using ConWep [30]. The ConWep algorithm is an empirical formula for calculating explosive loads. By inputting parameters, such as the type, mass, initiation method, distance, and height of the explosive, various aspects of the explosive load can be calculated [31]. The pressure–impulse criterion is used to evaluate the blast loading, and it is found that the blast loading closely matches the pressure and impulse of the impact loading. This comparison shows that both equivalence criteria are suitable for simulating blast loading. However, when using the pressure–impulse criterion for conversion, the resulting impact is equivalent to a blast condition; when using the impulse criterion, the results are not unique and the charge must be assumed before the corresponding blast parameters can be calculated.

Table 3. Equivalent conversion of impact loading and blast loading.

Q	Impact Environment			Explosive Environment				
	v m/s	P MPa	I MPa·ms	W kg	R m	Z m/kg$^{1/3}$	P MPa	I MPa·ms
①	12.76	6.457	8.869	2690.679	12.728	0.915	6.457	8.869
②	12.76	6.457	8.869	216	2.964	0.494	31.206	8.869

Note: Q—Criteria for equivalence; v—Velocity; T—Time; P—Peak pressure; I—Impulse; W—Spherical charge of weight; R—Standoff distance; Z—Scaled distance; ①—Pressure–impulse criterion; ②—Impulse criterion.

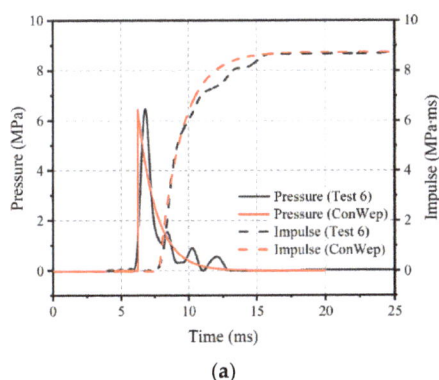
(a)

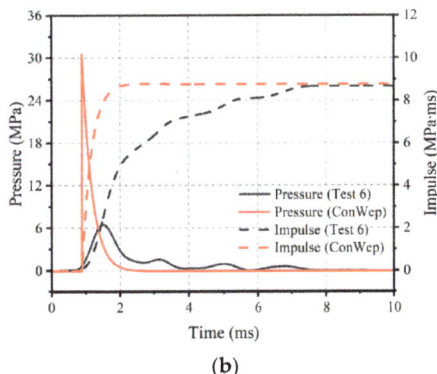
(b)

Figure 9. Comparison of impact loading and ideal blast loading. (a) Pressure–impulse criterion; (b) Impulse criterion.

3.3. Analysis and Discussion

The results of tests 1 to 5 are shown in Figure 10. As the thickness of the planar rubber sheet increases, the peak pressure decreases and the impulse also tends to increase gradually. However, there is a significant difference in the curve profile between the impact loading and the ideal blast loading. For example, at a cushion thickness of 50 mm, the impact loading first rises rapidly, then falls rapidly to zero, then rises again to around 2 MPa and finally falls slowly. This phenomenon is due to the oscillations of the rubber. The phenomenon of secondary peaks may be related to the compression of the rubber. After the initial contact, the rubber is compressed and removed from the specimen surface, at which point the contact force is almost zero; as the rubber reaches the densification stage, the load increases again and the curve becomes smoother. Therefore, the shock loads generated under the above conditions cannot be converted to blast loading using the pressure–impulse criterion, thus, the impulse criterion should be used.

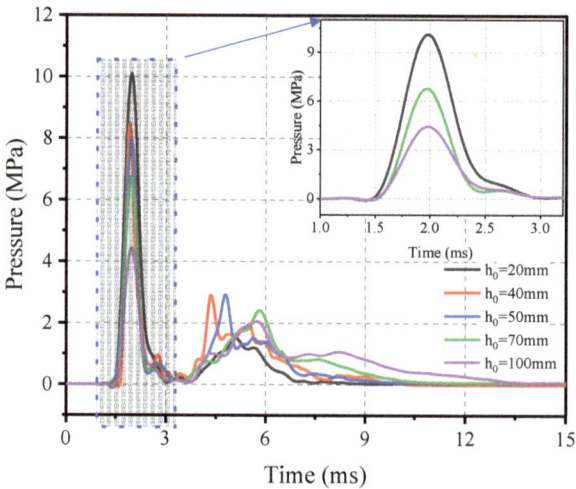

Figure 10. Pressure–time history versus thickness.

The results of tests 6 to 9 are shown in Figure 11. The shape of the pressure profile is characterized by a steep increase in pressure followed by a rapid decay for a duration of approximately 3 to 5 s, which is normal for the equivalent blast loading. From the peak trend, it is evident that pressure and impulse increase as the shock velocity increases. The impact loading is smoother due to the pyramidal shape of the rubber. Compared to the planar rubber case, the pressure tends to fall more gently after reaching its peak, although the difference is not significant initially. This makes it possible to apply both equivalence criteria to simulate blast loading when pyramidal rubber is used as the impact cushion.

Table 4 presents the impact force, loading time, pressure, and impulse for three repeated tests, along with the average and variance of the data. In addition, Figure 12 shows the pressure and impulse time history curves of the three sets of repeated experiments. The results indicate that the equipment has high loading accuracy and the data collection reliability of the test is also high, meeting the requirements of load repeatability for mechanical impact simulation explosion tests.

Table 4. Experimental results of three repeated tests.

Tests No.	F (kN)	T (ms)	P (MPa)	I (MPa·ms)
3	1991.483	8.164	7.966	9.119
10	2103.832	7.862	8.415	9.779
11	2064.585	8.258	8.258	9.224
Average	2053.3	8.095	8.213	9.374
Variance	2167.392	0.029	0.035	0.084

Notes: F—Force; T—Time; P—Peak pressure; I—Impulse.

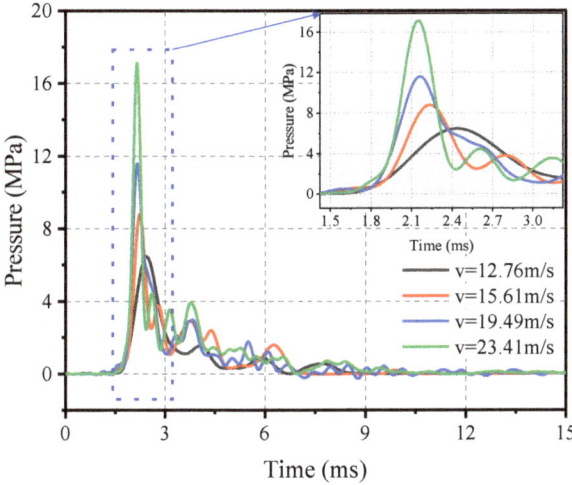

Figure 11. Pressure–time history versus impact velocity.

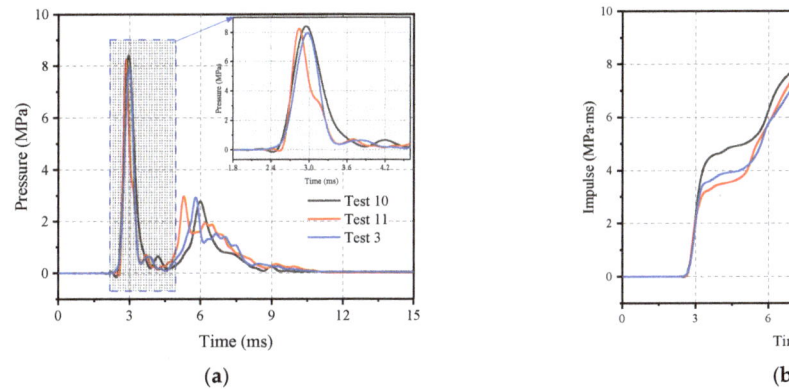

Figure 12. Comparison of impact loading of three repeated tests (**a**) Pressure–time curve; and (**b**) Impulse–time curve.

4. Numerical Simulations

In this study, we utilized the non-linear dynamic analysis software LS-DYAN to simulate the impact loading caused by RC plates when subjected to a blast simulator. Through a comparison of the numerical simulations and experimental test data, we were able to verify the accuracy of the numerical model and the reliability of the test method. Additionally, we investigated the effect of rubber shape, the impact velocity, the bottom thickness, and the upper thickness on the impact loading.

4.1. Material Models
4.1.1. Concrete

The CSCM CONCRETE (MAT_159) material model, which is available in LS-DYNA, is used to simulate the dynamic performance of reinforced concrete protection structures during vehicle collisions [32]. This material model was developed by the Federal Highway Administration and its parameters are defined based on the results of cubic compression tests. Table 5 shows the parameters of this material model that are used in the present study to model concrete. It is important to note that this material model is specifically designed to simulate the behavior of roadside reinforced concrete protection structures and has been validated for this purpose.

4.1.2. Steel

The steel of the slabs in the present study is modeled using the material model Plastic Kinematic (MAT_003) in Ls-Dyna [33], which is an elastic-plastic model with kinematic and isotropic hardening. Reports of material property tests provided by steel producers are used in the numerical simulation of test cases. The expression for the dynamic yield strength of steel, taking into account the effect of strain rate on the intrinsic structure relationship of the material, is as follows:

$$\sigma_y = \left[1 + (\dot{\varepsilon}/C)^{1/P}\right]\left(\sigma_0 + \beta E_P \varepsilon_P^{eff}\right) \quad (1)$$

where, σ_y is the dynamic yield strength of the steel, $\dot{\varepsilon}$ is the strain rate, C and P are the parameters of the strain rate, σ_0 is the initial yield strength of the steel, β is the hardening parameter, E_P is the hardening modulus, and ε_P^{eff} is the effective plastic strain. The input material parameters of steel in the current study are tabulated in Table 5. The Plastic Kinematic material model in Ls-Dyna is capable of accurately capturing the complex material behavior of steel under such extreme loading conditions.

4.1.3. Rubber

Blatz–Ko rubber is a combination of Blatz and Ko [34] defined by a hyper-elastic rubber model using type II Piola–Kirchoff stresses. The Blatz–Ko strain energy density function is a powerful tool for modeling compressible types of rubber, and it can be expressed in a precise mathematical form.

$$W = \frac{1}{2}G\left(\frac{I_2}{I_3} + 2\sqrt{I_3} - 5\right) \quad (2)$$

where, G is the shear modulus at infinitesimal deformation, E is the Young's modulus of elasticity and υ is the Poisson's ratio. $I(n=1,2,3)$ is the invariant of the Cauchy–Green deformation tensor. Equation (2) contains only one material constant, G. The material parameters are shown in Table 5.

Table 5. Input parameters for concrete, steel and rubber material models.

Material	Parameter	Value	Comments
Concrete	RO (Density)	2400 kg/m^3	Material test data
	FPC (Uniaxial compression strength)	45.6 MPa	
	NPLOT	1	According to [32,35,36]
	INCRE	0	
	IRATE (Rate effects options)	1	
	Elements erode	1.1	
	RECOV	0	
	IRETRC (Cap retraction option)	0	
	Pre-existing damage	0	
	DAGG (Maximum aggregate size)	24 mm	
	UNITS (Units options)	4	
Steel	Density	7800 kg/m^3	Material test data
	Young's modulus	2.09 × 10^5 MPa	
	Poisson's ratio	0.3	
	Yield stress	435.3 or 450.1 MPa	
Rubber	Density	1.27 kg/m^3	According to [34,37]
	Poisson's ratio	0.463	
	Shear modulus	24 MPa	

4.2. Model Calibration and Validation

4.2.1. Numerical Model

The test results clearly indicate that using flat rubber as an impact cushion leads to a secondary peak in the impact load, which is distinct from blast loading and thus unsuitable for simulating them. Therefore, a parametric analysis of pyramidal rubber was conducted to investigate the impact loading characteristics, taking into account the effects of impact velocity and rubber thickness. Figure 13 illustrates the numerical model used in this study. Initially, the original design was a 500 mm long and 500 mm wide pyramidal rubber during the experiment, which required a large mold for processing. Considering the cost and processing time, it was divided into 25 small pyramidal rubbers measuring 100 mm in length and width, each requiring only a small mold. In the numerical calculations, since the dispersed small pyramidal shapes needed to be considered for contact, we simplified the model accordingly. Based on the comparison between the numerical and experimental results, this simplification was found to be feasible.

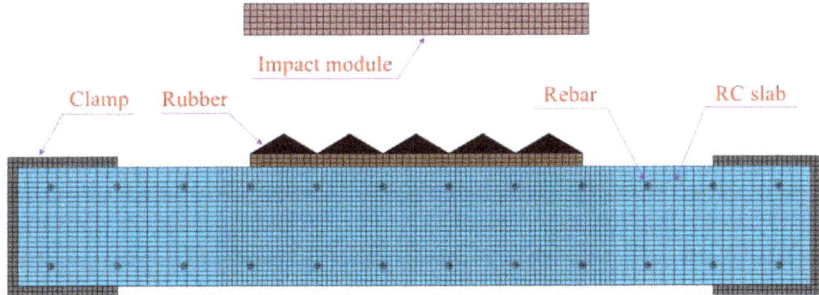

Figure 13. Finite element model.

The model comprises an RC plate, rubber, steel impact module, and fixture. Meshing was performed using 8-node solid hexahedral cells, and mesh convergence analysis was conducted to determine the appropriate cell size. Based on the convergence analysis, a concrete mesh size of 7.5 mm was used, which was doubled outside the range of ±600 mm from the center of the slab to reduce calculation time. The impact module had a cell size of 10 mm in both the side length and thickness direction. The upper half of the rubber was pyramidal and tangentially treated to achieve a mesh size of approximately 10 mm. The grid division details are presented in Figure 14. Further mesh refinement was found to yield similar simulation results, but would significantly increase calculation time. Details of the mesh refinement analysis will be presented in Section 4.2.3.

4.2.2. Boundary Conditions

Reasonable boundary conditions are crucial for obtaining accurate numerical results. In this study, we assume the supporting structure of the RC slab to be a rigid body that is fixed, and thus surface-to-surface contact was used, which uses a penalty function to determine the contact force to define the contact between the specimens and the supporting structure to constrain the specimens. To simplify the calculation and save computing resources, we directly define the impact velocity using *INITIAL_VELOCITY_GENERATION, applying a downward vertical initial velocity to the impact module. Additionally, a surface-to-surface contact was used to define the contact between the impact module, rubber, and concrete.

4.2.3. Grid Refinement Analysis

The numerical model's grid size was determined by conducting five analyses with varying grid resolutions, and the results are presented in Table 6. The grid convergence tests involved five cell sizes, namely 5 mm, 7.5 mm, 10 mm, 20 mm, and 30 mm. The peak pressures calculated for the models with the five grid sizes were found to be very similar, with a maximum error of only 3.41%. This indicates that reducing the grid size has little impact on the numerical results, but it significantly increases the computational time. Therefore, a grid size of 7.5 mm was selected for this study to strike a balance between accuracy and computational efficiency.

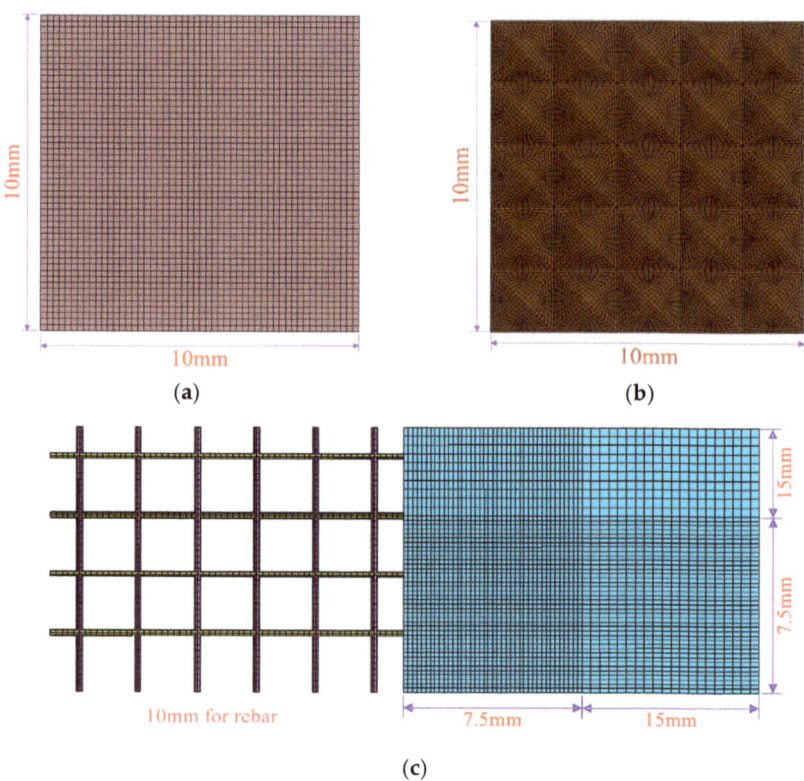

Figure 14. Detail of the grid division. (**a**) Impact module; (**b**) Rubber; (**c**) and RC slab.

Table 6. Mesh sizes and results for mesh refinement analysis.

Unit Size (mm)	5	7.5	10	20	30
Pressure (MPa)	26.420	26.505	26.107	25.934	25.631
Impulse (MPa·ms)	17.058	17.358	17.445	17.322	17.203
Computational time (min)	34	13	7	2	1

4.2.4. Comparison of Experimental and Numerical Results

The numerical model was calibrated by comparing the results of numerical calculations with the experimental test results. In Figure 15a,b, the pressure–time histories of Test 9 and the peak pressure and impulse of Tests 6 to 9 are presented, respectively. Additionally, the percentage differences between the experimental results and the numerical results are presented in Table 7. The comparison results indicate that the pressure curve obtained via numerical simulation is in good agreement with the measurement results.

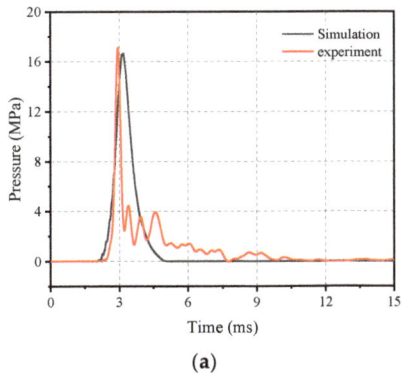
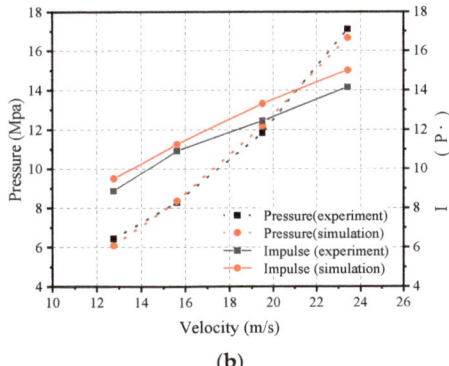

Figure 15. Comparison of numerical and experimental results. (**a**) Pressure profile; and (**b**) Peak pressure and impulse.

Table 7. Comparison of numerical calculations with experimental test.

Tests No.	Experimental Test		Simulation		Deviation	
	Pressure (MPa)	Impulse (MPa·ms)	Pressure (MPa)	Impulse (MPa·ms)	Pressure (%)	Impulse (%)
6	6.457	8.870	6.083	9.489	−5.602%	6.979%
7	8.289	10.912	8.35	11.242	0.736%	3.024%
8	11.577	11.912	12.18	13.307	5.209%	11.711%
9	17.108	14.152	16.646	14.997	−2.700%	5.971%

4.3. Parametric Studies

4.3.1. Effect of Rubber Shapes

The effect of rubber shape on the impact loading was investigated by comparing the results of six rubber shapes that are shown in Figure 16. The impact velocity was 20 m/s and the values of h_1 and h_2 are 20 mm and 30 mm, respectively. Figure 17a shows the impact response of six rubber shapes of rubber as an impact cushion for a reinforced concrete slab with a thickness of 180 mm. As expected, the pressure–time curve is smoothest when the upper side l_a is equal to 0. As l_a gradually increases, the pressure–time curve begins to oscillate and reaches a maximum when l_a is 100 mm. In addition, Figure 15b, highlights the peak pressure and impulse of the impact response for l_a from 0 to 90 mm and the fitted curve, excluding the case where the upper side l_a is 100 mm. The peak pressure and impulse of the upper side between 0 and 90 mm can be calculated according to the following equation.

$$P_{l_a} = 22.124 + 4.409 \times \sin(\pi \times (l_a - 9.254)/9.72), \ 0 \leq l_a \leq 90 \text{ mm} \tag{3}$$

$$I_{l_a} = 22.668 + 2.543 \times \sin(\pi \times (l_a + 0.498)/6.49), \ 0 \leq l_a \leq 90 \text{ mm} \tag{4}$$

4.3.2. Effect of Impact Velocities

To further characterize the effect of velocity on the impact loading characteristics, velocities ranging from 10 to 50 m/s were set, and numerical calculations were carried out. Figure 18 shows the load profile and the relationship between peak pressure and impulse versus velocity. As the speed increases, the peak pressure gradually increases, while the time of the load decreases accordingly, a connection can be established using

Equation (5). The impulse also increases with velocity, unlike the pressure, which increases at a progressively slower rate and can be described using Equation (6).

$$P_v = -21.275 + 17.135 e^{-v/28.489} \tag{5}$$

$$I_v = 565.256 - 13.415 e^{(-v/9.804)} - 555.419 e^{(-v/1972.268)} \tag{6}$$

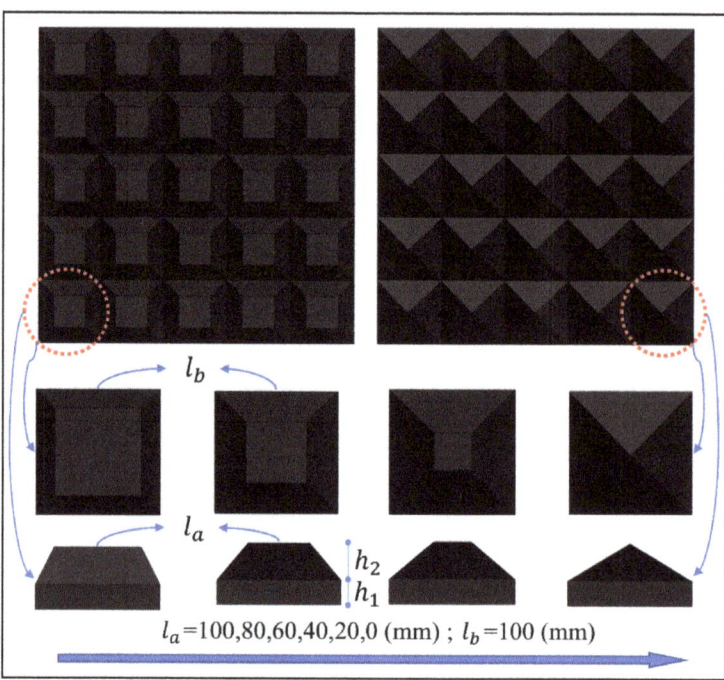

Figure 16. Schematic of rubber shapes variation of the upper side.

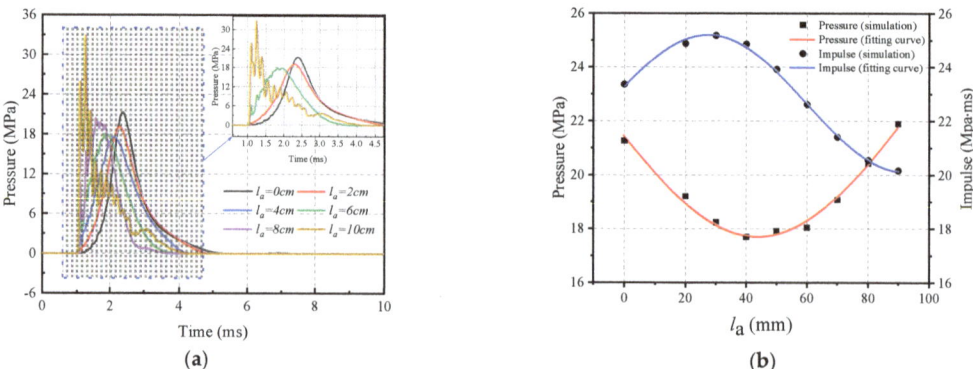

Figure 17. Influence of rubber shapes on the impact loading. (**a**) Pressure–time history; and (**b**) Peak pressure and impulse.

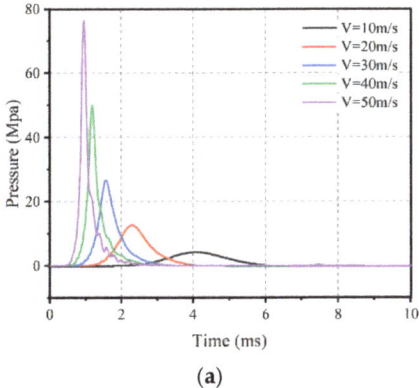 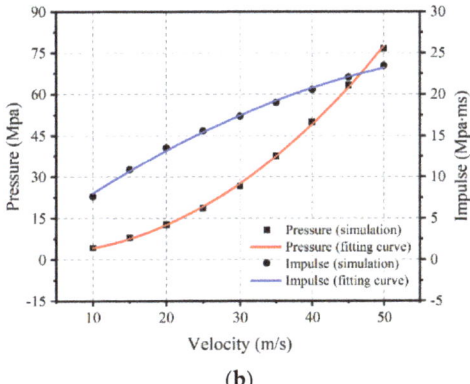

(a) (b)

Figure 18. Influence of impact velocity on the impact loading. (**a**) Pressure–time history; and (**b**) Peak pressure and impulse.

4.3.3. Effect of Bottom Thicknesses

The effect of bottom thickness h_1 on the impact loading was investigated by comparing the results of ten bottom thicknesses. At an impact velocity of 20 m/s, the h_2 value was held constant at 30 mm while the h_1 varied from 20 mm to 170 mm, as illustrated in Figure 19. The peak pressure initially decreases significantly as h_1 increases, and then gradually decreases while impulse increases linearly with the h_1, as shown in Figure 20. To simplify the calculations, the results are fitted with the peak pressure calculated as shown in Equation (7) and with the impulse calculated using Equation (8).

$$P_{h_1} = 5.454 + 10.794 e^{-h_1/4.983}, \ 2 \text{ cm} \leq h_1 \leq 17 \text{ cm} \tag{7}$$

$$I_{h_1} = 61.479 - 48.185 e^{-h_1/310.853}, \ 2 \text{ cm} \leq h_1 \leq 17 \text{ cm} \tag{8}$$

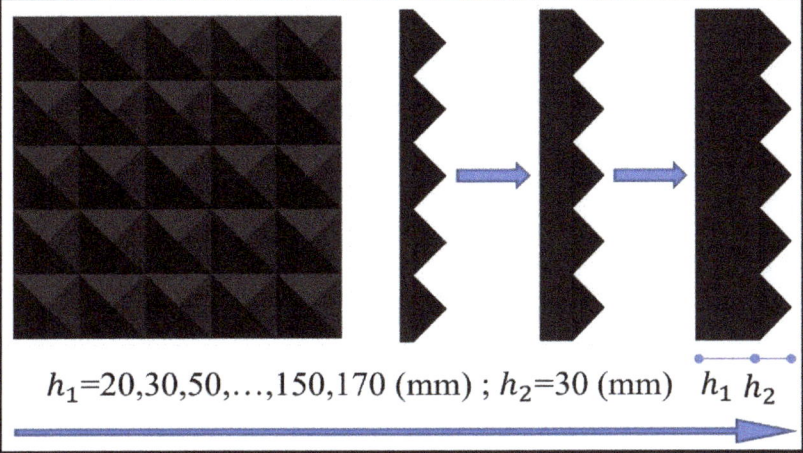

Figure 19. Schematic of the variation of the rubber bottom thickness h_1.

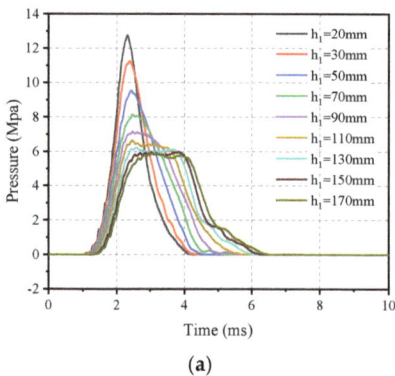 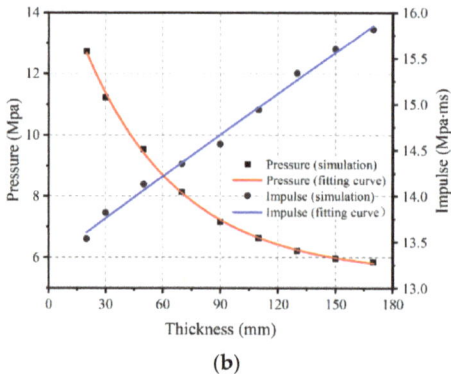

(a)   (b)

Figure 20. Influence of bottom thickness h_1 on the impact loading. (**a**) Pressure–time history; and (**b**) Peak pressure and impulse.

4.3.4. Effect of Upper Thicknesses

Similarly, the effect of upper thickness on the impact loading was investigated by comparing the results of nine upper thicknesses of pyramidal rubber. At an impact velocity of 20 m/s, the bottom thickness h_1 was held constant at 20 mm while the h_2 thickness varied from 10 mm to 130 mm, as detailed in Figure 21. The impact response of nine upper thicknesses of rubber as impact cushion for a reinforced concrete slab with a thickness of 180 mm is shown in Figure 22.

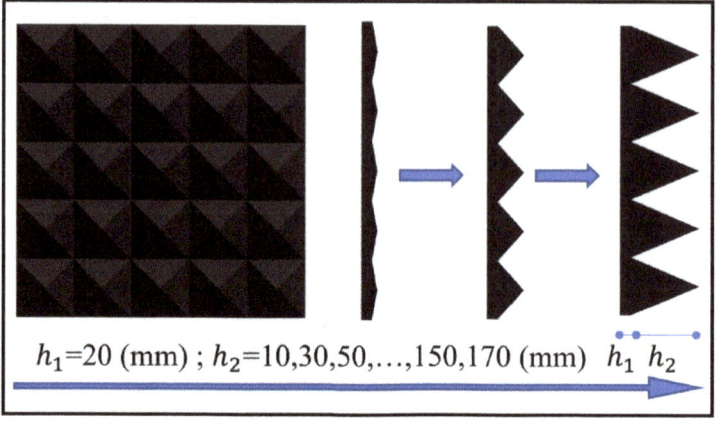

Figure 21. Schematic of the variation of the rubber upper thickness h_2.

It can be seen that the pressure gradually decreases while the time of load and impulse increase accordingly as h_2 gradually increases. It is worth noting that when h_2 is between 70 mm and 130 mm, the impulse is almost unaffected, accompanied by a significant reduction as h_2 continues to increase. To simplify the calculations, the results are fitted with the peak pressure calculated as shown in Equation (9) and with the impulse calculated using Equation (10).

$$P_{h_2} = 2.664 + 45.495 e^{-h_2/1.14} + 9.17 e^{-h_2/10.12}, \ 1 \text{ cm} \leq h_2 \leq 17 \text{ cm} \tag{9}$$

$$I_{h_2} = 7.811 - 2.927 \times h_2 - 0.41 \times h_2{}^2 + 0.026 \times h_2{}^3 - 6.441^{-4} \times h_2{}^4, \ 1 \text{ cm} \leq h_2 \leq 13 \text{ cm} \tag{10}$$

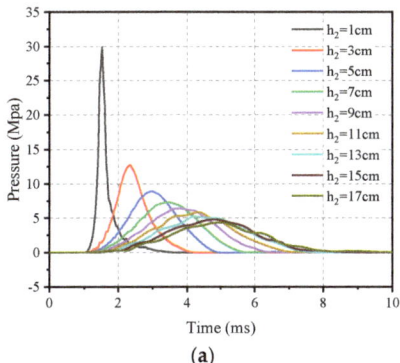 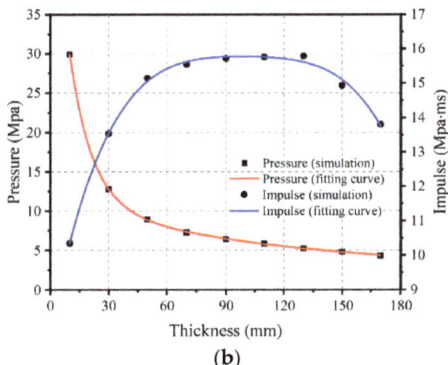

Figure 22. Influence of upper thickness h_2 on the impact loading. (**a**) Pressure–time history; (**b**) Peak pressure and impulse.

5. Summary and Conclusions

In this study, experimental tests were performed on four specimens to demonstrate the feasibility of the VMLH Blast Simulator for simulating blast loading. A numerical model was also developed to predict the impact loading using LS-DYNA. The validity of the model has been calibrated against experimental test results. Using the calibrated model, further studies are carried out to investigate the effect of different parameters on the impact loading of RC plates. The parameters investigated within the scope of this study were rubber shapes, impact velocities and the bottom thicknesses and the top thicknesses of the pyramidal rubber. The following conclusions can be drawn from the detailed experimental and numerical studies presented in this paper.

(1) The use of pyramidal rubber with a 0 mm upper side was more effective in regulating the peak pressure and impulse of impact loading compared to planar rubber with a 100 mm upper side. This was evident in the pressure–time curve, which closely resembled ideal blast loading. However, when l_a was between 40 mm and 100 mm, the pressure profile oscillated significantly, making it unsuitable for the pressure–impulse criterion.

(2) The impact velocity was found to have a significant effect on the pressure and impulse of impact loading. For a pyramidal rubber thickness of 50 mm, both pressure and impulse increased rapidly with increasing velocity. When the speed increases from 12.76 m/s to 23.41 m/s, the corresponding range of peak pressure is from 6.457 to 17.108 MPa, with an increase of 164.22%. The corresponding range of impulse is from 8.573 to 14.151 MPa·ms, with an increase of 65.07%.

(3) Variations in the upper thickness of the pyramidal rubber have a more positive effect on the impact loading than the bottom thickness. Notably, increasing the top thickness from 30 mm to 130 mm resulted in a 59.01% decrease in peak pressure and a 16.64% increase in impulse. Conversely, increasing the bottom thickness from 30 mm to 130 mm resulted in a 44.59% decrease in peak pressure and an 11.01% increase in impulse. As the bottom thickness increases, it takes longer for the rubber to compress and become compressed. When bottom thickness is 110 mm, the pressure reaches its peak and then barely changes in a time of approximately 2 ms, which is unsuitable for using the pressure–impulse criterion for modeling blast loading. Increasing the upper thickness only allows the pressure to rise and fall more smoothly without changing its shape characteristics, making it possible to adopt both criteria for simulating blast loading.

(4) The impact loading and blast loading can be converted using the "pressure-impulse" and "impulse" criterion. By obtaining the peak pressure and impulse of the impact loading, a corresponding explosive environment can always be found. These criteria can be widely applied in simulating blast loading using non-explosive methods.

Author Contributions: Conceptualization, G.Y. and L.W.; methodology, W.W.; investigation, J.M.; resources, G.Y.; data curation, W.Z.; writing—original draft preparation, Z.X.; writing—review and editing, G.Y. and W.W.; supervision, L.W.; project administration, G.Y.; funding acquisition, G.Y. and L.W. All authors have read and agreed to the published version of the manuscript.

Funding: This research received no external funding.

Institutional Review Board Statement: Not applicable.

Informed Consent Statement: Not applicable.

Data Availability Statement: The data presented in this study are available on request from the corresponding author.

Conflicts of Interest: The authors declare no conflict of interest.

References

1. Liao, Q.; Yu, J.; Xie, X.; Ye, J.; Jiang, F. Experimental study of reinforced UHDC-UHPC panels under close-in blast loading. *J. Build. Eng.* **2022**, *46*, 103498. [CrossRef]
2. Wang, L.; Cheng, S.; Liao, Z.; Yin, W.; Liu, K.; Ma, L.; Wang, T.; Zhang, D. Blast Resistance of Reinforced Concrete Slabs Based on Residual Load-Bearing Capacity. *Materials* **2022**, *15*, 6449. [CrossRef] [PubMed]
3. Wang, W.; Yang, G.; Yang, J.; Wang, J.; Wang, X. Experimental and numerical research on reinforced concrete slabs strengthened with POZD coated corrugated steel under contact explosive load. *Int. J. Impact Eng.* **2022**, *166*, 104256. [CrossRef]
4. Schenker, A.; Anteby, I.; Gal, E.; Kivity, Y.; Nizri, E.; Sadot, O.; Michaelis, R.; Levintant, O.; Ben-Dor, G. Full-scale field tests of concrete slabs subjected to blast loads. *Int. J. Impact Eng.* **2008**, *35*, 184–198. [CrossRef]
5. Dharmasena, K.P.; Wadley, H.N.; Xue, Z.; Hutchinson, J.W. Mechanical response of metallic honeycomb sandwich panel structures to high-intensity dynamic loading. *Int. J. Impact Eng.* **2008**, *35*, 1063–1074. [CrossRef]
6. Stapczynski, J.S. Blast injuries. *Ann. Emerg. Med.* **1982**, *11*, 687–694. [CrossRef] [PubMed]
7. Bartyczak, S.; Mock, J. Versatile gas gun target assembly for studying blast wave mitigation in materials. In *AIP Conference Proceedings*; AIP Publishing: Long Island, NY, USA, 2012; pp. 501–504. [CrossRef]
8. Schleyer, G.K.; Lowak, M.J.; Polcyn, M.A.; Langdon, G.S. Experimental investigation of blast wall panels under shock pressure loading. *Int. J. Impact Eng.* **2007**, *34*, 1095–1118. [CrossRef]
9. Thiagarajan, G.; Kadambi, A.V.; Robert, S.; Johnson, C.F. Experimental and finite element analysis of doubly reinforced concrete slabs subjected to blast loads. *Int. J. Impact Eng.* **2015**, *75*, 162–173. [CrossRef]
10. Rodrígueznikl, T. Experimental Simulations of Explosive Loading on Structural Components: Reinforced Concrete Columns with Advanced Composite Jackets. Ph.D. Thesis, University of California, San Diego, CA, USA, 2006.
11. Oesterle, M.G. *Blast Simulator Wall Tests: Experimental Methods and Mitigation Strategies for Reinforced Concrete and Concrete Masonry*; University of California, San Diego: La Jolla, CA, USA, 2009.
12. Gram, M.M.; Clark, A.J.; Hegemier, G.A.; Seible, F. Laboratory simulation of blast loading on building and bridge structures. In *Structures Under Shock and Impact Ix*; WIT Press: Billerica, MA, USA, 2006; Volume 87, pp. 33–44. [CrossRef]
13. Freidenberg, A. Advancements in Blast Simulator Analysis Demonstrated on a Prototype Wall Structure. Ph.D. Thesis, University of California, San Diego, CA, USA, 2013.
14. Huson, P. Experimental and Numerical Simulations of Explosive Loading on Structural Components: Composite Sandwich Connections. Ph.D. Thesis, University of California, San Diego, CA, USA, 2012.
15. Wolfson, J.C. Blast Damage Mitigation of Steel Structures from Near-Contact Charges. Ph.D. Thesis, University of California, San Diego, CA, USA, 2008.
16. Paul, S.C.; Lv, P.; Xiao, Y.-J.; An, P.; Liu, S.-Q.; Luo, H.-S. Thalidomide in rat liver cirrhosis: Blockade of tumor necrosis factor-alpha via inhibition of degradation of an inhibitor of nuclear factor-kappaB. *Pathobiology* **2006**, *73*, 82–92. [CrossRef] [PubMed]
17. Peroni, M.; Solomos, G.; Caverzan, A.; Larcher, M.; Valsamos, G. Assessment of dynamic mechanical behaviour of reinforced concrete beams using a blast simulator. *EPJ Web Conf.* **2015**, *94*, 1010. [CrossRef]
18. Marco, P.; George, S.; Pierre, P.; Alessio, C. *Electrical Blast Simulator (e-BLAST): Design, Development and First Operational Tests*; European Union: Brussels, Belgium, 2015.
19. Hetherington, J.; Smith, P. *Blast and Ballistic Loading of Structures*; CRC Press: Boca Raton, FL, USA, 2014; ISBN 9780429077982.
20. Krauthammer, T. *Modern Protective Structures*; CRC Press: Boca Raton, FL, USA, 2008.
21. Tekalur, S.A.; Bogdanovich, A.E.; Shukla, A. Shock loading response of sandwich panels with 3-D woven E-glass composite skins and stitched foam core. *Compos. Sci. Technol.* **2009**, *69*, 736–753. [CrossRef]
22. Kinney, G.F.; Graham, K.J. *Explosive Shocks in Air*, 2nd ed.; Springer: Berlin/Heidelberg, Germany, 1985; ISBN 9783642866845.
23. Eswaran, M.; Parulekar, Y.M.; Reddy, G.R. Introduction to Structural Dynamics and Vibration of Single-Degree-of-Freedom Systems. In *Textbook of Seismic Design: Structures, Piping Systems, and Components*; Reddy, G.R., Muruva, H.P., Verma, A.K., Eds.; Springer Singapore: Singapore, 2019; pp. 61–93, ISBN 978-981-13-3175-6.

24. U.S. Departments of the Army, the Navy and the Air Force. *Structures to Resist the Effects of Accidental Explosions (TM 5-1300/NAVFAC P-397/AFR 88-22, Revision 1)*; Departments of the Army, Air Force, and Navy and the Defense Special Weapons Agency: Washington, DC, USA, 1990.
25. Chopra, A. *Dynamics of Structures*; Pearson Education: Upper Saddle River, NJ, USA, 2005.
26. Zhou, X.Q.; Kuznetsov, V.A.; Hao, H.; Waschl, J. Numerical prediction of concrete slab response to blast loading. *Int. J. Impact Eng.* **2008**, *35*, 1186–1200. [CrossRef]
27. Rasouli, A.; Toopchi-Nezhad, H. The influence of confined water on blast response of reinforced concrete slabs: Experimental investigation. *J. Build. Eng.* **2020**, *30*, 101285. [CrossRef]
28. Wu, J.; Zhou, Y.; Zhang, R.; Liu, C.; Zhang, Z. Numerical simulation of reinforced concrete slab subjected to blast loading and the structural damage assessment. *Eng. Fail. Anal.* **2020**, *118*, 104926. [CrossRef]
29. Wang, W.; Zhang, D.; Lu, F.; Tang, F.; Wang, S. Pressure-impulse diagram with multiple failure modes of one-way reinforced concrete slab under blast loading using SDOF method. *J. Cent. South Univ.* **2013**, *20*, 510–519. [CrossRef]
30. Departments of the Army, Air Force, and Navy and the Defense Special Weapons Agency. *Design and Analysis of Hardened Structures to Conventional Weapons Effects, TM 5-855-1/AFPAM 32-1147(I)/NAVFAC P-1080/DAHSCWEMAN-97*; Departments of the Army, Air Force, and Navy and the Defense Special Weapons Agency: Washington, DC, USA, 1997.
31. Shuaib, M.; Daoud, O. Numerical Modelling of Reinforced Concrete Slabs under Blast Loads of Close-in Detonations Using the Lagrangian Approach. *J. Phys. Conf. Ser.* **2015**, *628*, 12065. [CrossRef]
32. Trentacoste, M. *Users Manual for LS-DYNA Concrete Material Model 159*; U.S. Department of Transportation: San Francisco, CA, USA, 2007.
33. *LS-DYNA Keyword User's Manual (LS-DYNA R8.0)*; Livermore Software Technology Corporation: Livermore, CA, USA, 2015.
34. Blatz, P.J.; Ko, W.L. Application of Finite Elastic Theory to the Deformation of Rubbery Materials. *J. Rheol.* **1962**, *6*, 223–252. [CrossRef]
35. Murray, Y.D. Theory and Evaluation of Concrete Material Model 159. In Proceedings of the 8th International LS-DYNA Users Conference, Detroit, MI, USA, 2–4 May 2004.
36. Parfilko, Y. Study of Damage Progression in CSCM Concretes Under Repeated Impacts. Master's Thesis, Kate Gleason College of Engineering, Rochester, NY, USA, 2017.
37. ANSYS. *ANSYS LS-DYNA User's Guide*; ANSYS Inc.: Canonsburg, PA, USA, 2008.

Disclaimer/Publisher's Note: The statements, opinions and data contained in all publications are solely those of the individual author(s) and contributor(s) and not of MDPI and/or the editor(s). MDPI and/or the editor(s) disclaim responsibility for any injury to people or property resulting from any ideas, methods, instructions or products referred to in the content.

Article

The Strain Rate Effects of Coral Sand at Different Relative Densities and Moisture Contents

Kai Dong [†], Kun Jiang [*,†] and Wenjun Ruan

School of Energy and Power Engineering, Nanjing University of Science and Technology, Nanjing 210094, China; dongkai@njust.edu.cn (K.D.)
* Correspondence: jkeddy@163.com
† These authors contributed equally to this work.

Abstract: A 37-mm-diameter split Hopkinson pressure bar (SHPB) apparatus was used for impact loading tests to determine the effects of the relative density and moisture content on the dynamic properties of coral sand. The stress–strain curves in the uniaxial strain compression state were obtained for different relative densities and moisture contents under strain rates between 460 s^{-1} and 900 s^{-1}. The results indicated that with an increase in the relative density, the strain rate becomes more insensitive to the stiffness of the coral sand. This was attributed to the variable breakage-energy efficiency at different compactness levels. Water affected the initial stiffening response of the coral sand, and the softening was correlated with the strain rate. Strength softening due to water lubrication was more significant at higher strain rates due to the higher frictional dissipation. The volumetric compressive response of the coral sand was investigated by determining the yielding characteristics. The form of the constitutive model has to be changed to the exponential form, and different stress–strain responses should be considered. We discuss the effects of the relative density and water content on the dynamic mechanical properties of coral sand and clarify the correlation with the strain rate.

Keywords: coral sand; strain rate; moisture content; relative density; volumetric compressive

Citation: Dong, K.; Jiang, K.; Ruan, W. The Strain Rate Effects of Coral Sand at Different Relative Densities and Moisture Contents. *Materials* 2023, *16*, 4217. https://doi.org/10.3390/ma16124217

Academic Editor: Wei Wang

Received: 11 May 2023
Revised: 1 June 2023
Accepted: 5 June 2023
Published: 7 June 2023

Copyright: © 2023 by the authors. Licensee MDPI, Basel, Switzerland. This article is an open access article distributed under the terms and conditions of the Creative Commons Attribution (CC BY) license (https://creativecommons.org/licenses/by/4.0/).

1. Introduction

Reef flats and lagoons in tropical coastal areas of the world are generally covered by coral sand, which is also called calcareous sand because of its high calcium carbonate content [1]. During the past years, due to the development and utilization of marine resources, many countries have established wharves, airports, oil depots, and other infrastructure on island reefs. Researchers have conducted many studies on the mechanical properties of coral sand as related to engineering needs and the results provided valuable technical support for human exploration and utilization of marine resources [2–5].

Coral sand is porous and brittle due to a large number of inherent defects in the interior of the particles [6]. It originates from dead coral and shellfish, and is widely used in reef construction. As an engineering material, sand has strong adaptability and should not only be able to support the designed static load but also withstand dynamic loads such as strong shocks, penetrations, and accidental or man-made explosions [7,8]. For a large number of impact engineering problems and accidental emergencies, whether dealing with specific engineering problems or conducting experimental research, the most common three-dimensional stress state of the material is the one-dimensional strain state due to the instantaneousness of the load [9]. Under strong dynamic loads, the mechanical properties of the coral sand have to be determined at a high strain rate (HSR) for safe use in reef development [10].

Coral sand has complex mechanical behaviors under HSR loading and these behaviors differ from those under static loading. Many external factors affect the dynamic mechanical properties of coral sand, such as the relative density, moisture content, particle gradation, and sampling location [11]. These factors, together with the intrinsic strain rate, determine the dynamic mechanical behavior of coral sand. In the moist ocean environment, the water saturation of coral sand and the compactness depend on the location [12,13]. Researchers conducted numerous studies on terrigenous silica bedrock soils and analyzed the influence of compaction and saturation on quartz sand [14–22]. However, the mechanical properties of coral sand are substantially different from those of terrigenous sand; thus, the results obtained from quartz sand are not applicable to biogenic coral sand [23,24]. Few studies have been conducted on the HSR characteristics of coral sand. Xiao, Lv, Wu, and Wei investigated the particle breakage characteristics and influence of the moisture content; the study focused on the comparison of coral sand and quartz sand to explain the different dynamic mechanical characteristics attributed to the fragile particles of coral sand [25–30]. However, the effects of different strain rates of coral sand have not been comprehensively considered, especially the relationship between the inherent strain rate sensitivity and external natural or artificial conditions, such as compactness and water content.

In this work, one-dimensional strain impact loading tests at different strain rates were performed to determine the dynamic responses of coral sand for different relative densities and moisture contents. Based on previous work [31], the effect of the strain rate and stiffness of coral sand was investigated in detail at different relative densities. In addition, impact loading tests on coral sand with different moisture contents under various strain rates were conducted by a split Hopkinson pressure bar. The effect of the loading rate on partially saturated coral sand was evaluated by determining the mechanism of the softening and yielding response. A novel compressive equation of state (EOS) model describing the relationship between the average pressure and volumetric strain was established by considering different stress–strain responses. The results of this study have great significance for theoretical calculations related to coral reef engineering.

2. Introduction of the SHPB Test

2.1. Test Device and Test Method

A dynamic mechanical test was conducted with a Φ37 mm split Hopkinson pressure bar (SHPB) made of aluminum alloy with a density of 2.85 g/cm^3, an elastic modulus of 72 GPa, and an elastic wave speed of about 5026 m/s. The test device is illustrated in Figure 1. The lengths of the incident bar and the transmission bar are 2000 mm, respectively, and the length of the striker bar is 400 mm. Through the instantaneous release of high-pressure nitrogen, the striker bar is driven to impact the incident bar at high speed to produce a stress pulse, a pulse shaper made of rubber (Φ10 mm × 1 mm, Young's modulus of 7.83 MPa) is attached to the front face of the incident bar. A sleeve made of high-strength steel (Young's modulus 210 GPa, Poisson's ratio 0.29) with an inner diameter of 37 mm (a tolerance of 0~0.01 mm) and an outer diameter of 43 mm is used to restrict the lateral displacement of the sample. A pair of high-precision semiconductor strain gauges are symmetrically attached to the outer walls of the incident bar, the transmission bar, and the sleeve to measure the strain during loading.

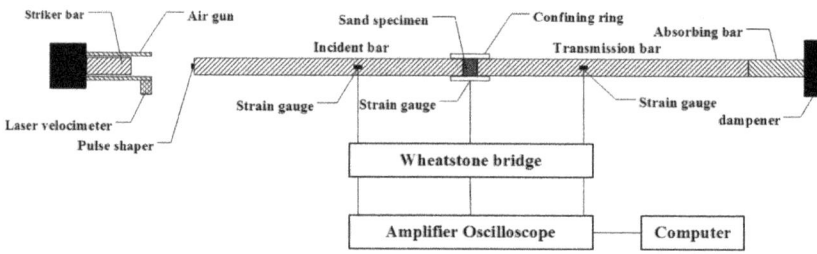

Figure 1. Schematic diagram of modified SHPB.

The strain gauge is a semiconductor strain gauge with the type of SB-3.8-120-p-2 made from Avic Zhonghang Electronic Measuring Instruments Co., Ltd. of Hanzhong, Hanzhong, China, with a resistance value of 120 ohms, and the sensitivity coefficient k = 110. The Elsys TranNET FE data acquisition system produced in Switzerland is used for data acquisition, where the sampling frequency is set as 2 MHz, so the data interval obtained is 0.5 μs. The metric signals are transmitted through the Wheatstone bridge and amplifier oscilloscope and converted to voltage signals that are stored in a computer. The strain value at the measured position is calculated by Equation (1) using the parameters of the strain gauge and the amplifier oscilloscope.

$$\varepsilon = \frac{V_m}{\zeta \cdot k \cdot V_a / \eta} \tag{1}$$

where, V_m is the measured voltage signal, ζ is the factor of Wheatstone bridge by using the quarter-bridge, half-bridge and full-bridge, and ζ = 0.25, 0.5, and 1.0, respectively. The half-Wheatstone bridge was used for the tests in this paper, k is the sensitivity coefficient of the strain gauge, V_a is the input voltage of the Wheatstone bridge, η is the amplification factor of the amplifier in the data acquisition system.

In addition, when the stress wave propagates in the bars, the tensile fracture occurs at the connection between the strain gauge and the conductor. As shown in Figure 2, soft foam can be filled between the conductor and the bar, so that the conductor has a flexible buffer when under axial tension, which greatly avoids a conductor fracture.

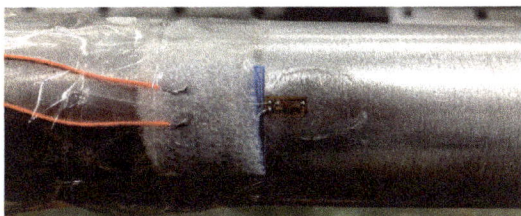

Figure 2. Strain gauge pasting and protection.

Assuming that the sample is exposed to uniform stress with uniform deformation during the loading process, the following equation applies: $\varepsilon_i(t) + \varepsilon_r(t) = \varepsilon_t(t)$. According to the one-dimensional stress wave theory, the strain rate $\dot{\varepsilon}(t)$, axial strain $\varepsilon_x(t)$ and axial stress $\sigma_x(t)$ of the sample during the testing process are obtained using Equation (2) [16,17,28,32].

$$\left.\begin{array}{l} \dot{\varepsilon}(t) = -2\frac{C_0}{L_s}\varepsilon_r(t) \\ \varepsilon_x(t) = -2\frac{C_0}{L_s}\int_0^t \varepsilon_r(t)dt \\ \sigma_x(t) = \frac{A_0}{A_s}E_0\varepsilon_t(t) = E_0\varepsilon_t(t) \end{array}\right\} \tag{2}$$

where A_s and A_0 are the cross-sectional areas of the specimen and bar, respectively, $\varepsilon_i(t)$, $\varepsilon_r(t)$, and $\varepsilon_t(t)$ are the strain of the incident, reflected, and transmitted signals, respectively, L_s is the length of the specimen.

When the sample is compressed and deformed under impact loading, the radial expansion is restrained by the elasticity of the sleeve. The circumferential strain of the sleeve is calculated according to the pulse measured by the strain gauge on the outer wall of the sleeve. The pressure on the inner wall of the cylinder σ_{rr} and the radial displacement of the inner wall of the sleeve ε_{rr} are obtained according to the dimensions of the thick-walled cylinder [17,28]. Since the sample is closely attached to the sleeve during compression, based on the interfacial equilibrium condition, the confining pressure and radial strain at the center of the sample are obtained using Equation (3).

$$\begin{cases} \sigma_{rr} = \sigma_{\theta\theta} = 0.5(\alpha^2 - 1)E_{sl}\varepsilon_{sl} \\ \varepsilon_{rr} = \varepsilon_{\theta\theta} = 0.5[(1-\nu_{sl}) + (1+\nu_{sl})\alpha^2]\varepsilon_{sl} \end{cases} \tag{3}$$

where α is the ratio of the outer diameter to the inner diameter of the sleeve; E_{sl}, ν_{sl}, and ε_{sl} are Young's modulus, Poisson's ratio, and the measured strain of the sleeve, respectively. The three principal stress components of the stress tensor are obtained from the measured axial and circumferential pulses. The average pressure P and the volumetric strain ε_v of the sample are defined in Equation (4).

$$P = \frac{1}{3}(\sigma_x + 2\sigma_{rr}), \quad \varepsilon_v = \varepsilon_x + 2\varepsilon_{rr} \approx \varepsilon_x \qquad (4)$$

2.2. Coral Sand Samples

The coral sand investigated in this study (ECS) was obtained from the Hainan province in China near the location where the sand used by Lv (LCS) was obtained [28]. The content of $CaCO_3$ is over 90%. For the analysis of the mechanical properties, particles with a diameter larger than 2.23 mm and smaller than 0.15 mm were excluded; the mass of these particles was less than 8% of the total. The physical properties of the dry ECS and LCS are shown in Table 1; the ASTM standard was used [33]. The specific gravity of the ECS and LCS is 2.81. The particle size distribution of the sieved sand is shown in Figure 3, and the scanning electron microscopy (SEM) micrographs are shown in Figure 4. The ECS has superior grading than the LCS.

The preparation of sand specimens has been described in detail in previous studies [19,28,31]. Certain discreteness in the mechanical properties of geotechnical materials requires that the sampling error is strictly controlled during the experiment. First, sieving was conducted prior to sampling using nine different sieve hole sizes and weighing was conducted by particle group using an electronic scale with an accuracy of 0.01 g. This was followed by uniform mixing with a measurement uncertainty of the total mass of ±0.03 g. Second, the length and flatness of the test device were strictly controlled, and the error was within 0.04 mm to ensure the uniform size of the specimens in repeated tests. The test device after assembly is shown in Figure 5. Coral sand is located between two platens which are the same material and diameter as the bar, and have a length of 30 mm. The screws are used to secure the sleeve and SHPB system during assembly, and should be removed during the tests.

Table 1. Physical properties of the dry coral sand.

Sand Type	D_{50} Particle Size (mm)	Coefficient of Uniformity C_u	Coefficient of Curvature C_c	Maximum Dry Density ρ_{max} (g/cm³)	Minimum Dry Density ρ_{min} (g/cm³)
ECS	0.48	2.36	0.92	1.317	1.136
LCS	0.55	1.86	0.95	1.377	1.180

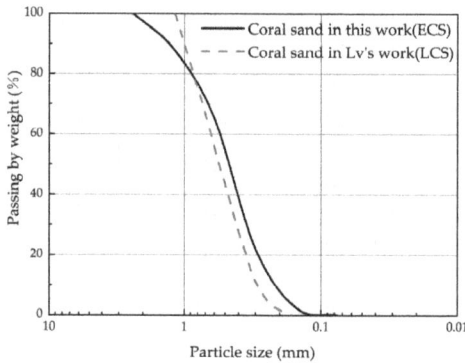

Figure 3. Particle size distribution of the coral sand.

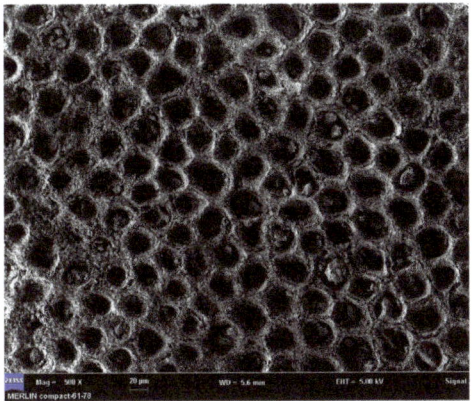

Figure 4. SEM image of the porous coral sand.

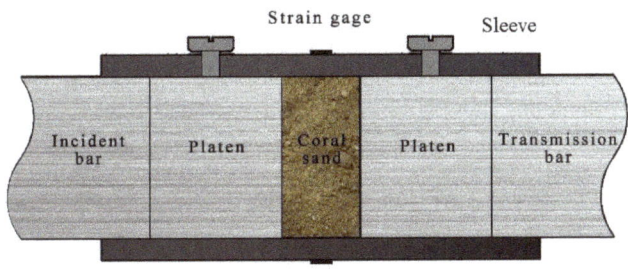

Figure 5. Test section for the coral sand sample.

After repeated tests and checks, three relative densities were selected, and the relative density Dr can be expressed as Equation (5). The experimental parameters of the specimens are shown in Table 2.

$$\mathrm{Dr} = \frac{(\rho_d - \rho_{\min}) \cdot \rho_{\max}}{(\rho_{\max} - \rho_{\min}) \cdot \rho_d} \cdot 100\% \tag{5}$$

Table 2. Specimen parameters in the experiments of different densities.

Relative Density Dr	Specimen Quality (g)	Specimen Parameters		Repeated Times
		Density ρ_d (g/cm^3)	Thickness (mm)	
30%		1.178	11.86	
60%	15	1.219	11.46	5~8
90%		1.260	11.08	

The impact loading tests of the moist coral sand were conducted at a relative density of 60%. The different moisture content conditions are shown in Table 3. The dry density of the coral sand was 1.219 g/cm^3, and the void ratio e was 1.306. The maximum moisture content in this study was 30%, and the maximum saturation was 64.54%. To ensure the uniform distribution of water in the sample, the dry sand is divided into 3–5 parts. Every part is placed in the sleeve and subsequently, a syringe is used to sprinkle the equal division of water into the sand. After 3~5 time repetitions, the thickness of the sample is adjusted using the platens.

Table 3. Experimental parameters of the specimens with different moisture contents.

Moisture Content	Saturation Degree	Density (g/cm^3)
0%	0%	1.219
2%	4.30%	1.242
4%	8.61%	1.266
6%	12.91%	1.290
8%	17.21%	1.315
10%	21.51%	1.339
20%	43.03%	1.461
30%	64.54%	1.583

3. Test Results and Discussion

3.1. Pulses and Stress Analysis

In the SHPB test, how to handle the contact fit between the sleeve and the sample, bar, and the platen has a significant influence on the accuracy of the test results. The friction force on the inner wall of the sleeve is a key factor that needs to be avoided. Most scholars use vaseline or lubricating oil to reduce the friction effect of the contact during the test, but many scholars still have a very large initial oscillation in the transmission curve obtained. Martin [16] stated that the vibration of the transmission pulse in the initial stage of loading was a problem that is difficult to explain. This problem can be neglected in high-strength quartz sand but coral sand has low particle strength, and multiple peaks caused by the vibration in the initial stage may result in the misinterpretation of the mechanical properties. Various confining sleeves have been used and shown that the initial vibration may be caused by the asymmetric contact friction between the sleeve and the platen or bar. As shown in Figure 6, the wavy vibration caused by friction is eliminated by ensuring strict processing accuracy, polishing, and grinding of the inner surface of the sleeve, and an application of a thin layer of lubricating oil prior to the experiment.

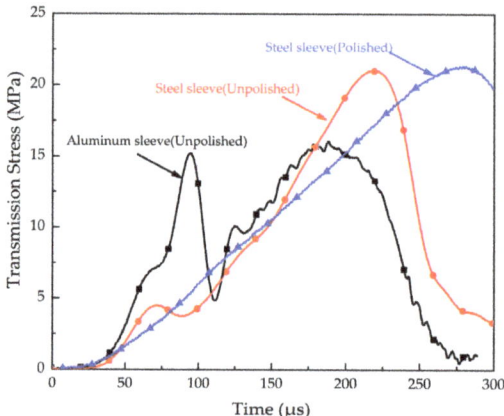

Figure 6. Transmission signals obtained using different sleeves.

The sample is connected to the end of the Hopkinson bar through the platen. The incident and transmission strain signals without the specimen (validation test) but assembly sleeve and platens are shown in Figure 7. The results show that the platen and the sleeve have little influence on the test accuracy, and the one-dimensional propagation of the stress pulse is ensured. Typical signals recorded from the strain gauges with the specimen including the incident pulse, reflect pulse, transmission pulse, and strain signal of the sleeve during the test are shown in Figure 8.

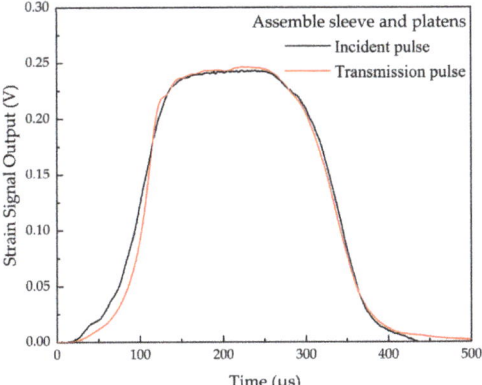

Figure 7. Incident and transmission strain signals without the specimen.

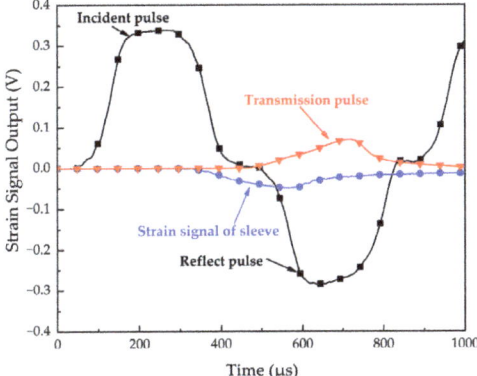

Figure 8. Typical signals recorded from the strain gauges with the specimen.

The different strain rates of the specimen are obtained by adjusting the velocity of the striker bar. As the velocity increases, it becomes challenging to ensure uniform loading of the sample under HSR loading. The stress equilibrium at the front and back ends of the sample is the key standard for determining the effectiveness of the test. The back end can be directly measured using transmitted waves of the transmission bar, while the front end requires the use of incident wave subtract reflected measured from the incident bar, i.e., $\sigma_i - \sigma_r$. The most common method to achieve stress equilibrium is using pulse shaping technology, which increases the rise time of the incident pulse. Figure 9 shows the stress–time history curves of the front and back ends for the HSR and the lowest sample density in the test. It is observed that the stress values are similar at the front and back ends of the sample, indicating that the specimen is under uniform stress during dynamic loading.

Test reproducibility is an important aspect of geotechnical material testing. The strain pulses recorded by the dynamic strain gauges and the stress–strain curves under the same conditions were obtained using Equation (1). As shown in Figure 10, the consistency of multiple tests demonstrates high reliability and good reproducibility of impact loading. The maximum value of the circumferential strain recorded on the sleeve is in the range of $10^{-5} \sim 10^{-4}$ and the axial strain value of the sample is in the range of 0.08~0.18. This result demonstrates that the sample is in a state of one-dimensional deformation during impact loading. Since the sample deformation is constrained by the steel sleeve, its pressure is very high although the circumferential strain is small and the pressure can be calculated by

Equation (2). The larger axial deformation is due to the free compression of the bars on the specimen, which is determined by the impact velocity and specimen properties.

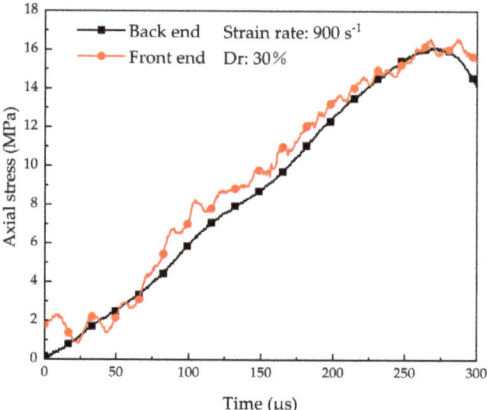

Figure 9. Dynamic stress equilibrium check of the sand sample.

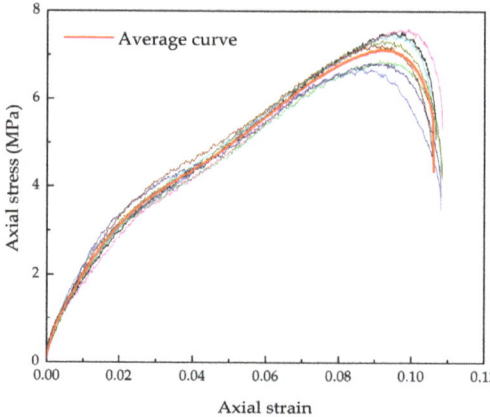

Figure 10. Reproducibility of the axial stress–strain of the test results.

3.2. Strain Rate Effected by Compaction of Dry Coral Sand

The stress–strain curves of the dry coral sand for the three densities at the strain rates of 460 s^{-1}, 650 s^{-1}, 800 s^{-1}, and 900 s^{-1} were obtained in the literature [31]. The mechanism of the strain rate effect was analyzed, but the relationship between compactness and strain rate was not analyzed in detail in the literature [31]. The curves represent the average of multiple tests (Figure 11). For dynamic compression, the stress–strain responses in this study were significantly different from the experimental results of Lv [28]. An inflection point (yielding point) [10] was observed in the initial deformation stage; this was not observed by Lv. Different yielding characteristics of the stress–strain curves of sand have been reported in the literature, but few scholars have explained the underlying reasons [10]. Lin compared the mechanical properties of Ottawa sand and distinguished two types of responses, i.e., fluid-like and solid-like behaviors; it was concluded that yielding was related to the particle size distribution [20]. The yielding mechanism of coral sand during initial loading is related to the sudden collapse of the specimen skeleton caused by extensive particle breakage. The ECS grading was better than the LCS grading, and the average particle size was smaller; therefore, the ECS particles are more difficult to breakage due to the initial strong skeleton support. When the loading pressure of the specimen exceeded the initial strength (i.e., the yield stress), many were crushed, resulting in a solid mass; therefore, yielding occurred rapidly. However, the LCS particles that were crushed during the entire compression and the curve exhibited fluid-like characteristics.

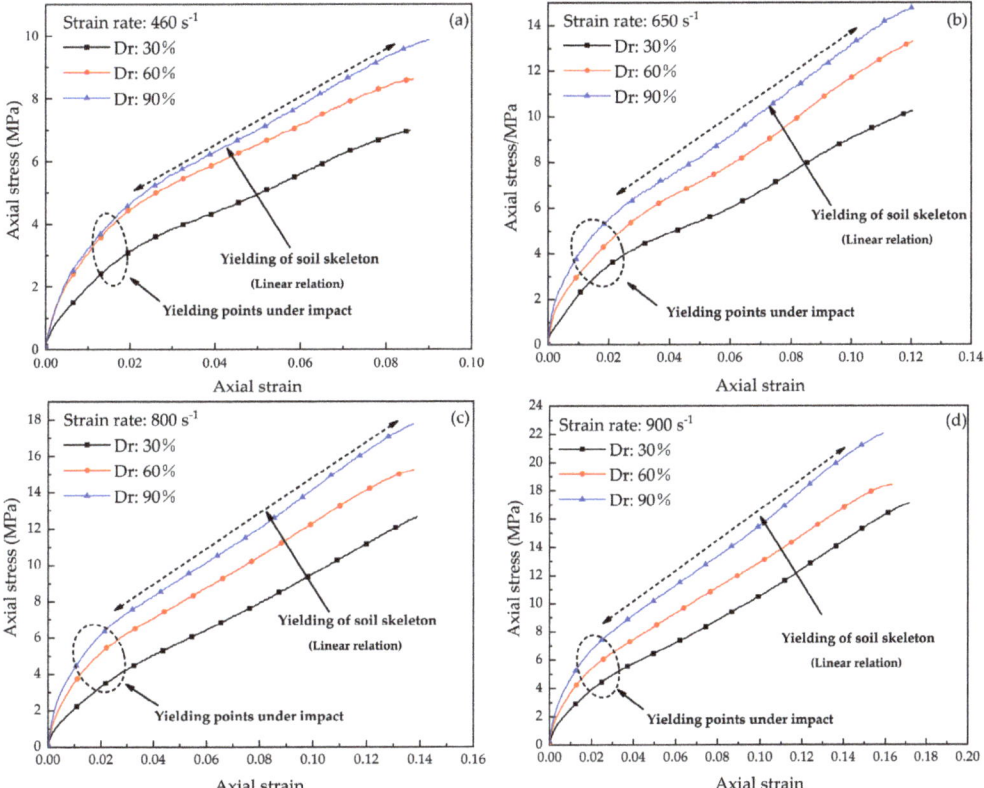

Figure 11. Stress–strain curves of dry coral sand under different compactness: (**a**) 460 s^{-1} (**b**) 650 s^{-1} (**c**) 800 s^{-1} (**d**) 900 s^{-1}.

Static compression tests were conducted on the coral sand with the three relative densities using a conventional material test system (MTS) [31]. The yielding points occurred at a strain of 0.02 (Figure 12). This result is similar to the yield strain under HSR loading. With the increase in the compaction level and the increase in the strain rate, the coral sand exhibited increasing stiffness. The dimensionless normalized stress [10] was determined by the ratio of the HRS stress from the dynamic uniaxial compression test and the static stress from a conventional quasi-static test to evaluate the increase in strength due to HSR loading at different strains. As shown in Figure 13, the normalized stress is almost constant during the yielding of the soil skeleton. At the same HSR loading, as the density decreases, the normal stress level increases, indicating that the strain rate sensitivity is closely related to the relative density. This phenomenon is related to the breakage-energy efficiency. A decrease in the compaction level results in higher particle degrees of freedom, thereby increasing the proportion of frictional dissipation [34]. Therefore, the ratio of the crushing energy to the total input energy (i.e., breakage-energy efficiency) decreases. The lower the breakage-energy efficiency during compression, the larger the strain rate effect is [18,20]. This explains why the coral sand shows an increasing strain rate sensitivity when the compaction level decreases.

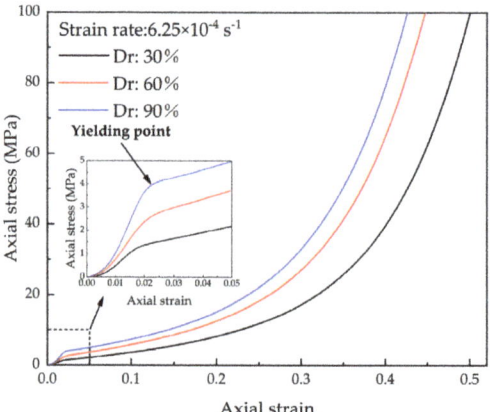

Figure 12. Axial stress–strain curves of dry coral sand under static loading [31].

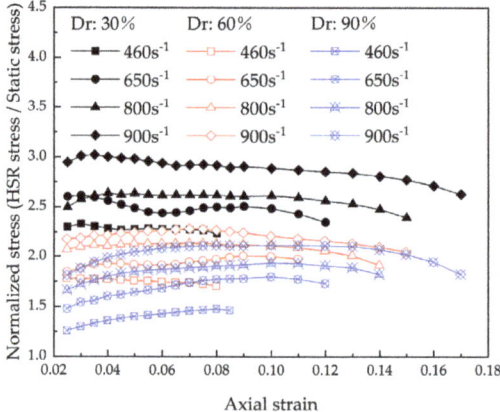

Figure 13. Normalized stress of dry coral sand at different strains in the HSR uniaxial compression tests.

3.3. Effect of Moisture Content on Dynamic Mechanical Properties of Coral Sand

The test results of different moisture contents at strain rates of 460 s^{-1}, 650 s^{-1}, and 800 s^{-1} are shown in Figure 14. The highest strength of the sample is observed at moisture contents between 6% and 8% and the lowest strength occurs at 20%. The moisture content has little influence on the strength of the coral sand under unsaturated conditions, but some observations regarding the mechanical properties can be made. Generally, the water has a softening effect on the strength of coral sand. Research has shown that water significantly reduces the strength of terrestrial soil, such as quartz sand and clay before reaching a high saturation state [10,16]. However, in this experiment, the influence of the water on the entire strength of the ECS is smaller than that of LCS [29]. Water reduces the friction of particles [10,16,29], and the difference between the responses of the two samples is related to the difference in the frictional dissipation of the particle motion during compression. The frictional dissipation results from the movement of the unbroken particles and the movement of the small sub-particles when the particles are crushed. The ECS has superior grading than the LCS; therefore, there is less particle breakage and the friction dissipation is lower in the ECS than the LCS during compression. As a result, the moist ECS does not decrease significantly due to the lower lubrication efficiency.

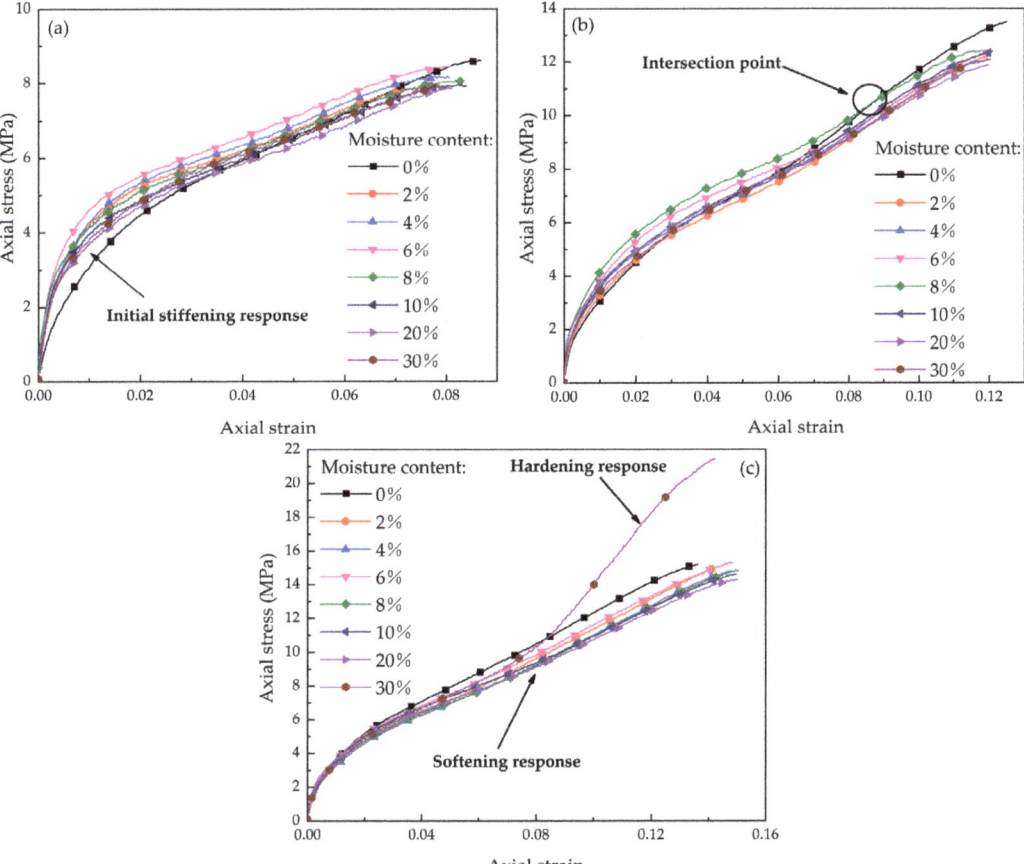

Figure 14. Stress–strain curves of the samples with different moisture contents at different strain rates: (**a**) 460 s^{-1}, (**b**) 650 s^{-1}, (**c**) 800 s^{-1}.

The strength of water-bearing sand is higher than that of dry sand during the initial compression process but it is slightly lower than that of dry sand with increasing deformation. This yielding phenomenon is observed at the strain rate of 460 s^{-1} and 650 s^{-1} (Figure 14a,b). These properties are related to the high porosity of coral sand. In this study, the increase in the initial modulus occurs because water is present in the cavities of the particles, resulting in an increase in the strength of the skeleton. However, the water in the supporting pores lubricates the secondary particles as the particles are crushed, causing a decrease in the modulus after yielding.

This effect is also related to the strain rate. It was interesting that when the strain rate increased, the position of the intersection point (where the stress is equal to that of dry sand) moves towards the origin of the coordinates on the abscissa. As shown in Figure 14, the intersection points are located between 0.04 and 0.08 for the strain rate of 460 s^{-1}, between 0.02 and 0.07 for 650 s^{-1}, and near 0.01 for 800 s^{-1}. That is, as the loading rate increases, the initial stiffening response for moist sand occurs at a different location on the curve. This phenomenon is also closely related to the strain rate effect of coral sand. The results in Section 3.2 and those of Huang provide the explanation [18]. As the loading speed increases, the friction dissipation ratio increases, and the water lubrication becomes more effective.

The stress increases sharply at the strain rate of 800 s^{-1} for the initial moisture content of 30%, indicating that the coral sand exhibits a hardening response in the compaction state. The saturation is about 77% upon reaching the compaction point. This demonstrates that coral sand reaches the compaction state earlier than quartz sand or clay, whose saturation is more than 90% [16]. Although there are few inter-particle pores in the hardened section, some unbroken coral sand still has a high internal porosity and does not reach a high saturation state. The strain still increases in the sample due to the release and compression of internal pores caused by particle breakage. However, the compression mechanism in the hardening response is different from that of the non-compaction stage due to the disappearance of the inter-particle pores. In the compaction state, the specimen cannot be easily compressed due to the absence of inter-particle movement. Therefore, the stress increases rapidly with a higher modulus.

The research on the mechanical properties of dry coral sand is relatively comprehensive. In order to quantitatively analyze the impact of water content on coral sand, the obtained regular conclusions are applied to previous studies. Based on the dimensionless stress ratio, the relationship between the stress ratio of water containing coral sand and dry sand at different strain rates is established, as shown in Figure 15. It can be more clearly seen that as the strain rate increases, the softening effect of water on coral sand becomes more pronounced.

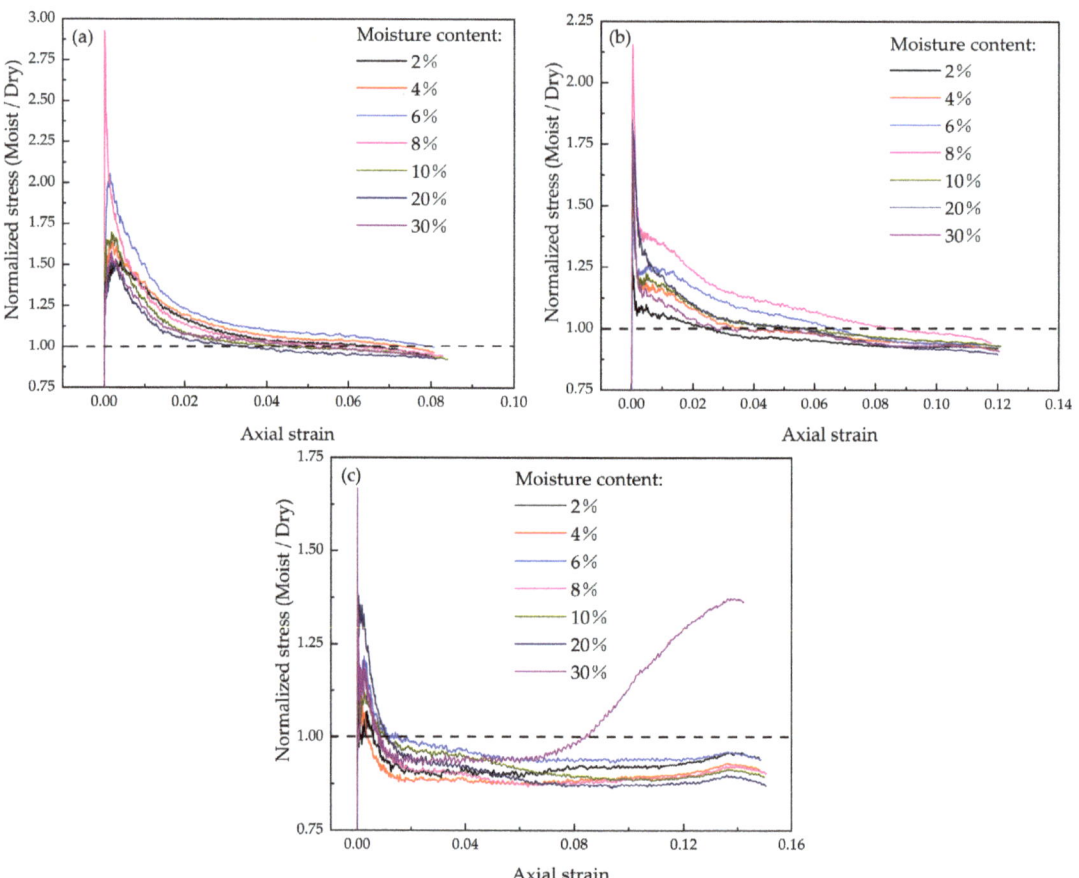

Figure 15. Normalized stress of moist coral sand to dry coral sand: (**a**) 460 s^{-1}, (**b**) 650 s^{-1}, (**c**) 800 s^{-1}.

3.4. Effect of Lateral Pressure and Equation of State

The circumferential strain of the sleeve at strain rates of 460 s^{-1}, 650 s^{-1}, 800 s^{-1} and 900 s^{-1} was obtained from the pulses recorded at the outer face of the sleeve. The signals were converted into the confining pressure of the sample using Equation (3). As shown in Figure 16, the duration of the pressure increases from zero to the peak value in about 290 μs, which is consistent with the axial loading duration of the sample. The confining pressure increases with the increase in the compaction level at the same strain rate.

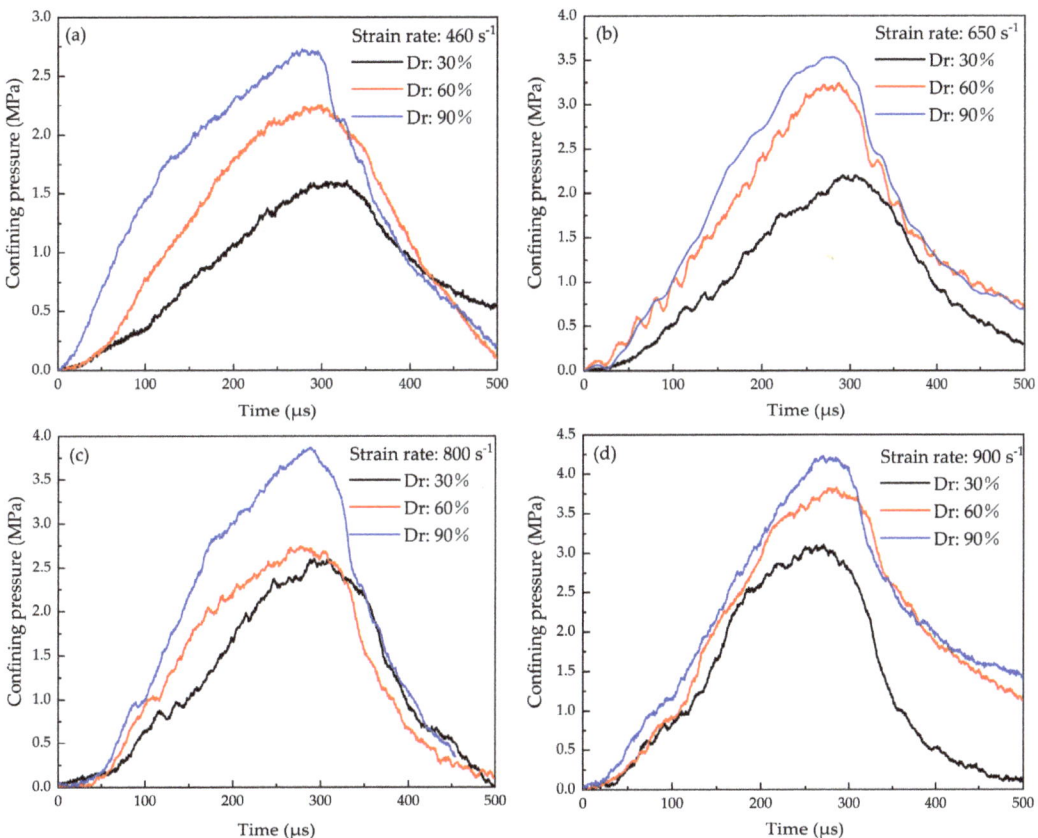

Figure 16. The confining pressure of the coral sand samples at different strain rates: (**a**) 460 s^{-1} (**b**) 650 s^{-1} (**c**) 800 s^{-1} (**d**) 900 s^{-1}.

The relationship between the average pressure and volumetric strain ($P - \varepsilon_v$) of the coral sand at different strain rates is calculated using Equation (4), as shown in Figure 17. However, in this study, in the initial compression stage of the coral sand, the slope of these curves of the coral sand decreases or even remains constant. Therefore, it is necessary to establish the constitutive model and fit the EOS for the solid-like response of the coral sand using exponential form.

The EOS is using the form of $P = a \times \varepsilon_v^b$. The fitting results of the EOS for the three relative densities at different strain rates are shown in Table 4. The value of the goodness of fit $R^2 > 0.95$ indicates that the power exponent form is suitable to describe the solid-like response of coral sand.

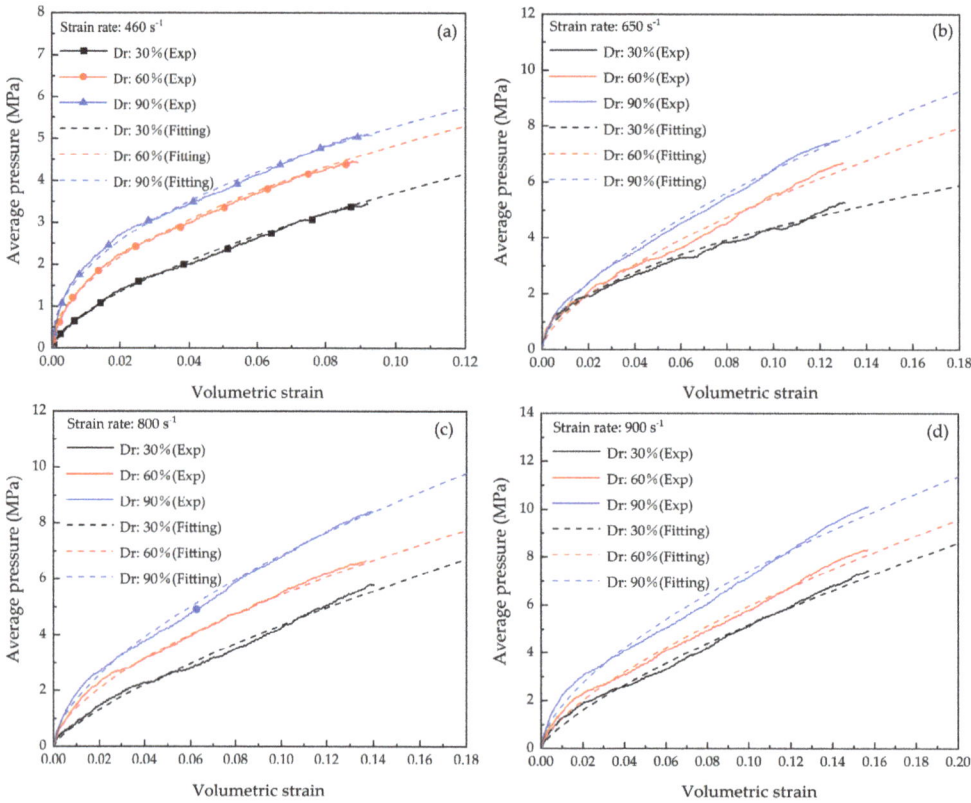

Figure 17. Relationship of average pressure and volumetric strain at different strain rates: (**a**) 460 s^{-1} (**b**) 650 s^{-1} (**c**) 800 s^{-1} (**d**) 900 s^{-1}.

Table 4. Fitting parameters of the EOS.

Strain Rate	Dr = 30%		Dr = 60%		Dr = 90%	
	a	b	a	b	a	b
460 s^{-1}	15.83	0.63	15.31	0.50	14.92	0.45
650 s^{-1}	13.85	0.50	23.80	0.64	26.80	0.62
800 s^{-1}	23.86	0.74	21.65	0.60	27.87	0.61
900 s^{-1}	27.79	0.73	28.51	0.68	30.86	0.62

4. Conclusions

Coral sand has high porosity, irregularly shaped particles, and strain-rate dependency, and exhibits complex mechanical properties. An understanding of the essential mechanical properties allows us to determine the influence of the relative density and the water content on the dynamic mechanical behavior of coral sand, thus providing scientific guidance for practical engineering design and applications. The following conclusions were determined based on the HRS impact experiments of coral sand:

(1) A significant correlation was observed between the strain rate and the stiffness with increasing relative density of coral sand. The breakage-energy efficiency decreases with an increase in the relative density, and the strain rate becomes more insensitive to the stiffness of the coral sand.

(2) The initial stiffening response of the moist coral sand decreased as the loading rate increased. Water had a softening effect on the strength of the coral sand after yielding. Due to the increase in the frictional dissipation of the coral sand with increasing strain rates, the lubrication effect of the water was more noticeable as the strength decreased.
(3) The internal porosity is an important factor affecting the compaction characteristics of coral sand at high saturated moisture content. The hardening state occurred when the inter-particle pores disappeared, and sand breakage became more difficult due to the restricted inter-particle movement.
(4) The compressive response of coral sand should be determined before establishing the pressure-volumetric strain equation. The exponential form of the EOS has to be used for solid-like coral sand.

Author Contributions: Conceptualization, K.D.; methodology, K.D. and K.J.; validation, K.D. and K.J.; formal analysis, K.D. and K.J.; investigation, K.D.; resources, K.J. and W.R.; data curation, K.D. and W.R.; writing—original draft preparation, K.D.; writing—review and editing, K.J.; project administration, K.J. and W.R.; funding acquisition, K.J. All authors have read and agreed to the published version of the manuscript.

Funding: This research received no external funding.

Institutional Review Board Statement: Not applicable.

Informed Consent Statement: Not applicable.

Data Availability Statement: Not applicable.

Conflicts of Interest: The authors declare no conflict of interest.

Abbreviations

The following abbreviations are used in this manuscript:

HRS	High strain rate
SHPB	Split Hopkinson pressure bar
ECS	Coral sand investigated in this study
LCS	Coral sand used by Lv
MTS	Material test system
EOS	Equation of State

References

1. Noorany, I. Classification of Marine Sediments. *J. Geotech. Eng.* **1989**, *115*, 23–37. [CrossRef]
2. Kong, D.; Fonseca, J. On the kinematics of shelly carbonate sand using X-ray micro tomography. *Eng. Geol.* **2019**, *261*, 105268. [CrossRef]
3. Ata, A.; Salem, T.N.; Hassan, R. Geotechnical characterization of the calcareous sand in northern coast of Egypt. *Ain Shams Eng. J.* **2018**, *9*, 3381–3390. [CrossRef]
4. Coop, M.R.; Sorensen, K.K.; Bodas Freitas, T.; Georgoutsos, G. Particle breakage during shearing of a carbonate sand. *Géotechnique* **2004**, *54*, 157–163. [CrossRef]
5. Yu, F. Particle breakage in triaxial shear of a coral sand. *Soils Found.* **2018**, *58*, 866–880. [CrossRef]
6. Li, H.; Chai, H.; Xiao, X.; Huang, J.; Luo, S. Fractal breakage of porous carbonate sand particles: Microstructures and mechanisms. *Powder Technol.* **2020**, *363*, 112–121. [CrossRef]
7. Rigby, S.E.; Fay, S.D.; Clarke, S.D.; Tyas, A.; Reay, J.J.; Warren, J.A.; Gant, M.; Elgy, I. Measuring spatial pressure distribution from explosives buried in dry Leighton Buzzard sand. *Int. J. Impact Eng.* **2016**, *96*, 89–104. [CrossRef]
8. Omidvar, M.; Malioche, J.D.; Bless, S.; Iskander, M. Phenomenology of rapid projectile penetration into granular soils. *Int. J. Impact Eng.* **2015**, *85*, 146–160. [CrossRef]
9. Field, J.E.; Walley, S.M.; Proud, W.G.; Goldrein, H.T.; Siviour, C.R. Review of experimental techniques for high rate deformation and shock studies. *Int. J. Impact Eng.* **2004**, *30*, 725–775. [CrossRef]
10. Omidvar, M.; Iskander, M.; Bless, S. Stress-strain behavior of sand at high strain rates. *Int. J. Impact Eng.* **2012**, *49*, 192–213. [CrossRef]
11. He, H.; Li, W.; Kostas, S. Small strain dynamic behavior of two types of carbonate sands. *Soils Found.* **2019**, *59*, 571–585. [CrossRef]

12. Wang, X.; Jiao, Y.; Wang, R.; Hu, M.; Meng, Q.; Tan, F. Engineering characteristics of the calcareous sand in Nansha Islands, South China Sea. *Eng. Geol.* **2011**, *120*, 40–47. [CrossRef]
13. Wang, X.; Wang, X.; Jin, Z.; Zhu, C.; Wang, R.; Meng, Q. Investigation of engineering characteristics of calcareous soils from fringing reef. *Ocean Eng.* **2017**, *134*, 77–86. [CrossRef]
14. Bragov, A.M.; Kotov, V.L.; Lomunov, A.K.; Sergeichev, I.V. Measurement of the Dynamic Characteristics of Soft Soils using the Kolsky Method. *J. Appl. Mech. Tech. Phys.* **2004**, *45*, 580–585. [CrossRef]
15. Bragov, A.M.; Lomunov, A.K.; Sergeichev, I.V.; Proud, W.; Tsembelis, K.; Church, P. A Method for Determining the Main Mechanical Properties of Soft Soils at High Strain Rates (10^3–$10^5 s^{-1}$) and Load Amplitudes up to Several Gigapascals. *Tech. Phys. Lett.* **2005**, *31*, 530–531. [CrossRef]
16. Martin, B.E.; Chen, W.; Song, B.; Akers, S.A. Moisture effects on the high strain-rate behavior of sand. *Mech. Mater.* **2009**, *41*, 786–798. [CrossRef]
17. Luo, H.; Lu, H.; Cooper, W.L.; Romanduri, R. Effect of Mass Density on the Compressive Behavior of Dry Sand Under Confinement at High Strain Rates. *Exp. Mech.* **2011**, *51*, 1499–1510. [CrossRef]
18. Huang, J.; Xu, S.; Hu, S. Effects of grain size and gradation on the dynamic responses of quartz sands. *Int. J. Impact Eng.* **2013**, *59*, 1–10. [CrossRef]
19. Luo, H.; Cooper, W.L.; Lu, H. Effects of particle size and moisture on the compressive behavior of dense Eglin sand under confinement at high strain rates. *Int. J. Impact Eng.* **2014**, *65*, 40–55. [CrossRef]
20. Lin, Y.; Yao, W.; Jafari, M.; Wang, N.; Xia, K. Quantification of the Dynamic Compressive Response of Two Ottawa Sands. *Exp. Mech.* **2017**, *57*, 1371–1382. [CrossRef]
21. Wang, S.; Shen, L.; Maggi, F.; El-Zein, A.; Nguyen, G.D.; Zheng, Y.; Zhang, H.; Chen, Z. Influence of dry density and confinement environment on the high strain rate response of partially saturated sand. *Int. J. Impact Eng.* **2018**, *116*, 65–78. [CrossRef]
22. Barr, A.D.; Clarke, S.D.; Tyas, A.; Warren, J.A. Effect of Moisture Content on High Strain Rate Compressibility and Particle Breakage in Loose Sand. *Exp. Mech.* **2018**, *58*, 1331–1334. [CrossRef]
23. Lv, Y.; Li, F.; Liu, Y.; Fan, P.; Wang, M. Comparative study of coral sand and silica sand in creep under general stress states. *Can. Geotech. J.* **2017**, *54*, 1601–1611. [CrossRef]
24. Van Impe, P.O.; Van Impe, W.F.; Manzotti, A. Compaction control and related stress–strain behavior of off-shore land reclamations with calcareous sands. *Soils Found.* **2015**, *55*, 1474–1486. [CrossRef]
25. Xiao, Y.; Liu, H.; Xiao, P.; Xiang, J. Fractal crushing of carbonate sands under impact loading. *Géotech. Lett.* **2016**, *6*, 199–204. [CrossRef]
26. Xiao, Y.; Yuan, Z.; Lv, Y.; Wang, L.; Liu, H. Fractal crushing of carbonate and quartz sands along the specimen height under impact loading. *Constr. Build. Mater.* **2018**, *182*, 188–199. [CrossRef]
27. Wei, H.; Xu, R.; Ma, L.; Xin, L.; Li, Z.; Meng, Q. Effects of water saturation and salinity on particle crushing of single coral sand. *Powder Technol.* **2023**, *426*, 118666. [CrossRef]
28. Lv, Y.; Liu, J.; Xiong, Z. One-dimensional dynamic compressive behavior of dry calcareous sand at high strain rates. *J. Rock Mech. Geotech. Eng.* **2019**, *11*, 192–201. [CrossRef]
29. Lv, Y.; Liu, J.; Zuo, D. Moisture effects on the undrained dynamic behaviour of calcareous sand at high strain rates: Split Hopkinson pressure bar tests. *Geotech. Test. J. ASTM* **2019**, *42*, 725–746.
30. Wu, Q.; Liu, Q.; Zhuang, H.; Xu, C.; Chen, G. Experimental investigation of dynamic shear modulus of saturated marine coral sand. *Ocean Eng.* **2022**, *264*, 112412.
31. Dong, K.; Ren, H.; Ruan, W.; Ning, H.; Guo, R.; Huang, K. Study on strain rate effect of coral sand. *Explos. Shock Waves* **2020**, *40*, 093102. (In Chinese)
32. Lee, O.S.; Lee, J.Y.; Kim, G.H.; Hwang, J.S. High Strain-Rate Deformation of Composite Materials Using a Split Hopkinson Bar Technique. *Key Eng. Mater.* **2000**, *183–187*, 307–312. [CrossRef]
33. ASTM D4253-16; Standard Test Methods for Maximum Index Density and Unit Weight of Soils Using a Vibratory Table. ASTM International: West Conshohocken, PA, USA, 2016.
34. De Cola, F.; Pellegrino, A.; Globner, C.; Penumadu, D.; Petrinic, N. Effect of Particle Morphology, Compaction, and Confinement on the High Strain Rate Behavior of Sand. *Exp. Mech.* **2017**, *58*, 223–242. [CrossRef]

Disclaimer/Publisher's Note: The statements, opinions and data contained in all publications are solely those of the individual author(s) and contributor(s) and not of MDPI and/or the editor(s). MDPI and/or the editor(s) disclaim responsibility for any injury to people or property resulting from any ideas, methods, instructions or products referred to in the content.

Article

A Research Investigation into the Impact of Reinforcement Distribution and Blast Distance on the Blast Resilience of Reinforced Concrete Slabs

Yangyong Wu [1,2], Jianhui Wang [1,*], Fei Liu [1,*], Chaomin Mu [2], Ming Xia [1] and Shaokang Yang [1]

1. Institute of Defense Engineering, Academy of Military Sciences, People's Liberation Army, Beijing 100850, China; yywu_aust@163.com (Y.W.)
2. School of Safety Science and Engineering, Anhui University of Science and Technology, Huainan 232001, China; chmmu@mail.ustc.edu.cn
* Correspondence: 13598184300@163.com (J.W.); 13525944181@163.com (F.L.)

Abstract: Reinforcement is one of the important factors affecting the anti-blast performance of reinforced concrete (RC) slabs. In order to study the impact of different reinforcement distribution and different blast distances on the anti-blast performance of RC slabs, 16 model tests were carried out for RC slab members with the same reinforcement ratio but different reinforcement distribution and the same proportional blast distance but different blast distances. By comparing the failure patterns of RC slabs and the sensor test data, the impact of reinforcement distribution and blast distance on the dynamic response of RC slabs was analyzed. The results show that, under contact explosion and non-contact explosion, the damage degree of single-layer reinforced slabs is more serious than that of double-layer reinforced slabs. When the scale distance is the same, with the increase of distance, the damage degree of single-layer reinforced slabs and double-layer reinforced slabs increases first and then decreases, and the peak displacement, rebound displacement and residual deformation near the center of the bottom of RC slabs gradually increase. When the blast distance is small, the peak displacement of single-layer reinforced slabs is smaller than that of double-layer reinforced slabs. When the blast distance is large, the peak displacement of double-layer reinforced slabs is smaller than that of single-layer reinforced slabs. No matter how large the blast distance, the rebound peak displacement of the double-layer reinforced slabs is smaller, and the residual displacement is larger. The research in this paper provides a reference for the anti-explosion design, construction and protection of RC slabs.

Keywords: reinforced concrete slabs; blast resistance; reinforcement distribution; blast distance; anti-blast performance; model tests

1. Introduction

In today's world, peace and development have become the main theme of the times, and the world is in a relatively peaceful and stable situation. However, violent terrorist attacks and local wars caused by religious and racial discrimination and hegemonism have emerged in an endless stream, posing serious challenges to the safety protection design of buildings and structures around the world [1–3]. In addition, in the process of industrial production and processing, the use of various flammable and explosive dangerous goods is inevitable, which will bring potential risks to the safety of buildings and structures and seriously threaten the safety of people's lives and property [4,5]. The explosion load is ignored in the design and construction of traditional buildings. Attention should be paid to the anti-explosion design of important buildings such as stations, schools, business centers and other places with dense traffic. It is necessary to perform a good job in the prevention of violent terrorist attacks [6–8].

The basic structural elements of a building include beams, slabs, columns and walls. Studying the impact of blast load on individual structural elements is of great significance [9–12]. When studying the overall building, various structural elements are coupled with each other, and it is difficult to distinguish the contribution of a certain structural element to the blast resistance of the building.

Slabs are important load-bearing components in buildings, so it is important to study the anti-explosion performance of slabs for the design and protection of buildings [13,14]. Traditional research is mainly based on typical RC slabs. By changing the reinforcement ratio [15–25], strength of the concrete [16–18], span of the slab [16], thickness of the slab [17,19,20], strength of the reinforcement [18] and scale distance [20,21], the data on the failure form, blast pit, displacement and reflected overpressure of the reinforced concrete slab after the explosion are obtained. The damage degree and anti-explosion performance of the RC slab are evaluated by the bearing angle [18], displacement [22], residual bearing capacity [23] and P-I curve [16,24]. Among them, the reinforcement ratio is the most influential factor.

Recent research on slabs mainly includes the application of new materials, reinforcement technology and the introduction of new failure prediction methods. In terms of new materials, on the one hand, RC slabs are made of carbon-fiber-reinforced polymer [26], low ductility reinforcement [27] and basalt-fiber-reinforced plastic bars [28], so the loss of slabs under an explosion load is smaller.

On the other hand, by changing the material of concrete, the RC slabs are made of ultra-high-strength concrete [29], ultra-high-performance fiber-reinforced concrete [30], superabsorbent polymer honeycomb concrete [31], 200 MPa ultra-high-performance fiber-reinforced concrete [32] and ultra-high-ductility concrete mixed with ultra-high-performance concrete [33,34], which have a better anti-explosion performance than ordinary RC slabs.

In terms of reinforcement technology, Mendonca et al. [35] studied the use of foam to strengthen RC slabs through experiments and concluded that the slabs strengthened with foam had different pressure modes compared with ordinary slabs, and that the displacement and acceleration increased instead. Maadoun et al. [26] strengthened the RC slab by bonding carbon-fiber-reinforced polymer (CFRP) and concluded that the strengthened slab has a better flexural bearing capacity and stiffness under an explosive load. Gao et al. [36] verified the finite element model of the ultra-high-performance concrete slab strengthened with polyurea based on the experiment. Through changing the reinforcement ratio and scale distance to carry out an anti-explosion numerical simulation of the slab, they obtained the prediction formula of the end rotation angle of the ultra-high-performance concrete slab strengthened with polyurea under a near-field explosion. Gao et al. [37] carried out an explosion resistance experiment of an RC slab with a porous energy-absorbing material foam aluminum protective layer, verified the finite element model based on the experimental data, studied the damage rule of the foam aluminum density and longitudinal reinforcement ratio on reinforced concrete and concluded that the greater the reinforcement ratio, the better the explosion resistance effect of the RC slab. Thiagarajan and Reynolds [38] studied the anti-explosion performance of high-strength concrete slabs strengthened with high-strength vanadium steel through an explosion simulator, concluded that slabs with a larger spacing of steel bars have a smaller ductility and gave the damage mode of the panel.

In terms of introducing new failure prediction methods, Almustafa et al. [39] studied the influence of 10 input characteristics on the maximum displacement of RC slabs under an explosive load based on the random forest algorithm. This method has achieved good results in predicting the maximum displacement of RC slabs, and is more efficient and accurate than the existing numerical calculation methods. Shishegaran et al. [40] evaluated various models based on normalized square error and fractional deviation and concluded that the best model for predicting the maximum deflection of the plate is multiple Ln equation regression.

In the latest research, both the application of new materials and the reinforcement of slabs will increase the construction cost of buildings. In previous studies, increasing the reinforcement ratio can enhance the anti-explosion performance of RC slabs, which will also lead to increased costs. In this paper, by fixing the reinforcement ratio and changing the distribution of reinforcement in the slabs, the difference in anti-explosion performance between single-layer reinforced slabs and double-layer reinforced slabs with the same reinforcement ratio under contact explosion and non-contact explosion was studied. This can determine which type of reinforcement distribution in slabs has a better blast resistance without increasing costs. In previous studies, the conditions for changing the scale distance to change the blast load were discussed. By using a fixed scale distance, the influence of blast distance on the blast resistance of RC slabs was studied in this paper, which can verify whether the load conditions determined by scale distance are reliable.

2. Test Overview
2.1. Design of Specimens

The size of slabs was 2000 mm × 2000 mm × 100 mm, and they were HRB400E-reinforced and had a diameter of 8 mm. The concrete strength grade was C40, and the thickness of concrete protective layer was 20 mm. Single-layer two-way reinforcement and double-layer two-way reinforcement were adopted for the components. The spacing of single-layer reinforcement slabs was 100 mm, the spacing of double-layer two-way reinforcement was 200 mm, and the spacing of layers was 600 mm. The number of single-layer reinforced slabs was S1–S8, and the number of double-layer reinforced slabs was D1–D8. The information of RC slab specimens is shown in Table 1, and the reinforcement diagram is shown in Figure 1.

Table 1. Information of RC slab specimens.

Types of RC Slabs	Model of Reinforcement	Model of Concrete	Reinforcement Ratio (%)	Number
Single-layer reinforced slab	HRB400E	C40	0.45	8
Double-layer reinforced slab	HRB400E	C40	0.45	8

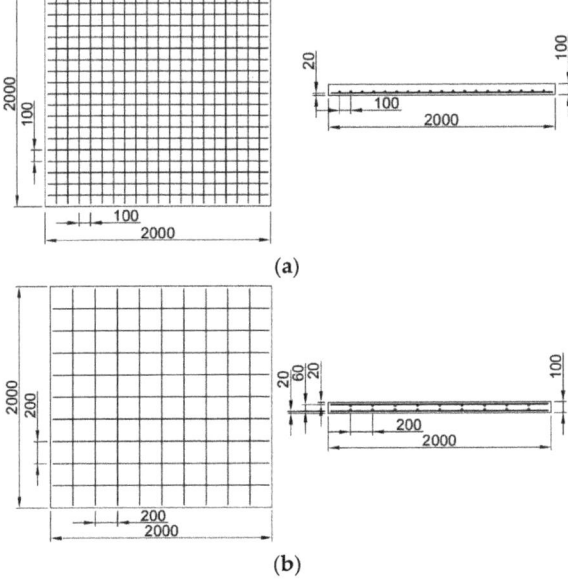

Figure 1. Reinforcement diagram of RC slabs; unit is mm. (a) Single-layer reinforced slab; (b) double-layer reinforced slab.

2.2. Test Conditions

The working conditions of contact explosion are shown in Table 2, and the working conditions of non-contact explosion are shown in Table 3.

Table 2. Working conditions of contact explosion.

Types of RC Slabs	Specimen	Charge (kg)
Single-layer reinforced slab	S1	0.2
	S2	0.4
	S3	0.8
	S4	1.6
Double-layer reinforced slab	D1	0.2
	D2	0.4
	D3	0.8
	D4	1.6

Table 3. Working conditions of non-contact explosion.

Types of RC Slabs	Specimen	Charge (kg)	Blast Distance (m)	Scale Distance (m·kg$^{-1/3}$)
Single-layer reinforced slab	S1	0.2	0.25	0.43
	S2	0.4	0.32	0.43
	S3	0.8	0.40	0.43
	S4	1.6	0.50	0.43
Double-layer reinforced slab	D1	0.2	0.25	0.43
	D2	0.4	0.32	0.43
	D3	0.8	0.40	0.43
	D4	1.6	0.50	0.43

2.3. Material Properties

2.3.1. Material Properties of Concrete

The concrete specimens were poured at the same time. Six concrete cubes with a size of 150 mm × 150 mm × 150 mm were retained for compression test when pouring the specimens. They were cured in the same environment as the components. The compressive strength of the six concrete cubes was 45.8, 48.2, 47.4, 46.8, 46.4 and 47.6 MPa, respectively, and the average compressive strength of the cubes was 47.0 MPa.

2.3.2. Material Properties of Reinforcement

The model of reinforcement was HRB400E, and the diameter was 8 mm. Its mechanical properties are shown in Table 4.

Table 4. Mechanical properties of reinforcement.

Model of Reinforcement	Elastic Modulus (GPa)	Yield Strength (MPa)	Tensile Strength (MPa)	Yield Strain (%)	Elongation (%)
HRB400E	200	455	587.5	0.23	21

2.4. Arrangement of Test

The sample of reinforced concrete slab was fixed onto a rigid frame made of I-beam, which adopts one-way support. Two clamps were installed on each side of the slab through bolt fastening. The contact explosion test installs explosives in the center of the slab, and the non-contact explosion test lifts explosives directly above the center of the slab. The explosives used in the test were stacked by standard TNT explosive blocks. The mass of standard TNT explosive block is 200 g, and the structural dimension is 100 mm × 50 mm × 25 mm, detonated with digital detonator. A displacement sensor with model DH5G107 was arranged in the center of bottom face to measure the dynamic displacement of the mid-span of the slab. Due to the fact that the concrete at the center of the bottom face may fall due to collapse, placing the displacement sensor here will damage. Therefore, move the displacement sensor RC board side by 30 cm. The experimental layout is shown in Figure 2.

Figure 2. Experimental layout diagram. (**a**) Overall layout of the experiment; (**b**) arrangement of displacement sensor.

3. Test Results and Analysis of Contact Explosion

3.1. Damage Form of Contact Explosion

Under the contact explosion, the dynamic response law of the RC slab was studied by changing the charge. The contact explosion experimental results are shown in Table 5 and Figure 3.

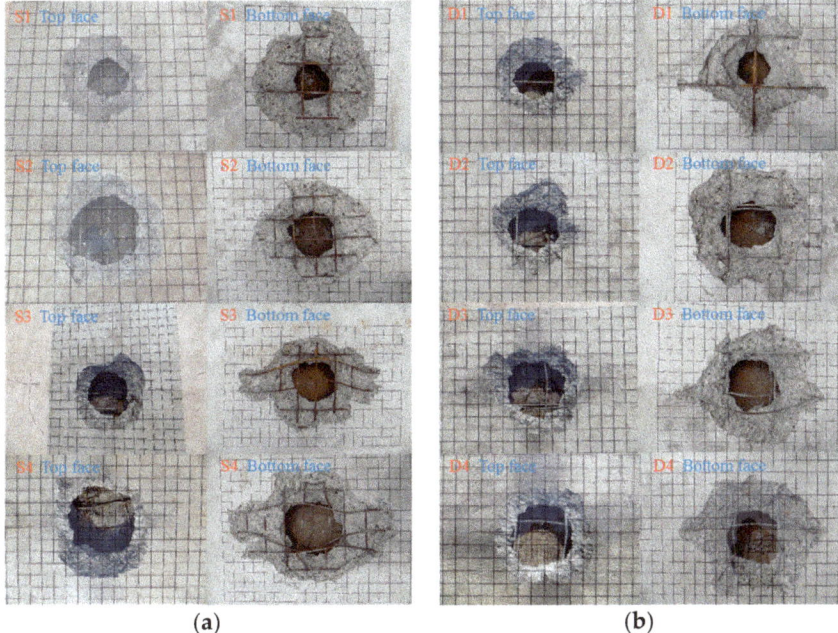

Figure 3. Contact explosion failure mode of RC slabs. (**a**) Single-layer reinforced slab; (**b**) Double-layer reinforced slab.

Table 5. Test results of contact explosion of RC slabs.

Types of RC Slabs	Specimen	Description of Phenomenon
Single-layer reinforced slabs	S1	The reinforced concrete slab has a through hole, with a hole diameter of 15 cm. Four bars are exposed transversely and two bars are exposed longitudinally. The bending value of the bars is 2 cm.
	S2	The reinforced concrete slab has a through hole, with a hole diameter of 21 cm. Five bars are exposed horizontally, four bars are exposed longitudinally and the bending value of the bars is 5 cm.
	S3	The reinforced concrete slab has a through hole, with a hole diameter of 23 cm. Five bars are exposed horizontally and longitudinally, and the bending value of the bars is 10 cm.
	S4	The reinforced concrete slab has a through hole, with a hole diameter of 27.5 cm. Five bars are exposed transversely and seven bars are exposed longitudinally. The bending value of the bars is 10.5 cm.
Double-layer reinforced slabs	D1	The reinforced concrete slab has a through hole, with a hole diameter of 14 cm, and two bars are exposed horizontally and longitudinally. The bending value of the first layer of reinforcement is 3 cm, and the bending value of the second layer of reinforcement is 4 cm.
	D2	The reinforced concrete slab has a through hole, with a hole diameter of 18 cm. Two bars are exposed horizontally and longitudinally. The bending value of the first layer of reinforcement is 4 cm, and the bending value of the second layer of reinforcement is 6 cm.
	D3	The reinforced concrete slab has a through hole, with a hole diameter of 22 cm. Two bars are exposed horizontally and longitudinally. The bending value of the first layer of reinforcement is 6.5 cm, and the bending value of the second layer of reinforcement is 10.5 cm.
	D4	The reinforced concrete slab has a through hole, with a hole diameter of 23.5 cm. Two bars are exposed horizontally and longitudinally. The bending value of the first layer of reinforcement is 7.5 cm, and the bending value of the second layer of reinforcement is 11.5 cm.

3.2. Failure Mode of Contact Explosion

The failure mode of the RC slab under the contact explosion load is mainly local failure, which can be summarized into four types, namely explosion pit, explosion collapse, explosion penetration and explosion punching. The schematic diagram of the four failure modes is shown in Figure 3. Within the range of charge in this paper, the single-layer reinforced slab and double-layer reinforced slab both show explosive penetration damage, and the concrete medium near the center of the top face is crushed and peeled off to form a blast hole. The compression stress wave caused by the explosion will be reflected on the bottom face, and the resulting reflection stretching effect will cause the bottom face concrete to crack and collapse, thus forming a collapse hole. In addition, the top face blast hole and bottom face collapse hole will penetrate up and down.

3.3. Analysis of Damage Area of Contact Explosion

The damage area of the top face, the damage area of the bottom face and the diameter of the through hole of RC slabs under contact explosion are shown in Table 6.

Table 6. Measurement data of damage form under contact explosion.

Types of RC Slabs	Specimen	Charge (kg)	The Damage Area of Top Face (cm^2)	The Damage Area of Bottom Face (cm^2)	Diameter of Through Hole (cm)
Single-layer reinforced slabs	S1	0.2	750	1400	15
	S2	0.4	1000	1950	21
	S3	0.8	1100	2225	23
	S4	1.6	1275	2850	27.5
Double-layer reinforced slabs	D1	0.2	650	1375	14
	D2	0.4	825	1500	18
	D3	0.8	975	1925	22
	D4	1.6	1125	2125	23.5

Under the contact explosion, the top face is compressed and destroyed to form a blasting pit. The parameters that affect the blasting damage form of the top face include the explosion source factors (such as charge and explosive density) and the medium factors (such as concrete strength, concrete density, reinforcement strength, reinforcement density and wave velocity). Due to the main compression failure of the top face, it is assumed that the damage area of top face A_1 is a function of charge M, the density of explosive ρ_1, compressive strength of concrete f_c, density of concrete ρ_2, compressive strength of reinforcement f_c', density of reinforcement ρ_3, wave velocity V and thickness of slabs H. It is assumed that the damage area of bottom face A_2 is a function of charge M, density of explosive ρ_1, tensile strength of concrete f_t, density of concrete ρ_2, tensile strength of reinforcement f_t', density of reinforcement ρ_3, wave velocity V and thickness of slabs H. It is assumed that the through hole diameter D is a function of charge M, density of explosive ρ_1, compressive strength of concrete f_c, tensile strength of concrete f_t, density of concrete ρ_2, compressive strength of reinforcement f_c', tensile strength of reinforcement f_t', density of reinforcement ρ_2, wave velocity V and thickness of slabs H. The dimension of each parameter is shown in Table 7. For the dimensional analysis of the damage area of top face A_1, it can be expressed as Equation (1):

$$A_1 = f(M, \rho_1, f_c, \rho_2, f_c', \rho_3, V, H) \tag{1}$$

Table 7. Dimension of blasting parameters.

Parameter	Symbol	Dimensions
Charge	M	M
Density of explosive	ρ_1	ML^{-3}
Compressive strength of concrete	f_c	$ML^{-1}T^{-2}$
Tensile strength of concrete	f_t	$ML^{-1}T^{-2}$
Density of concrete	ρ_2	ML^{-3}
Compressive strength of reinforcement	f_c'	$ML^{-1}T^{-2}$
Tensile strength of reinforcement	f_t'	$ML^{-1}T^{-2}$
Density of reinforcement	ρ_3	ML^{-3}
Wave velocity	V	LT^{-1}
The damage area of top face	A_1	L^2
The damage area of bottom face	A_2	L^2
The diameter of through hole	D	L
Thickness of slabs	H	L

Select M, f_c, ρ_2 as independent variables and list three dimensionless Π values as shown in Equation (2):

$$\Pi_1 = \frac{\rho_1}{M^{a_1} f_c^{a_2} \rho_2^{a_3}}, \Pi_2 = \frac{f_c'}{M^{b_1} f_c^{b_2} \rho_2^{b_3}}, \Pi_3 = \frac{\rho_3}{M^{c_1} f_c^{c_2} \rho_2^{c_3}},$$
$$\Pi_4 = \frac{V}{M^{d_1} f_c^{d_2} \rho_2^{d_3}}, \Pi_5 = \frac{A_1}{M^{e_1} f_c^{e_2} \rho_2^{e_3}}, \Pi_6 = \frac{H}{M^{f_1} f_c^{f_2} \rho_2^{f_3}} \tag{2}$$

The expression of each dimensionless quantity Π is shown in Equation (3):

$$\begin{aligned}
ML^{-3} &= M^{a_1}(ML^{-1}T^{-2})^{a_2}(ML^{-3})^{a_3} \\
ML^{-1}T^{-2} &= M^{b_1}(ML^{-1}T^{-2})^{b_2}(ML^{-3})^{b_3} \\
ML^{-3} &= M^{c_1}(ML^{-1}T^{-2})^{c_2}(ML^{-3})^{c_3} \\
LT^{-1} &= M^{d_1}(ML^{-1}T^{-2})^{d_2}(ML^{-3})^{d_3} \\
L^2 &= M^{e_1}(ML^{-1}T^{-2})^{e_2}(ML^{-3})^{e_3} \\
L &= M^{f_1}(ML^{-1}T^{-2})^{f_2}(ML^{-3})^{f_3}
\end{aligned} \tag{3}$$

Determine the indexes of Π according to the principle of dimensional consistency as shown in Equation (4):

$$\Pi_1 = \frac{\rho_1}{M^0 f_c^0 \rho_2^1}, \quad \Pi_2 = \frac{f_c'}{M^0 f_c^1 \rho_2^0}, \quad \Pi_3 = \frac{\rho_3}{M^0 f_c^0 \rho_2^1},$$
$$\Pi_4 = \frac{V}{M^0 f_c^{1/2} \rho_2^{1/2}}, \quad \Pi_5 = \frac{A_1}{M^{2/3} f_c^0 \rho_2^{2/3}}, \quad \Pi_6 = \frac{H}{M^{1/3} f_c^0 \rho_2^{-1/3}} \quad (4)$$

The dimensional function relationship can be obtained as shown in Equations (5) and (6):

$$\frac{A_1}{M^{2/3} \rho_2^{2/3}} = f\left(\frac{\rho_1}{\rho_2}, \frac{f_c'}{f_c}, \frac{\rho_3}{\rho_2}, \frac{V}{f_c^{1/2} \rho_2^{1/2}}, \frac{H}{M^{1/3} \rho_2^{-1/3}}\right) \quad (5)$$

$$A_1 = M^{2/3} \rho_2^{2/3} f\left(\frac{\rho_1}{\rho_2}, \frac{f_c'}{f_c}, \frac{\rho_3}{\rho_2}, \frac{V}{f_c^{1/2} \rho_2^{1/2}}, \frac{H}{M^{1/3} \rho_2^{-1/3}}\right) \quad (6)$$

The explosion and the material of media in the experiment are constant and the density of explosive ρ_1, compressive strength of concrete f_c, density of concrete ρ_2, compressive strength of reinforcement f_c', density of reinforcement ρ_3, wave velocity V and thickness of slabs H are constant, so the above function form can be simplified as Equation (7):

$$A_1 = k_1 M^{2/3} + a \quad (7)$$

In the same way, the functional form of the damage area of the bottom face can be simplified as Equation (8):

$$A_2 = k_2 M^{2/3} + b \quad (8)$$

The functional form of the diameter of through hole can be simplified as Equation (9):

$$D = k_3 M^{1/3} + c \quad (9)$$

Based on the test data of single-layer reinforced slabs and double-layer reinforced slabs, the fitting formula of the damage area of the top face A_1, the damage area of the bottom face A_2 and the diameter of through hole D are obtained by fitting the function form derived from dimensional analysis, as shown in Equations (10)–(15). The fitting curve is shown in Figures 4–6, and the determination coefficient R^2 of the fitting curve is greater than 0.8. For the explosion test, the fitting result is ideal. It can be seen from the formula that k_1, k_2 and k_3 are parameters reflecting the anti-explosion performance of the medium. The smaller their values, the better the anti-explosion performance of the RC slab.

Single-layer reinforced slabs: $A_1 = 0.047 M^{2/3} + 0.067 \quad R^2 = 0.902$ (10)

Double-layer reinforced slabs: $A_1 = 0.044 M^{2/3} + 0.055 \quad R^2 = 0.919$ (11)

Single-layer reinforced slabs: $A_2 = 0.132 M^{2/3} + 0.108 \quad R^2 = 0.942$ (12)

Double-layer reinforced slabs: $A_2 = 0.076 M^{2/3} + 0.114 \quad R^2 = 0.906$ (13)

Single-layer reinforced slabs: $D = 0.199 M^{1/3} + 0.046 \quad R^2 = 0.913$ (14)

Double-layer reinforced slabs: $D = 0.162 M^{1/3} + 0.055 \quad R^2 = 0.882$ (15)

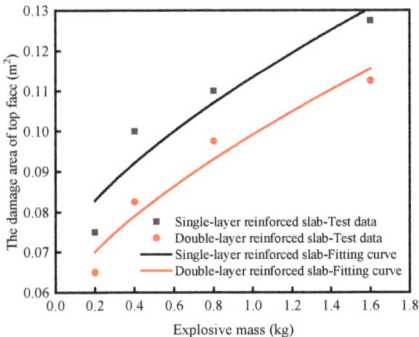

Figure 4. The change rule of the damage area of top face under contact explosion with the charge.

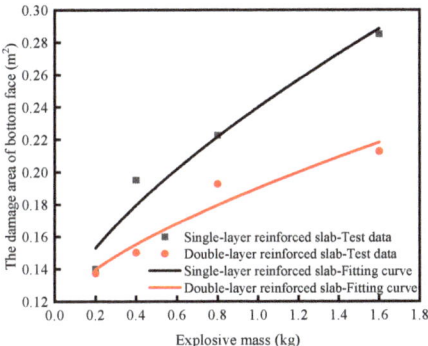

Figure 5. The change rule of the damage area of bottom face under contact explosion with the charge.

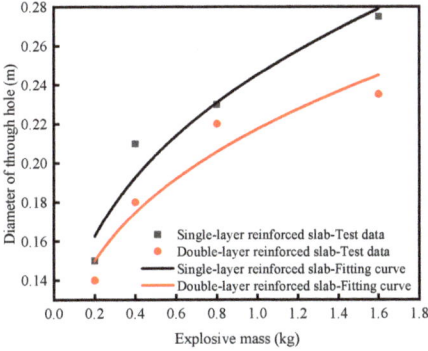

Figure 6. The change rule of the diameter of through hole under contact explosion with the charge.

4. Experimental Results and Analysis of Non-Contact Explosion
4.1. Damage Form of Non-Contact Explosion

Under the non-contact explosion, the dynamic response law of the RC slab is studied by changing the charge. The contact explosion experimental results are shown in Table 8 and Figure 7.

Table 8. Test results of non-contact explosion of RC slabs.

Types of RC Slabs	Specimen	Description of Phenomenon
Single-layer reinforced slabs	S5	There are traces of explosion on the top face, but no crater or crack is formed; there is a small crack in the center of the bottom face.
	S6	There are traces of explosion on the top face, but there are no explosion pits and tiny cracks; there is a seismic collapse pit on the bottom face with an area of 1075 cm^2 and a depth of 4.5 cm. Three bars are exposed horizontally and one bar is exposed longitudinally.
	S7	There are traces of explosion on the top face, but there are no explosion pits and tiny cracks; there is a collapse pit on the bottom face with an area of 975 cm^2 and a depth of 4 cm. Two bars are exposed horizontally and two bars are exposed longitudinally.
	S8	There are traces of explosion on the top face, but there are no explosion pits and tiny cracks; there are circumferential cracks and cracks emanating from the center to the periphery on the bottom face, and the diameter of circumferential cracks is 7 cm.
Double-layer reinforced slabs	D5	The reinforced concrete slab is free of damage and cracks.
	D6	There are traces of explosion on the top face, but there are no explosion pits and tiny cracks; there is a seismic collapse pit on the bottom face with an area of 1400 cm^2 and a depth of 3.5 cm. One steel bar is exposed horizontally and two steel bars are exposed longitudinally.
	D7	There are traces of explosion on the top face, but there are no explosion pits and tiny cracks; there are circumferential cracks and cracks emanating from the center to the periphery on the bottom face, and the diameter of circumferential cracks is 6 cm.
	D8	There are traces of explosion on the top face, but there are no explosion pits and tiny cracks; there are circumferential cracks and cracks emanating from the center to the periphery on the bottom face, and the diameter of circumferential cracks is 5 cm.

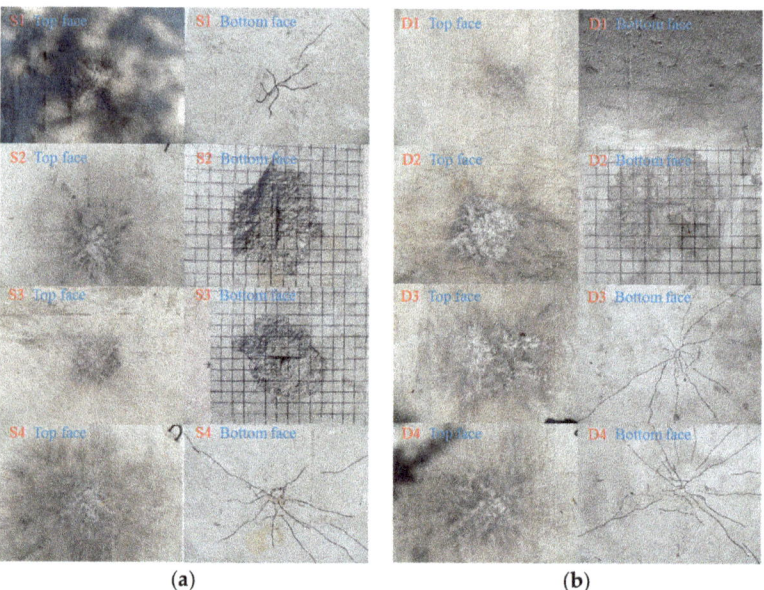

Figure 7. Non-contact explosion failure mode of RC slabs. (**a**) Single-layer reinforced slab; (**b**) double-layer reinforced slab.

4.2. Failure Mode of Contact Explosion

The failure modes of RC slabs under non-contact explosive loads are mainly local failure and overall bending failure. The explosion shock wave causes damage on the blast face of the slab, and there is no crack or fine crack on the blast face. When the compression stress wave propagates to the bottom face of the slab, it will reflect and transmit, and the compression stress wave will be transformed into a tensile wave. Because of the low tensile strength of the concrete, the center of the bottom face will crack due to bending, the concrete will crack or even peel off and the slab will be bent.

4.3. Analysis of Damage Area of Contact Explosion

The damage area of the bottom face is shown in Table 9.

Table 9. Measurement data of damage form under non-contact explosion.

Types of RC Slabs	Specimen	Explosive Mass (kg)	Blast Distance(m)	Scale Distance (m·kg$^{-1/3}$)	The Damage Area of the Bottom Face (cm^2)
Single-layer reinforced slab	S5	0.2	0.25	0.43	-
	S6	0.4	0.32	0.43	1075
	S7	0.8	0.40	0.43	975
	S8	1.6	0.50	0.43	-
Double-layer reinforced slab	D5	0.2	0.25	0.43	-
	D6	0.4	0.32	0.43	1400
	D7	0.8	0.40	0.43	-
	D8	1.6	0.50	0.43	-

It can be seen from Figure 7 and Table 9 that the damage of the single-layer reinforced slab is more serious than that of the double-layer reinforced slab. When the scale distance is constant, with an increase in the blast distance, there will be slight cracks on the top face of the slab. The blast face of the slab will gradually increase from only cracks to concrete falling off, and then the concrete falling off area will gradually decrease until only cracks appear. The severity of the damage form will first increase and then decrease. This is because the initial burst distance and charge quantity are relatively small. Due to the size effect, it is difficult for the smaller-size explosives to damage the larger-size reinforced concrete slab. With an increase in the blast distance and charge quantity, the shackles of the size effect are broken, and the explosive produces large local damage on the reinforced concrete slab, but the damage at this time is caused by the joint action of the blast load and explosive gas after the explosion. With a further increase in the blast distance and charge quantity, the damage caused by explosives on the reinforced concrete slab gradually changes from local damage to overall damage. Due to the increase in blast distance, the explosive gas after the explosion of explosives escapes into the air. At this time, only the blast load acts on the reinforced concrete slab, resulting in a gradual reduction in the damage caused by explosives on the reinforced concrete slab.

4.4. Response Analysis of Displacement

The comparison of displacement–time-history curves of single-layer reinforced slabs and double-layer reinforced slabs is shown in Figure 8. The displacement data of the test are shown in Table 10.

Table 10. The displacement data of test.

Specimen	The Peak of Displacement (mm)	The Peak of Rebound Displacement (mm)	Residual Deformation (mm)
S5	−3.99	1.85	−0.70
S6	−8.24	5.87	−1.99
S7	−13.19	11.64	−4.25
S8	−17.99	18.40	−5.61
D5	−4.10	−1.10	−1.97
D6	−10.10	4.28	−4.06
D7	−11.98	9.59	−4.36
D8	−13.91	13.12	−6.95

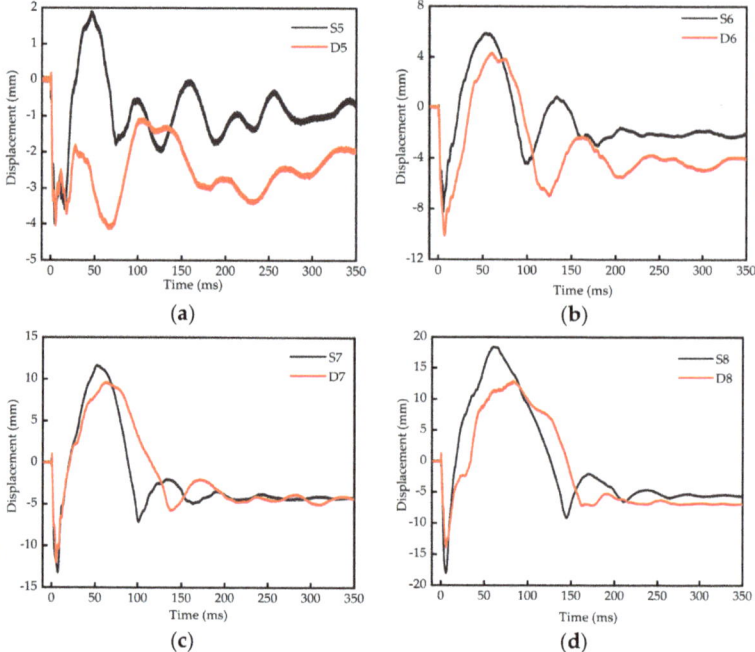

Figure 8. Comparison of displacement–time-history curves of single-layer and double-layer RC slabs. (**a**) 0.2 kg; (**b**) 0.4 kg; (**c**) 0.8 kg; (**d**) 1.6 kg.

It can be seen from Figure 8 and Table 10 that the peak displacement of a single-layer reinforced plate is 3.99 mm and 8.24 mm, respectively, the peak displacement of a double-layer reinforced slab is 4.10 mm and 10.10 mm, respectively, and the peak displacement of a single-layer reinforced slab is smaller when the blast distance and charge are relatively small; that is, the blast distance is 0.25 m and 0.32 m and the charge is 0.2 kg and 0.4 kg. When the blast distance and charge are relatively large—that is, when the blast distance is 0.40 m and 0.50 m and the charge is 0.8 kg and 1.6 kg—the displacement peak of the double-layer reinforced slab is 11.98 mm and 13.91 mm, respectively, the displacement peak of the single-layer reinforced slab is 13.19 mm and 17.99 mm, respectively, and the displacement peak of the double-layer reinforced slab is smaller. When the blast distance and charge are relatively small, the RC slab is mainly subject to local damage. At this time, the reinforcement of the single-layer reinforced slab is closer to the bottom face, and its displacement peak value is smaller. When the blast distance and charge are relatively large, the RC slab is mainly subject to overall damage. At this time, the overall structure of the double-layer reinforced slab is better, and its displacement peak value is smaller. The peak rebound displacement of a single-layer reinforced slab is 1.85, 5.87, 11.64 and 18.40 mm, respectively, and that of a double-layer reinforced slab is −1.10, 4.28, 9.59 and 13.12 mm, respectively. The peak rebound displacement of a single-layer reinforced slab is greater than that of a double-layer reinforced slab. The residual deformation of a single-layer reinforced slab is 0.70, 1.99, 4.25 and 5.61 mm, respectively, and that of a double-layer reinforced slab is 1.97, 4.06, 4.36 and 6.93 mm, respectively. The residual deformation of a single-layer reinforced slab is less than that of a double-layer reinforced slab.

The comparison of displacement–time-history curves of RC slabs with different blast distances is shown in Figure 9. It can be seen from Figure 9 that the peak displacement, rebound peak displacement and residual deformation of both single-layer and double-layer reinforced slabs increase with an increase in the blast distance and charge.

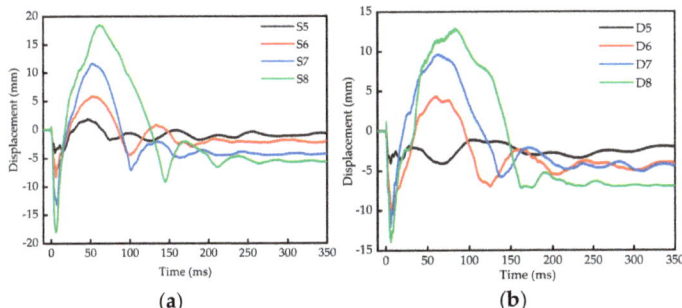

Figure 9. Comparison of displacement–time-history curves of RC slabs with different blasting distances. (**a**) Single-layer reinforced slab; (**b**) double-layer reinforced slab.

5. Discussion

In this paper, the damage degree of RC slabs can be determined through the failure patterns and test data. From the perspective of failure patterns, the degree of damage to reinforced concrete slabs increases with an increase in charge under contact explosion, and, under the same conditions, the damage degree of double-layer reinforced slabs is significantly smaller than that of single-layer reinforced slabs. The damage degree of RC slabs increases with an increase in charging under a non-contact explosion, and, under the same conditions, the damage degree of double-layer reinforced slabs is significantly smaller than that of single-layer reinforced slabs.

However, from the perspective of test data, the results of the test are not entirely consistent. The damage of RC slabs is often related to the dynamic response of the RC slabs. The parameter tested in this paper was displacement, and the maximum bearing rotation angle θ_{max} can be calculated through the peak displacement. The dimensionless parameter θ_{max} is a standard for evaluating the degree of damage to reinforced concrete slab components. The calculation formula for the maximum bearing rotation angle θ_{max} is shown in Equation (16) [41]:

$$\theta_{max} = \tan^{-1}\left(\frac{x_{max}}{L/2}\right) \qquad (16)$$

where L is the span of the RC slabs and x_{max} is the maximum displacement at the mid-span of the RC slabs.

In ref. [41], it is pointed out that the larger the maximum bearing rotation angle θ_{max}, the greater the degree of damage to the RC slabs. For RC slabs of the same size, the larger the peak displacement, the greater the degree of damage to the RC slabs. From the test data, it can be seen that, with an increase in the blast distance and charge, the peak displacement gradually increases, which means that the degree of damage to the RC slabs increases. When the blast distance and charge are small, the displacement of the double-layer reinforced slabs is greater than that of the single-layer reinforced slabs, which means that the damage degree of the double-layer reinforced slabs is greater than that of the single-layer reinforced slabs. When the blast distance and charge are large, the displacement of the single-layer reinforced slabs is greater than that of the double-layer reinforced slabs, which means that the damage degree of the single-layer reinforced slabs is greater than that of the double-layer reinforced slabs. This is not entirely consistent with the results obtained based on the failure patterns. When the degree of damage determined by different standards is inconsistent, it is necessary to comprehensively consider the degree of damage obtained by various standards. Usually, the standard with the most severe degree of damage can be used as the criterion.

Liu et al. [42] conducted experiments on arch structure with the same scale distance (0.5 m·kg$^{-1/3}$) and different blast distances and found that, the larger the blast distance, the more severe the damage to the arch structure and the larger the peak displacement.

However, in this article, an inconsistent conclusion was drawn, where, the larger the blast distance, the greater the failure of the slab, which first increases and then decreases, but the peak displacement always increases. The difference in scale distance between the two is not significant, and the reason for different conclusions may be due to the different range of blast distance. In the experiment of the former, the blast distance range was 0.5–1.0 m, whereas, in the experiment of this paper, the blast distance range was 0.25–0.50 m.

6. Conclusions

In this paper, field chemical explosion experiments were carried out on RC slabs with the same reinforcement ratio but different reinforcement distribution and the same blast distance but different scale distance. The influence of different reinforcement distribution and different blast distance on the anti-blast performance of RC slabs was studied and the damage form and test data of RC slabs were compared and analyzed.

1. The charge increased from 0.2 kg to 1.6 kg under contact explosion, and the damage of single-layer reinforced slabs was more serious than that of double-layer reinforced slabs. The fitting relationships of two RC slabs between the damage area of the top face, the damage area of the bottom face, the diameter of the through hole and charge were obtained. The damage area of the top face of single-layer reinforced slabs was 12.8–21.2% larger than that of double-layer reinforced slabs. The damage area of the bottom face of single-layer reinforced slabs was 1.8–34.1% larger than that of double-layer reinforced slabs. The diameter of the through hole of single-layer reinforced slabs was 4.5–17.0% larger than that of double-layer reinforced slabs.
2. When the scale distance was the same, the blast distance increased from 0.25 m to 0.50 m under non-contact explosion, the damage degree of single-layer reinforced slabs and double-layer reinforced slabs increased first and then decreased and the peak of displacement, the peak of rebound displacement and residual deformation near the center of the bottom face gradually increased.
3. When the blast distance was small, the peak displacement of single-layer reinforced slabs was 2.8–22.6% larger than that of double-layer reinforced slabs. When the blast distance was large, the peak displacement of double-layer reinforced slabs was 10.1–29.3% larger than that of single-layer reinforced slabs. No matter how the blast distance changed, the peak rebound displacement of the double-layer reinforced slabs was 17.6–27.1% smaller than that of the single-layer reinforced slabs, and the residual deformation of the double-layer reinforced slabs was 1.03–2.81 times that of the single-layer reinforced slabs.
4. When designing and constructing RC slabs, double-layer or multi-layer reinforcement can be considered to improve the blast resistance of RC slabs. When studying blast load conditions, not only scale distance but also blast distance should be considered.

Author Contributions: Conceptualization, J.W. and F.L.; methodology, C.M.; investigation, M.X.; resources, J.W.; data curation, S.Y.; writing—original draft preparation, Y.W.; writing—review and editing, J.W., F.L. and C.M.; supervision, F.L.; project administration, J.W.; funding acquisition, J.W. and C.M. All authors have read and agreed to the published version of the manuscript.

Funding: This research was funded by National Key Research and Development Program of China (grant number 2021YFC3100802).

Institutional Review Board Statement: Not applicable.

Informed Consent Statement: Not applicable.

Data Availability Statement: The data presented in this study are available on request from the corresponding author.

Conflicts of Interest: The authors declare no conflict of interest.

References

1. Sagaseta, J.; Olmati, P.; Micallef, K.; Cormie, D. Punching shear failure in blast-loaded RC slabs and panels. *Eng. Struct.* **2017**, *147*, 177–194. [CrossRef]
2. Kumar, V.; Kartik, K.; Iqbal, M. Experimental and numerical investigation of reinforced concrete slabs under blast loading. *Eng. Struct.* **2020**, *206*, 110125. [CrossRef]
3. Mendonca, F.; Urgessa, G.; Iha, K.; Rocha, R.; Rocco, J. Comparison of predicted and experimental behaviour of RC slabs subjected to blast using SDOF analysis. *Def. Sci. J.* **2018**, *68*, 138–143. [CrossRef]
4. Wang, W.; Zhang, D.; Lu, F.; Wang, S.; Tang, F. Experimental study and numerical simulation of the damage mode of a square reinforced concrete slab under close-in explosion. *Eng. Fail. Anal.* **2013**, *27*, 41–51. [CrossRef]
5. Lin, S.; Xu, Y.; Yang, P.; Gao, S.; Zhou, Y.; Xu, J. The research into the propagation law of the shock wave of a gas explosion inside a building. *Shock. Vib.* **2021**, *2021*, 4939014. [CrossRef]
6. Wang, W.; Zhang, D.; Lu, F.; Wang, S.; Tang, F. Experimental study on scaling the explosion resistance of a one-way square reinforced concrete slab under a close-in blast loading. *Int. J. Impact Eng.* **2012**, *49*, 158–164. [CrossRef]
7. Mendonca, F.; Urgessa, G.; Dutra, R.; Goncalves, R.; Iha, K.; Rocco, J. EPS foam blast attenuation in full-scale field test of reinforced concrete slabs. *Acta Sci. Technol.* **2020**, *42*, e40020. [CrossRef]
8. Lim, K.; Han, T.; Lee, J. Numerical simulation on dynamic behavior of slab-column connections subjected to blast loads. *Appl. Sci.* **2021**, *11*, 7573. [CrossRef]
9. Temsah, Y.; Jahami, A.; Khatib, J.; Sonebi, M. Numerical analysis of a reinforced concrete beam under blast loading. *MATEC Web Conf.* **2018**, *149*, 02063. [CrossRef]
10. Das Adhikary, S.; Chandra, L.; Christian, A.; Ong, K.C.G. SHCC-strengthened RC panels under near-field explosions. *Constr. Build. Mater.* **2018**, *183*, 675–692. [CrossRef]
11. Conrad, K.; Abass, B. Effects of transverse reinforcement spacing on the response of reinforced concrete columns subjected to blast loading. *Eng. Struct.* **2017**, *142*, 148–164.
12. Zhu, H.; Luo, X.; Ji, C.; Wang, X.; Wang, Y.; Zhao, C.; Zhang, L. Strengthening of clay brick masonry wall with spraying polyurea for repeated blast resistance. *Structures* **2023**, *53*, 1069–1091. [CrossRef]
13. Lyu, P.; Fang, Z.; Wang, X.; Huang, W.; Zhang, R.; Sang, Y.; Sun, P. Explosion test and numerical simulation of coated reinforced concrete slab based on blast mitigation polyurea coating performance. *Materials* **2022**, *15*, 2607. [CrossRef]
14. Studzinski, R.; Gajewski, T.; Malendowski, M.; Sumelka, W.; Al-Rifaie, H.; Peksa, P.; Sielicki, P. Blast test and failure mechanisms of soft-core sandwich panels for storage halls applications. *Materials* **2021**, *14*, 70. [CrossRef]
15. Zhao, C.; Wang, Q.; Lu, X.; Huang, X.; Mo, Y. Blast resistance of small-scale RCS in experimental test and numerical analysis. *Eng. Struct.* **2019**, *199*, 109610. [CrossRef]
16. Wang, W.; Zhang, D.; Lu, F.; Tang, F.; Wang, S. Pressure-impulse diagram with multiple failure modes of one-way reinforced concrete slab under blast loading using SDOF method. *J. Cent. South Univ.* **2013**, *20*, 510–519. [CrossRef]
17. Dua, A.; Braimah, A. Assessment of reinforced concrete slab response to contact explosion effects. *J. Perform. Constr. Fac.* **2020**, *34*, 04020061. [CrossRef]
18. Kee, J.; Park, J.; Seong, J. Effect of one way reinforced concrete slab characteristics on structural response under blast loading. *Adv. Concr. Constr.* **2019**, *8*, 277–283.
19. Gomathi, K.; Rajagopal, A. Dynamic performance of reinforced concrete slabs under impact and blast loading using plasticity-based approach. *Int. J. Struct. Stab. Dyn.* **2020**, *20*, 2043015. [CrossRef]
20. Zhao, C.; Ye, X.; Gautam, A.; Lu, X.; Mo, Y. Simplified theoretical analysis and numerical study on the dynamic behavior of FCP under blast loads. *Front. Struct. Civ. Eng.* **2020**, *14*, 983–997. [CrossRef]
21. Mendonca, F.; Urgessa, G.; Almeida, L.; Rocco, J. Damage diagram of blast test results for determining reinforced concrete slab response for varying scaled distance, concrete strength and reinforcement ratio. *An. Acad. Bras. Cienc.* **2021**, *93*, e20200511. [CrossRef] [PubMed]
22. Yao, S.; Zhang, D.; Chen, X.; Lu, F.; Wang, W. Experimental and numerical study on the dynamic response of RC slabs under blast loading. *Eng. Fail. Anal.* **2016**, *66*, 120–129. [CrossRef]
23. Wang, L.; Cheng, S.; Liao, Z.; Yin, W.; Liu, K.; Ma, L.; Wang, T.; Zhang, D. Blast resistance of reinforced concrete slabs based on residual load-bearing capacity. *Materials* **2022**, *15*, 6449. [CrossRef] [PubMed]
24. Thiagarajan, G.; Johnson, C. Experimental behavior of reinforced concrete slabs subjected to shock loading. *ACI Struct. J.* **2014**, *111*, 1407–1417. [CrossRef]
25. Krauthammer, T.; Astarlioglu, S. Direct shear resistance models for simulating buried RC roof slabs under airblast-induced ground shock. *Eng. Struct.* **2017**, *140*, 308–316. [CrossRef]
26. Maazoun, A.; Belkassem, B.; Reymen, B.; Matthys, S.; Vantomme, J.; Lecompte, D. Blast response of RC slabs with externally bonded reinforcement: Experimental and analytical verification. *Compos. Struct.* **2018**, *200*, 246–257. [CrossRef]
27. Gilbert, R.; Sakka, Z. Effect of reinforcement type on the ductility of suspended reinforced concrete slabs. *J. Struct. Eng.* **2007**, *133*, 834–843. [CrossRef]
28. Feng, J.; Zhou, Y.; Wang, P.; Wang, B.; Zhou, J.; Chen, H.; Fan, H.; Jin, F. Experimental research on blast-resistance of one-way concrete slabs reinforced by BFRP bars under close-in explosion. *Eng. Struct.* **2017**, *150*, 550–561. [CrossRef]

29. Li, J.; Wu, C.; Hao, H. An experimental and numerical study of reinforced ultra-high performance concrete slabs under blast loads. *Mater. Des.* **2015**, *82*, 64–76. [CrossRef]
30. Lin, X. Numerical simulation of blast responses of ultra-high performance fibre reinforced concrete panels with strain-rate effect. *Constr. Build. Mater.* **2018**, *176*, 371–382. [CrossRef]
31. Ren, J.; Cheng, H.; Yang, C.; Dai, R. Experimental study of blast resistance on BFRP bar-reinforced cellular concrete slabs fabricated with millimeter-sized saturated SAP. *Shock. Vib.* **2019**, *2019*, 5814172. [CrossRef]
32. Sherif, M.; Othman, H.; Marzouk, H.; Aoude, H. Design guidelines and optimization of ultra-high-performance fibre-reinforced concrete blast protection wall panels. *Int. J. Prot. Struct.* **2020**, *11*, 494–514. [CrossRef]
33. Liao, Q.; Xie, X.; Yu, J. Numerical investigation on dynamic performance of reinforced ultra-high ductile concrete-ultra-high performance concrete panel under explosion. *Struct. Concrete* **2022**, *23*, 3601–3615. [CrossRef]
34. Liao, Q.; Yu, J.; Xie, X.; Ye, J.; Jiang, F. Experimental study of reinforced UHDC-UHPC panels under close-in blast loading. *J. Build. Eng.* **2022**, *46*, 103498. [CrossRef]
35. Mendonca, F.; Urgessa, G.; Rocco, J. Experimental investigation of 50 MPa reinforced concrete slabs subjected to blast loading. *Ing. Investig.* **2018**, *38*, 27–33. [CrossRef]
36. Gao, B.; Wu, J.; Chen, Q.; Yu, J.; Yu, H. Effect of spraying polyurea on the anti-blast performance of the Ultra-High performance concrete slab. *Sensors* **2022**, *22*, 9888. [CrossRef]
37. Gao, H.; Liu, Z.; Yang, Y.; Wu, C.; Geng, J. Blast-resistant performance of aluminum foam-protected reinforced concrete slabs. *Explos. Shock. Waves* **2019**, *39*, 023101.
38. Thiagarajan, G.; Reynolds, K. Experimental behavior of High-Strength concrete one-way slabs subjected to shock loading. *ACI Struct. J.* **2017**, *114*, 611–620. [CrossRef]
39. Almustafa, M.; Nehdi, M. Machine learning model for predicting structural response of RC slabs exposed to blast loading. *Eng. Struct.* **2020**, *221*, 111109. [CrossRef]
40. Shishegaran, A.; Khalili, M.; Karami, B.; Rabczuk, T.; Shishegaran, A. Computational predictions for estimating the maximum deflection of reinforced concrete panels subjected to the blast load. *Int. J. Impact Eng.* **2020**, *139*, 103527. [CrossRef]
41. Wang, W. Study on Damage Effects and Assessments Method of Reinforced Concrete Structural Members under Blast Loading. Doctoral Thesis, National University of Defense Technology, Changsha, China, 2012.
42. Liu, G.; Liu, R.; Wang, W.; Wang, X.; Zhao, Q. Blast resistance experiment of underground reinforced concrete arch structure under top explosion. *Chin. J. Energetic Mater.* **2021**, *29*, 157–165.

Disclaimer/Publisher's Note: The statements, opinions and data contained in all publications are solely those of the individual author(s) and contributor(s) and not of MDPI and/or the editor(s). MDPI and/or the editor(s) disclaim responsibility for any injury to people or property resulting from any ideas, methods, instructions or products referred to in the content.

Article

Analytical Study of SH Wave Scattering by a Circular Pipeline in an Inhomogeneous Concrete with Density Variation

Zailin Yang [1], Chenxi Sun [1], Guanxixi Jiang [2], Yunqiu Song [1], Xinzhu Li [1] and Yong Yang [1,*]

[1] College of Aerospace and Civil Engineering, Harbin Engineering University, Harbin 150001, China
[2] School of Physical and Mathematical Sciences, Nanjing Tech University, Nanjing 211800, China
* Correspondence: yangyongheu@163.com

Abstract: In this paper, the shear horizontal (SH) wave scattering by a circular pipeline in an inhomogeneous concrete with density variation is studied. A model of inhomogeneous concrete with density variation in the form of a polynomial-exponential coupling function is established. By using the complex function method and conformal transformation, the incident and scattering wave field of SH wave in concrete are obtained, and the analytic expression of dynamic stress concentration factor (DSCF) around the circular pipeline is given. The results show that the inhomogeneous density parameters, the wave number of the incident wave and the angle of the incident wave in concrete are important factors affecting the distribution of dynamic stress around the circular pipe in concrete with inhomogeneous density. The research results can provide a theoretical reference and a basis for analyzing the influence of circular pipeline on elastic wave propagation in an inhomogeneous concrete with density variation.

Keywords: inhomogeneous concrete; SH wave; circular pipeline; complex function method; dynamic stress concentration factor

1. Introduction

In recent years, there has been significant interest in studying the propagation of elastic waves in solids, as it is crucial for understanding wave propagation mechanisms in engineering applications such as non-destructive testing of structures and the use of new materials [1–8]. Concrete, being a common and popular engineering material, has been extensively studied for its elastic wave propagation mechanism [9–14]. In many concrete structures, circular cavity structures such as pipelines are present, and understanding their response to dynamic loads is important for engineering purposes. At present, there are many different applications for elastic wave propagation in concrete. Many studies have focused on structural damage or defects in concrete. For instance, Ziaja [15] used elastic wave propagation to monitor the state of GFRP-reinforced concrete structural members. They used PZT (lead-zirconate-titanate) sensors to record the changing state of elastic waves caused by cracks and crack propagation in GFRP reinforced concrete structures, considering the material discontinuity caused by cracks and the influence of strain field on wave propagation. Yoon [16] analyzed the applicability of the elastic wave of impact echo (IE) and evaluated six types of prestressed concrete structures using multichannel analysis of surface waves, electromagnetic waves, and shear waves. A more accurate classification method for internal defects in pipelines was proposed by using electromagnetic wave, IE, and principal component analysis (PCA). Beniwal et al. [17] proposed two different ultrasonic imaging techniques designed to use more information contained in the scattered fields for concrete using scattered elastic compression and mode conversion shear wave field modes. Guo [18] established the basic equation of elastic wave propagation in damaged concrete media based on the basic principles of classical elastic dynamics and the damage mechanics model, and derived the fundamental solution of the system. Due to the existence of damage in the structure, the wave response of concrete, including the shape, amplitude, and propagation time of ultrasonic waves in the structure, will change obviously. Additionally, many scholars have used the impact-echo method to detect defects in concrete structures [19,20]. Ali [21] described the theoretical basis of a crack detection and location method for concrete samples based on the time for elastic waves generated by crack formation to reach a set of sensors located at the sample boundary, and presented a location method based on acoustic emission detection, and developed a discretization scheme for two-dimensional elastic equations. Uenishi [22] used a two-dimensional in-plane time-harmonic elastodynamics model to analyze the effects of P wave and SV wave incidence on a circular tunnel with lining located at a finite depth from a nearly flat free surface of a homogeneous isotropic linear elastic medium. They also discussed the influence of wavelength and incidence angle, covering layer thickness, and relative compliance on the relative compliance of the linear elastic lining. The results of spalling of lining concrete, buckling of side wall reinforcement, and disengagement of subgrade from invert were given.

In practical engineering structures, the uneven density of concrete can significantly impact the mechanical properties of materials and structures. For example, Lu and Liu [23] analyzed the maximum first principal stress and mid-span deflection increment of density gradient concrete continuous rigid frame bridges under shrinkage and creep effect. Their research aimed to address the problems of excessive mid-span deflection and box girder cracking. The results showed that the effect of shrinkage and creep was reduced by a continuous rigid frame bridge with density gradient concrete. This research provided a theoretical basis for the successful application of continuous rigid-frame bridge with density gradient concrete. Not only do inhomogeneous concrete structures appear in practical projects, but the 3D printing technology of concrete is also becoming increasingly mature, making it possible to prepare concrete materials with gradually functional gradients [24]. Foamed concrete with functional gradients also plays a role in protecting the structure. For example, Strieder [25] used a simplified model to study the influence of gradient concrete material distribution on crack reduction in mass concrete structures. It also demonstrated that graded concrete may help reduce the confinement stress and weaken the risk of cracks during the hardening of concrete. Wang [26] proposed a layered graded foamed concrete-filled tensile honeycomb structure, which achieved multi-level structural protection by adjusting its overall compression deformation mode to layer-by-layer deformation mode. For example, the coagulative density of the new material UR50 ultra-early-strength concrete could reach 2600 kg/m^3. This concrete is characterized by ultra-high strength, ultra-high toughness, ultra-impact resistance, and ultra-high durability. Wei et al. [27,28] also researched the mechanical properties and penetration resistance of this high-density concrete. The density of foamed concrete was relatively low, and the density span was large, ranging from 300 kg/m^3 to 1600 kg/m^3. Therefore, some scholars investigated the performance and structure of foamed concrete with different densities [29]. Foamed concrete backfill has been proven to be an ideal material for improving the bearing capacity of underground engineering structures, except for buried pipelines. Wang [30] studied the effect of foamed concrete backfill on improving the anti-knock performance of buried pipelines. In addition, the study of elastic wave propagation in heterogeneous concrete is of great significance. The propagation characteristics of elastic waves can be used to analyze the structural characteristics of heterogeneous concrete. Metais [31] investigated the impact of multiple scattering on the dispersion curve of phase dry surface waves by considering elastic circular inclusions in an elastic matrix. The dispersion curves were calculated using the global neighborhood algorithm and were inverted to obtain a solution for layered media with linear uniform and isotropic elastic layers. The study quantified the effect of multiple scattering on the results. As the phase velocity of surface waves does not change with frequency, a solution consisting of uniform layers was obtained through inversion. Many scholars have studied the dynamic stress response of homogeneous concrete with defects by numerical simulation or experiment. The dynamic stress response of inhomogeneous concrete with defects is rarely reported. For example, variations in density and shear modulus in inhomogeneous concrete, and structural forms of defects in concrete. These factors will affect the dynamic stress response of concrete to elastic waves. Even less research has been done on the analytical solutions to such problems. This paper aims to investigate the scattering of SH waves in inhomogeneous concrete containing a circular pipeline using complex function theory and conformal transformation proposed in Refs. [32–36]. In Section 2, the model of density inhomogeneous concrete and the wave field model are introduced. The governing equation is given in Section 3. In Section 4, the stress and displacement fields of the concrete are derived. In Section 5, unknown factors are solved according to boundary conditions, and the expression of dynamic stress concentration factor (DSCF) around the circular pipeline is obtained. In Section 6, the influence of reference wave number and two kinds of density inhomogeneous parameters on DSCF around the circular pipeline is discussed. Finally, Section 7 summarizes the work of this paper.

2. Concrete and Waves Field Model
2.1. Concrete Model

The density of concrete will change in the actual engineering structure [23]. In addition, long and thin pipelines such as drainage pipelines will also exist in concrete structures. In view of the above possible situations, this paper assumes that the density of concrete is inhomogeneous. We propose a concrete structure model in which a large volume of concrete contains a relatively small circular pipeline. It is assumed that concrete media is infinite in a two-dimensional plane, and the shear modulus of the concrete μ is considered to be a constant value of μ_0. The density of the concrete is expressed as a polynomial and exponential coupling function, which changes with x and y in two directions simultaneously and continuously. This problem model is shown in Figure 1. The expressions of concrete density are given by the following equations.

$$\rho(x,y) = \rho_0 \cdot A(x,y) \cdot B(x), \tag{1}$$

$$A(x,y) = \beta_1^2 \beta_2^2 (x^2 + y^2), \tag{2}$$

$$B(x) = \exp(2\beta_2 x), \tag{3}$$

where ρ_0 is the reference density of the concrete, β_1 and β_2 are the inhomogeneous parameters of density; The expression of function A is the polynomial structure of x and y, the expression of function B is the exponential form.

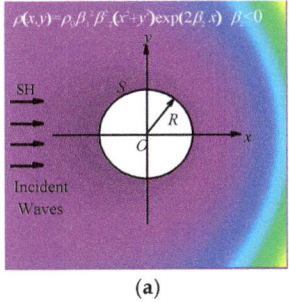
(a)

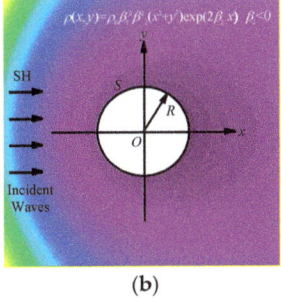
(b)

Figure 1. Model of inhomogeneous concrete containing a circular pipeline.

Since the numerical value of the density of the concrete should exist and be real, neither the inhomogeneous parameters β_1 or β_2 in the density distribution function expression can be equal to 0. Moreover, the values of inhomogeneous parameters β_1 and β_2 affect the variation form of concrete density. The variation of inhomogeneous parameters β_1 affects the density value in the infinite concrete, while the variation of inhomogeneous parameters β_2 not only affects the density value, but also directly affects the density distribution in the concrete.

Then, the expression of wave number k is given by the following formula:

$$k(x,y) = k_0\sqrt{A(x,y) \cdot B(x)}, \qquad (4)$$

$$k_0 = \frac{\omega_0}{\sqrt{\frac{\mu_0}{\rho_0}}}, \qquad (5)$$

where k_0 is the reference wave number, ω_0 is the circular frequency.

2.2. SH Waves Field Model

The scattering field model is shown in Figure 1. The radius of the circular pipeline is R, and the center of the circular pipeline coincides with the coordinate origin O. The density inhomogeneous parameter β_2 in Figure 1a is a positive number, whereas in Figure 1b, the density inhomogeneous parameter β_2 is negative. The change from yellow to purple represents that the concrete density changes from large to small. It can be seen from Figure 1 that the density changes in the 2D direction, and is symmetrically distributed along the x-axis. Based on the symmetry of density in the concrete, the incident SH wave is assumed to incident horizontally along the x-axis. When β_2 is positive, SH waves are incident from the low density to the high density of the concrete; when β_2 is negative, SH waves are incident from the high density to the low density of the concrete. The incident direction of the two cases is completely opposite.

3. Governing Equation

In the Cartesian coordinate system, the wave equation in the concrete with inhomogeneous density is given by the following equation:

$$\frac{\partial^2 \varphi(x,y)}{\partial x^2} + \frac{\partial^2 \varphi(x,y)}{\partial y^2} + k^2(x,y) \cdot \varphi(x,y) = 0, \qquad (6)$$

where $\varphi(x,y)$ is the displacement in the wave field, which is the function of x and y.

Based on the complex function theory, a set of complex variables, $z = x + iy$ and $\bar{z} = x - iy$, where introduced to transform the wave Equation (6) into the following equation:

$$\frac{\partial^2 \varphi}{\partial z \partial \bar{z}} + \frac{1}{4}k^2(z, \bar{z}) \cdot \varphi = 0, \qquad (7)$$

where $k(z, \bar{z})$ is expressed as $k(z, \bar{z}) = k_0 \beta_1 \beta_2 |z| \exp[0.5\beta_2(z + \bar{z})]$ in the coordinates of the complex variables.

To solve wave Equation (5), we need to introduce a new set of variables ζ and $\bar{\zeta}$

$$\zeta = w(z) = \beta_1(z - \frac{1}{\beta_2})\exp(\beta_2 z), \quad \bar{\zeta} = w(\bar{z}) = \beta_1(\bar{z} - \frac{1}{\beta_2})\exp(\beta_2 \bar{z}). \qquad (8)$$

By introducing a new set of variables, the Helmholtz equation with variable coefficients can be transformed into the standard one, allowing for the easy derivation of analytic solutions for displacement and stress fields using the separation of variables method. The standard form of the Helmholtz equation is expressed as:

$$\frac{\partial^2 \varphi}{\partial \zeta \partial \bar{\zeta}} + \frac{1}{4}k_0^2 \varphi = 0. \qquad (9)$$

4. Fields of Displacements and Stresses

The propagation direction of the incident wave is horizontal and the incident angle is $0°$. In the ζ-plane, the displacement of the wave field can be obtained by using the Helmholtz equation in the standard form, and the expression of the incident waves $\varphi^{(i)}$ is as follows.

$$\varphi^{(i)}(\zeta,\overline{\zeta}) = \varphi_0 \exp\left[\frac{ik_0}{2}(\zeta+\overline{\zeta})\right], \tag{10}$$

where φ_0 is the amplitude of the incident wave.

In addition, a circular pipeline exists in an infinite inhomogeneous concrete, and the scattering waves $\varphi^{(s)}$ from the circular pipeline is,

$$\varphi^{(s)}(\zeta,\overline{\zeta}) = \sum_{n=-\infty}^{\infty} C_n H_n^{(1)}(k_0|\zeta|)\left(\frac{\zeta}{|\zeta|}\right)^n, \tag{11}$$

where C_n are undetermined coefficients and $H_n^{(1)}$ is the first kind Hankel function of nth order.

In the concrete, the displacement field should be the superposition of incident waves and scattering waves displacement, so the displacement fields in the concrete with inhomogeneous density can be expressed as:

$$\varphi_t = \varphi^{(i)} + \varphi^{(s)}, \tag{12}$$

where φ_t represents the total displacement field.

In the complex plane, the expression of hoop stresses and radial stresses in the concrete with inhomogeneous density is given by the following equation,

$$\tau_{rz} = \mu_0\left(\frac{\partial\varphi}{\partial z}e^{i\theta} + \frac{\partial\varphi}{\partial \overline{z}}e^{-i\theta}\right), \tag{13}$$

$$\tau_{\theta z} = i\mu_0\left(\frac{\partial\varphi}{\partial z}e^{i\theta} - \frac{\partial\varphi}{\partial \overline{z}}e^{-i\theta}\right), \tag{14}$$

where r, θ, and z are the cylindrical coordinates.

By introducing variables ζ and $\overline{\zeta}$, Equations (13) and (14) can be transformed into the following,

$$\tau_{rz} = \mu_0\left(\frac{\partial\varphi}{\partial \zeta}\frac{d\zeta}{dz}e^{i\theta} + \frac{\partial\varphi}{\partial \overline{\zeta}}\frac{d\overline{\zeta}}{d\overline{z}}e^{-i\theta}\right), \tag{15}$$

$$\tau_{\theta z} = i\mu_0\left(\frac{\partial\varphi}{\partial \zeta}\frac{d\zeta}{dz}e^{i\theta} - \frac{\partial\varphi}{\partial \overline{\zeta}}\frac{d\overline{\zeta}}{d\overline{z}}e^{-i\theta}\right). \tag{16}$$

By substituting the displacement of incident waves and scattering waves into Equations (15) and (16), the stresses field in the infinite inhomogeneous concrete in the ζ plane can be obtained,

$$\tau_{rz}^{(i)} = \frac{1}{2}i\mu_0 k_0 \varphi_0 \beta_1 \beta_2 [z\exp(\beta_2 z + i\theta) + \overline{z}\exp(\beta_2 \overline{z} - i\theta)]\exp\left[\frac{ik_0}{2}(\zeta+\overline{\zeta})\right], \tag{17}$$

$$\tau_{\theta z}^{(i)} = -\frac{1}{2}\mu_0 k_0 \varphi_0 \beta_1 \beta_2 [z\exp(\beta_2 z + i\theta) - \overline{z}\exp(\beta_2 \overline{z} - i\theta)]\exp\left[\frac{ik_0}{2}(\zeta+\overline{\zeta})\right], \tag{18}$$

$$\tau_{rz}^{(s)} = \frac{\mu_0 k_0 \beta_1 \beta_2}{2}\sum_{n=-\infty}^{\infty} C_n\left\{H_{n-1}^{(1)}(k_0|\zeta|)\left(\frac{\zeta}{|\zeta|}\right)^{n-1} z\exp(\beta_2 z + i\theta) - H_{n+1}^{(1)}(k_0|\zeta|)\left(\frac{\zeta}{|\zeta|}\right)^{n+1} \overline{z}\exp(\beta_2 \overline{z} - i\theta)\right\}, \tag{19}$$

$$\tau_{\theta z}^{(s)} = \frac{i\mu_0 k_0 \beta_1 \beta_2}{2}\sum_{n=-\infty}^{\infty} C_n\left\{H_{n-1}^{(1)}(k_0|\zeta|)\left(\frac{\zeta}{|\zeta|}\right)^{n-1} z\exp(\beta_2 z + i\theta) + H_{n+1}^{(1)}(k_0|\zeta|)\left(\frac{\zeta}{|\zeta|}\right)^{n+1} \overline{z}\exp(\beta_2 \overline{z} - i\theta)\right\}. \tag{20}$$

5. Boundary Conditions and Dynamic Stress Concentration Factor of Circular Pipeline

According to the relationship between the infinite concrete and the circular pipeline, it can be determined that the stress freedom should be satisfied on the boundary of the circular pipeline. Therefore, the radial stresses should be zero on this boundary ($|z| = R$) as

$$\tau_{rz}^{(t)} = \tau_{rz}^{(i)} + \tau_{rz}^{(s)} = 0, \ |z| = R. \tag{21}$$

According to Equation (21), the radial stresses expressions of the incident and scattered waves are substituted into Equation (19), and the following expressions can be obtained,

$$\sum_{n=-\infty}^{\infty} C_n \xi_n = \xi, \tag{22}$$

where

$$\xi_n = H_{n-1}^{(1)}(k_0|\zeta|) \left(\frac{\zeta}{|\zeta|}\right)^{n-1} z \exp(\beta_2 z + i\theta) - H_{n+1}^{(1)}(k_0|\zeta|) \left(\frac{\zeta}{|\zeta|}\right)^{n+1} \overline{z} \exp(\beta_2 \overline{z} - i\theta), \tag{23}$$

$$\xi = -i\varphi_0 [z \exp(\beta_2 z + i\theta) + \overline{z} \exp(\beta_2 \overline{z} - i\theta)] \exp\left[\frac{ik_0}{2}(\zeta + \overline{\zeta})\right]. \tag{24}$$

To solve the unknown coefficient C_n, multiply both sides of Equation (22) by $e^{-im\theta}$e, and perform integration over the interval of 2π to obtain Equation (25).

$$\sum_{n=-\infty}^{\infty} \frac{C_n}{2\pi} \int_{-\pi}^{\pi} \xi_n e^{-im\theta} d\theta = \frac{1}{2\pi} \int_{-\pi}^{\pi} \xi e^{-im\theta} d\theta, \ (m = 0, \pm 1, \pm 2, \ldots, \pm n). \tag{25}$$

Then the dynamic stress concentration factor around the circular pipeline in the concrete can be obtained, which is given by the following formula,

$$\tau_{\theta z}^* = \left| \frac{\tau_{\theta z}^{(t)}}{\tau_0} \right|, \tag{26}$$

$$\tau_{\theta z}^{(t)} = -[z \exp(\beta_2 z + i\theta) - \overline{z} \exp(\beta_2 \overline{z} - i\theta)] \exp\left[\frac{ik_0}{2}(\zeta + \overline{\zeta})\right]$$
$$+ \frac{i}{\varphi_0} \sum_{n=-\infty}^{\infty} C_n \left\{ H_{n-1}^{(1)}(k_0|\zeta|) \left(\frac{\zeta}{|\zeta|}\right)^{n-1} z \exp(\beta_2 z + i\theta) + H_{n+1}^{(1)}(k_0|\zeta|) \left(\frac{\zeta}{|\zeta|}\right)^{n+1} \overline{z} \exp(\beta_2 \overline{z} - i\theta) \right\}, \tag{27}$$

where $\tau_{\theta z}^{(t)} = \tau_{\theta z}^{(i)} + \tau_{\theta z}^{(s)}$, $\tau_0 = \mu_0 k_0 \varphi_0 \beta_1 \beta_2$.

6. Numerical Results and Discussion

After establishing the scattering model of elastic waves by circular pipeline in the infinite inhomogeneous concrete, the DSCFs around the circular pipeline can be obtained when SH waves propagate in the horizontal direction. The distribution and variation rule of the DSCFs around the circular pipeline are analyzed and discussed. The dimensionless variables used in the analysis are reference wave number $k_0 R$, density inhomogeneous parameters β_1 and β_2, respectively. When other variables are the same as each other and the values of β_2 are opposite number to each other, the density distribution in the concrete is symmetric about the y-axis. When β_2 is positive, it can be considered that SH waves are incident horizontally from low density to high density in the concrete. When β_2 is negative, SH waves are incident horizontally from high density to low density in the concrete.

To analyze the impact of different values of inhomogeneous parameters on DSCF, the values of inhomogeneous parameters adopted in this paper were based on the selected variables in the study of reference [37]. The values of variables in reference [37] can be found in Appendix A. The distribution of DSCFs around the circular pipeline is given in Figure 2 when the density inhomogeneous parameter $\beta_1 = 0.5$. The values of $k_0 R$ are 0.5, 1.0 and

2.0, respectively. The density inhomogeneous parameter β_2 in Figure 2a,b is 0.1 and 0.5. As can be seen from the figure, since the distribution form of the inhomogeneous concrete is symmetric about the x-axis, and SH waves are also positively incident along the x-axis, the distribution of the DSCFs around the circular pipeline is also symmetric. When the reference wave numbers $k_0 R$ become two times larger, the amplitude of the DSCF around the circular pipeline increases. It can be inferred that higher DSCF values will be obtained in the case of high-frequency waves at this density. When β_2 increases, the maximum value of DSCF also increases, and the extreme point of DSCF shifts toward the back wave surface, and the distribution of DSCF around the circular pipeline becomes more regular. Figure 2c,d shows that when β_2 is negative, the direction of density distribution in the concrete changes concerning the symmetry of the y-axis. Contrary to the results in Figure 2a,b cases, the amplitude of DSCF decreases with the increase of reference wave number $k_0 R$, and higher DSCF values are obtained in the case of low-frequency waves. Compared with the case when β_2 is positive, the maximum value of DSCF when β_2 is negative is smaller. However, in the same way, that β_2 is positive, the extreme points of DSCF will shift toward the back wave surface of the circular pipeline with the increase of the absolute values of β_2. Thus, it can be inferred that when SH waves are incident into the concrete from two opposite directions, the distribution law and magnitude of DSCF will be changed.

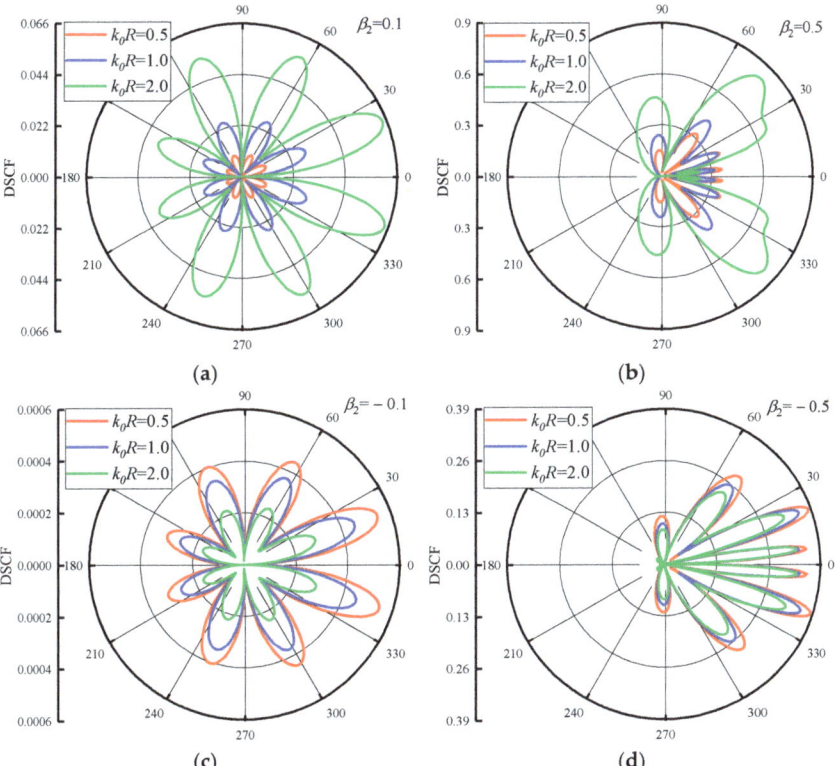

Figure 2. DSCF distribution around the circular pipeline with different reference wave numbers $k_0 R (\beta_1 = 0.5)$.

By comparing the results of the DSCFs around the circular pipeline in the radial and linear terms of the density given by Jiang [37], the distribution of DSCFs around the circular pipeline given in Figure 2 shows that when the density of the concrete is distributed in the form of a polynomial-exponential coupling, more extreme points appear in the DSCF curves. The amplitudes of DSCF all appear on the surface, and there is no obvious offset with the increase of reference wave number or inhomogeneous parameters. It shows that when the density of the concrete is distributed in this form, the influence on the distribution of DSCF around the circular pipeline is more severe.

The values of DSCFs under different reference wave numbers $k_0 R$ are given in Figures 3 and 4. The values of $k_0 R$ are set as 0.5, 1.0 and 2.0, respectively.

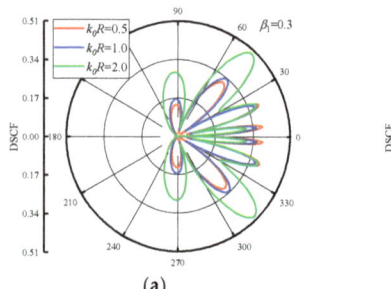
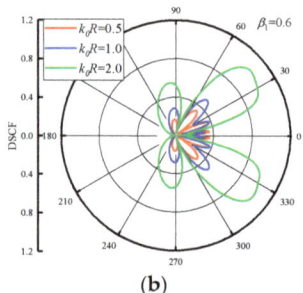
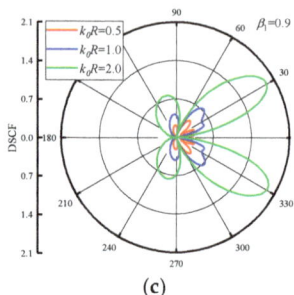

Figure 3. DSCF distribution around circular pipeline with different reference wave numbers $k_0 R$ ($\beta_2 = 0.5$).

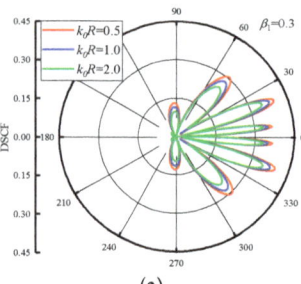
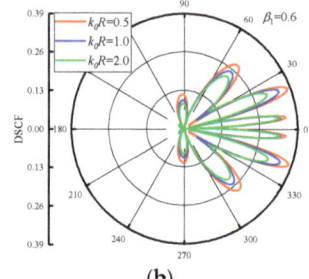
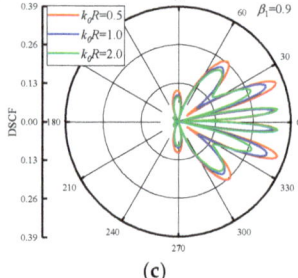

Figure 4. DSCF distribution around circular pipeline with different reference wave numbers $k_0 R$ ($\beta_2 = -0.5$).

Figure 3 shows the distribution of DSCF around the circular pipeline when the $\beta_2 = 0.5$ and $\beta_1 = 0.3, 0.6, 0.9$. The DSCFs around the circular pipeline increase with larger reference wave number $k_0 R$. When β_1 increases, it can be seen that the distribution of DSCF around the circular pipeline (20~40° and 320~340°) becomes regular, and the number of extreme points of DSCF around here decreases. When $k_0 R$ and β_1 change, the distribution of extreme points of DSCF is dominated by the back wave surface. Figure 4 shows the DSCF distribution around the circular pipeline when the $\beta_2 = -0.5$. It can be seen that when β_2 is negative, the maximum value of DSCF decreases with the increases of $k_0 R$, but its maximum value decreases slightly. The DSCF around the circular pipeline is complex, and there is no obvious distribution change with the increase of β_1, and the maximum value of DSCF is much smaller than the result in Figure 3. Thus, it can be found that when the incident direction of SH waves is from low density to high density, the distribution of DSCF in the process is more regular.

In the discussion of the influence of reference wave number $k_0 R$ on DSCF around the circular pipeline, it is obvious that the inhomogeneous parameters β_1 and β_2 will have a significant influence on DSCF around the circular pipeline. Figure 5 shows the variation of DSCF distribution around the circular pipeline with the values of β_2, when the $k_0 R = 0.5$, 1.0 and 1.5. We set the $\beta_1 = 2.0$ and changed the $\beta_2 = 0.2, 0.3$ and 0.4. As can be seen from Figure 5, when the β_2 increases, the maximum value of the DSCF around the circular pipeline increases along with it. Although the values of β_2 are small and the increment of each change is small, the amplitude of the DSCFs around the circular pipeline is very obvious, because the change of β_2 has a drastic impact on the density distribution and value in the concrete. In addition, as the reference wave number k_0R increases into an arithmetic sequence, it is found that the distribution of DSCF near 40° and 320° of the circular pipeline becomes more regular, and the number of extreme points of DSCF decreases. Figure 6 shows DSCF around the circular pipeline with the values of β_2 = 0.8, 1.0 and 1.2. When the value of β_2 is large, the maximum value of the DSCF around the circular pipeline will not change significantly with the values of k_0R variation. Therefore, it can be inferred that when β_2 increases to a certain value range, the change of reference wave number has a small impact on the amplitude of DSCF around the circular pipeline. In addition with the increase of β_2, the distribution of DSCF near 20° and 340° of the circular pipeline gradually becomes complex, and the number of extreme points of DSCF tends to increase. These changes are completely contrary to the changes when β_2 is small.

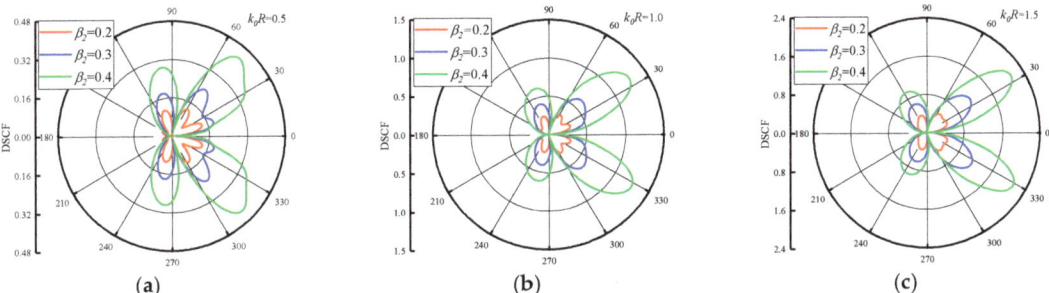

Figure 5. Distribution of DSCF around circular pipeline with different inhomogeneous parameter β_2 (β_2 = 0.2, 0.3, 0.4), (β_1 = 2.0).

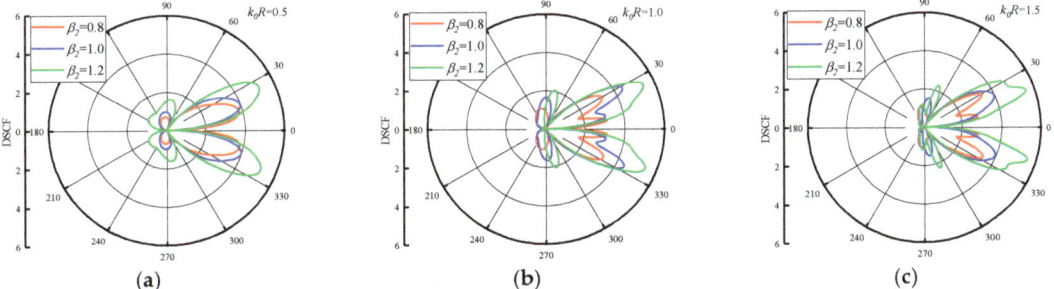

Figure 6. Distribution of DSCF around circular pipeline with different inhomogeneous parameter β_2 (β_2 = 0.8, 1.0, 1.2), (β_1 = 2.0).

The distribution of DSCFs around the circular pipeline with β_1 is given in Figures 7 and 8. The reference wave number k_0R = 0.5, 1.0 and 1.5. Figure 7 set the β_2 = 0.5, and change the β_1 = 1.2, 1.6 and 2.0. As can be seen from Figure 7, when the β_1 increases, the maximum value of DSCF around the circular pipeline increases along with it. When k_0R increases, it can be found that the distribution of DSCF near 30° and 330° of the circular pipeline changes gradually, and the number of extreme points of DSCF decreases. In addition, k_0R increases in an arithmetic sequence, and the maximum value of DSCF increases in an arithmetic sequence. Figure 8 shows the distribution of DSCF around the circular pipeline at β_2 = −0.5. SH waves are incident horizontally from high density to low density in the concrete. It can be found that the distribution of DSCF around the circular pipeline is more complex than in Figure 7, in which multiple extreme points appear on the back wave surface, and when β_1 increases, it has little influence on the amplitude of DSCF. When k_0R increases, the peak values of DSCF almost do not change, remaining within 0.4, which is smaller than the maximum value of DSCF in Figure 7. Only the distribution of DSCF near 30° and 330° of the circular pipeline changed slightly. It can be inferred that when SH waves are incident from high density to low density, the changes of k_0R and β_1 have a very limited impact on the amplitude and distribution of DSCF around the circular pipeline, and no obvious changes will occur.

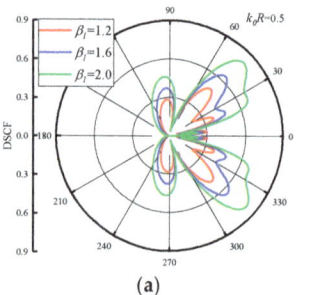 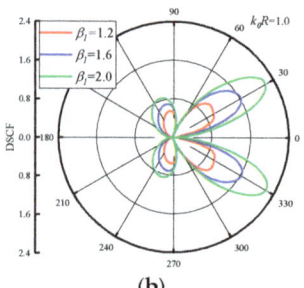 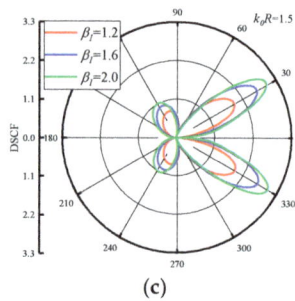

Figure 7. Distribution of DSCF around circular pipeline with different inhomogeneous parameters β_1 ($\beta_2 = 0.5$).

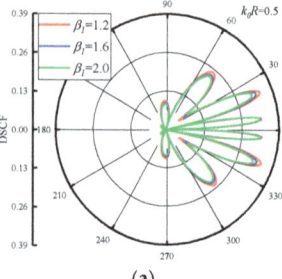 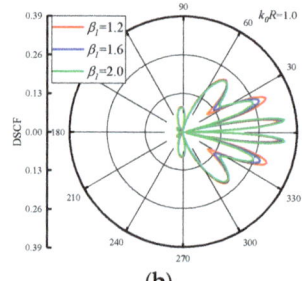 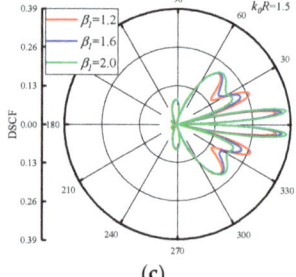

Figure 8. Distribution of DSCF around circular pipeline with different inhomogeneous parameters β_1 ($\beta_2 = -0.5$).

The distribution of DSCF around a circular pipeline is shown in Figures 2–8, and in most cases, there are significant differences in the distribution of DSCF between 20~40° and 320~340°. In order to analyze and summarize the distribution characteristics of DSCF around the circular pipeline more intuitively, this paper selects three observation points on the circular pipe at $\theta = 20°$, 30°, and 40°, and provides the changes in DSCF values under different variable influences in Tables 1–3.

The results of the values of DSCF at positions $\theta = 20°$, 30°, and 40° on a circular pipeline with $\beta_1 = 2.0$ and $k_0R = 0.5$, 1.0, 1.5 are shown in Tables 1–3. It can be visually observed that the values of DSCF at the position of $\theta = 30°$ on the circular pipeline are relatively more sensitive to changes in β_2 and k_0R compared to the other two observation points, based on both horizontal and vertical comparisons of the data in the three tables. It can be inferred that changes in the parameters of inhomogeneous density and reference wavenumber have the most significant impact on the values of DSCF at the position of $\theta = 30°$ on the circular pipeline. Therefore, the position $\theta = 30°$ is selected as the observation point. The continuous variation of DSCF with β_1, β_2 and k_0R at the observation point is analyzed and discussed.

Table 1. The values of DSCF at different positions of circular pipeline with $k_0R = 0.5$ ($\beta_1 = 2.0$).

k_0R	β_2	DSCF ($\theta = 20°$)	DSCF ($\theta = 30°$)	DSCF ($\theta = 40°$)
0.5	0.2	0.1349	0.0971	0.0562
	0.3	0.1846	0.1558	0.1810
	0.4	0.2520	0.3571	0.3985
	0.5	0.6116	0.6885	0.7974

Table 2. The values of DSCF at different positions of circular pipeline with $k_0R = 1.0$ ($\beta_1 = 2.0$).

k_0R	β_2	DSCF (DSCF ($\theta = 20°$))	DSCF ($\theta = 30°$)	DSCF ($\theta = 40°$)
1.0	0.2	0.3018	0.2555	0.2124
	0.3	0.4973	0.5548	0.6070
	0.4	0.8678	1.2262	1.2294
	0.5	1.6774	2.2196	1.9365

Table 3. The values of DSCF at different positions of circular pipeline with $k_0R = 1.5$ ($\beta_1 = 2.0$).

k_0R	β_2	DSCF ($\theta = 20°$)	DSCF ($\theta = 30°$)	DSCF ($\theta = 40°$)
1.5	0.2	0.4653	0.4670	0.4509
	0.3	0.8378	1.0856	1.1452
	0.4	1.4758	2.0635	1.9871
	0.5	2.0386	2.9974	2.4527

In Figure 9, DSCF changes continuously with β_1 at 30° of the circular pipeline with $k_0R = 1.0$, 2.0 and 3.0. The β_2 of Figure 9a–c are 0.5, 1.0 and 1.5, respectively. It can be seen that with the continuous increase of β_1, the DSCF curves fluctuate significantly at the 30° of the circular pipeline. As the reference k_0R increases, the oscillation frequency of DSCF curves increases. Another interesting point can be found in Figure 9b,c. Although the vibration frequencies of the DSCF curves in the subgraph are different, the maximum and minimum values in the same vibration cycle are approximately the same. It can be inferred that changing the density of the concrete with β_1 can compensate for the influence of the reference wavenumber on the amplitude of DSCF. In addition, it can be observed that when the values of β_2 increases, the oscillation frequency of DSCF curves will increase, but the oscillation amplitude of DSCF curves will decrease. It can be inferred that with the increase of concrete density and change of distribution, the amplitude of DSCF tends to be stable, and the fluctuation gradually decreases. The DSCF continuously changes curves with β_1 is given in Figure 9d–f, $\beta_2 = -0.5$, -1.0 and -1.5. SH waves incident horizontally from high density to low density of the concrete. It can be seen that compared with Figure 9a–c, the oscillation frequency of DSCF curves is relatively fast when the values of β_2 are negative, and the minimum value of each vibration period is approaching 0. Similarly, the maximum and minimum values of the three DSCF curves under different reference wave numbers are approximately the same in the same vibration period.

In Figures 10 and 11, the DSCF curves continuously change with β_2 at 30° of the circular pipeline $k_0R = 1.0$, 2.0 and 3.0 are given. The values of β_1 are 0.1, 0.3, 0.5. The values of β_2 in Figure 10 are the positive of number of continuous changes with a range of 0.1~2.0, while β_2 in Figure 11 are the negative number of a continuous range of -2.0~-0.1.

It can be observed in Figure 10 that the overall variation trend of DSCF curves increases with the increase of β_2. When $\beta_1 = 0.1$, it can be found that within the range of $0.1 < \beta_2 < 1.5$, the DSCF curves change with different values of k_0R basically coincide without significant difference. When $\beta_2 > 0.5$, the DSCF curves have obvious differences. When $\beta_1 = 0.3$, the DSCF curves have obvious differences after $\beta_2 > 0.5$. When $\beta_1 = 0.5$, the DSCF curves have obvious differences after $\beta_2 > 0.2$. Figure 11 shows the horizontal incident of SH waves from high density to low density of concrete. It can be found that as the absolute value of β_2 increases, the DSCF curves at 30° of the circular pipeline have an obvious oscillation phenomenon. When the values of k_0R and β_1 increase, the frequency of oscillation of DSCF curves will be higher, which is obviously different from the corresponding situation in Figure 10. When the values of β_1 and β_2 are both small, different k_0R has little influence on the DSCF results at 30° of the circular pipeline. As the values of β_1 increases the absolute value of β_2, which causes differences between DSCF curves under different k_0R conditions, will decrease, which is the same as the situation in Figure 10. This indicates that the density of the concrete with this density form will increase with the increases of β_1, so that the DSCF curves with different wave numbers appear in advance.

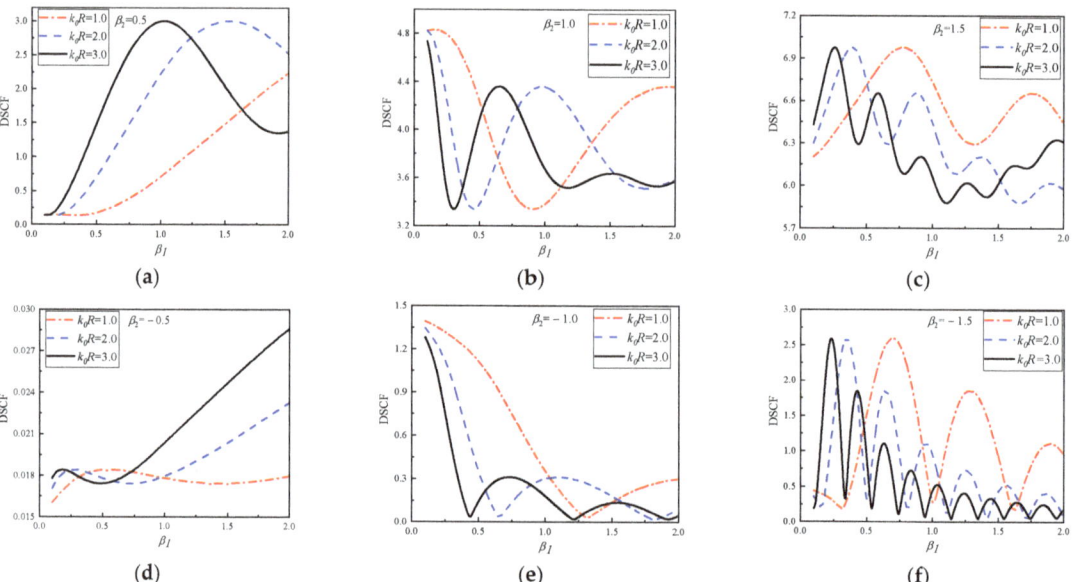

Figure 9. The change of DSCF at the 30° position of the circular pipeline with the inhomogeneous parameter β_1.

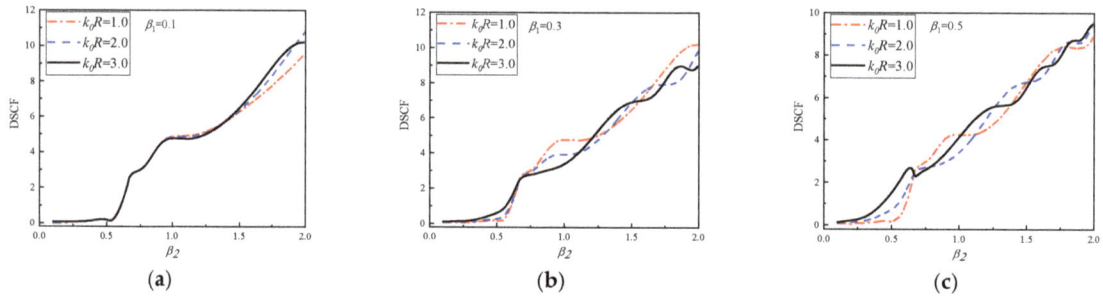

Figure 10. The change of DSCF at the 30° position of the circular pipeline with the inhomogeneous parameter β_2 ($0.1 \leq \beta_2 \leq 2.0$).

In Figure 12, the continuous change of DSCF curves with k_0R at 30° of circular pipeline is given. In Figure 12a–c, $\beta_1 = 1.0$, 1.5 and 2.0, $\beta_2 = 0.8$, 1.0, 1.2. It can be observed that SH waves are incident horizontally from low density to high density of the concrete. It can be observed in Figure 12 that with the continuous change of k_0R, the vibration amplitude of DSCF curves decreases. When the values of β_1 and β_2 increase, the oscillation frequency of DSCF curves will increase. In Figure 12d–f, the values of $\beta_2 = -0.8$, -1.0 and -1.2, SH waves are incident horizontally from the high-density to the low-density direction of the concrete. In this case, compared with the case in Figure 12a–c, the maximum value of DSCF curves is less than, and the minimum points on the DSCF curves are all approaching 0. It can be found that similar to the continuous change of DSCF with β_1 the maximum and minimum values of different DSCF curves in the same vibration period remain the same regardless of the values of β_1 in each subgraph of Figure 12.

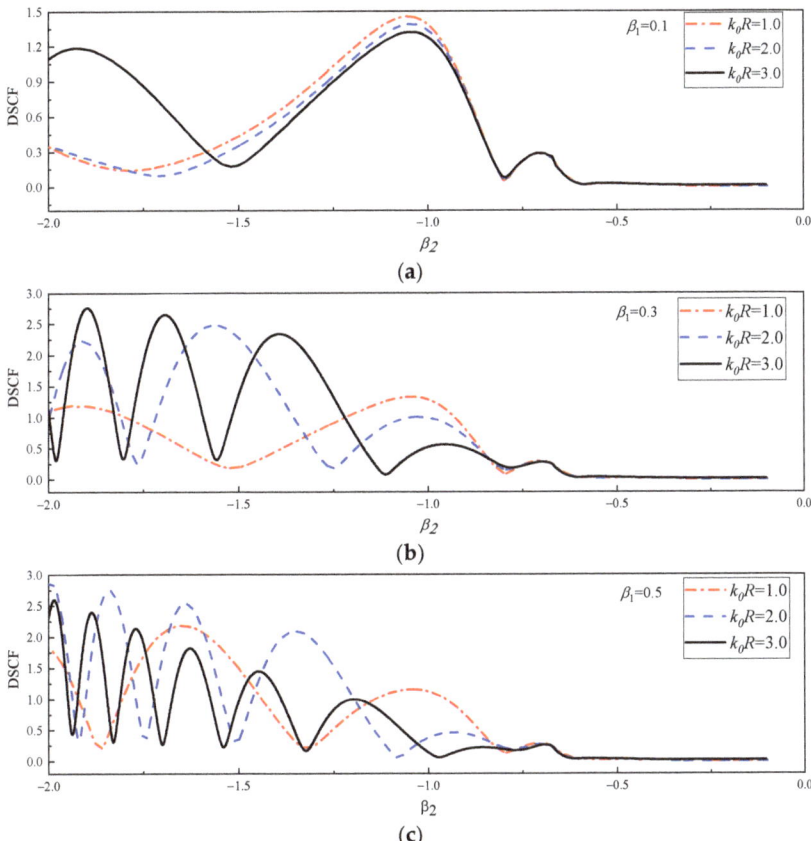

Figure 11. The change of DSCF at the 30° position of the circular pipeline with the inhomogeneous parameter β_2 ($-2.0 \leq \beta_2 \leq -0.1$).

In Figure 13, the continuous changes of DSCF curves with k_0R at 30° of the circular pipeline are given. $\beta_2 = 1.2$, 1.6, and 2.0, $\beta_1 = 0.1$, 1.0 and 2.0. It can be observed that with the increase of β_2, the peak value of DSCF also increases, the oscillation of DSCF curves will appear earlier and the oscillation frequency will be higher. At the same time, the larger β_2 is, the closer the occurrence time of the maximum values of DSCF at the position of 30° is to the quasi-static condition ($k_0R = 0.1$). With the continuous increase of k_0R, the amplitude of DSCF curves decreases. In addition, larger β_1 causes faster oscillation frequency of DSCF curves. Interestingly, even when the value of β_1 changes, the DSCF curves with $\beta_2 = 1.2$ always have an amplitude range of 4 to 6, the DSCF curves with $\beta_2 = 1.6$ always have an amplitude range of 6 to 8, and the DSCF curves with $\beta_2 = 2.0$ always have an amplitude range of 8 to 11.

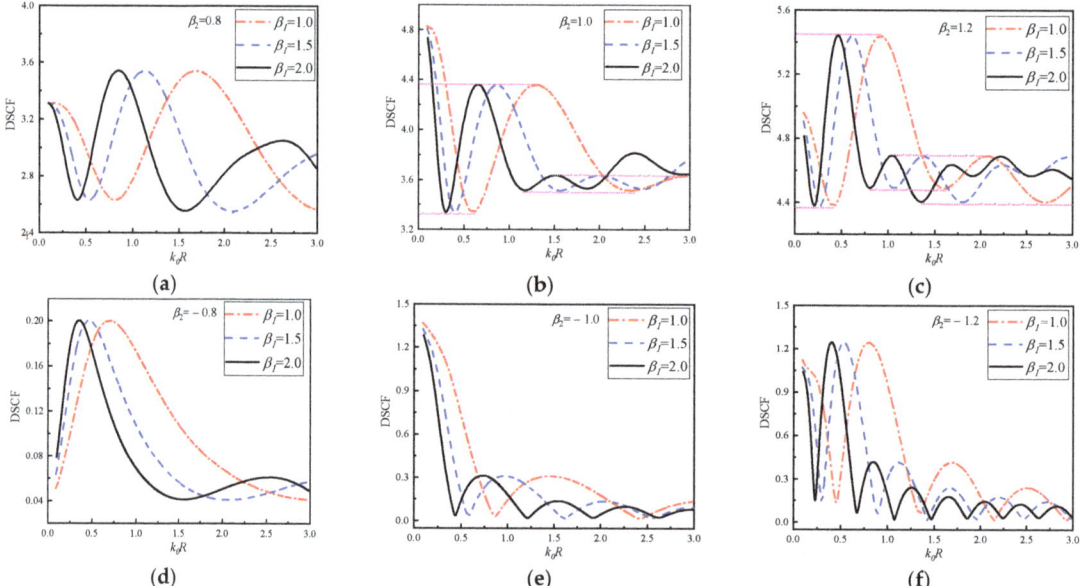

Figure 12. The change of DSCF at the 30° position of the circular pipeline with the reference wave number k_0R ($\beta_2 = \pm 0.8, \pm 1.0, \pm 1.2$).

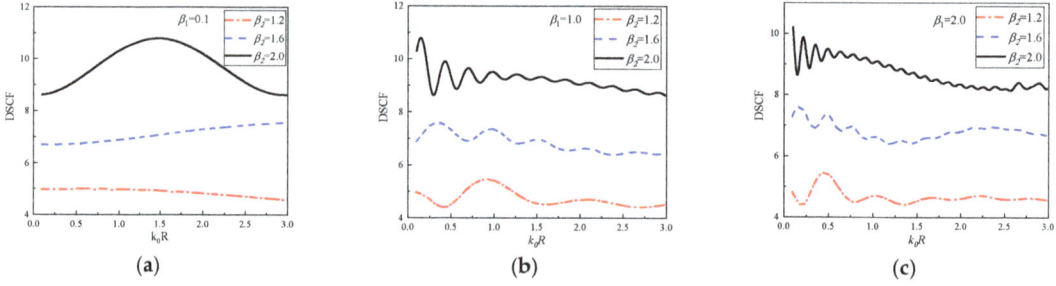

Figure 13. The change of DSCF at the 30° position of the circular pipeline with the reference wave number k_0R ($\beta_1 = 0.1, 1.0, 2.0$).

7. Conclusions

In this paper, the scattering of SH waves by circular pipeline in inhomogeneous concrete with polynomial-exponential coupling density distribution is studied based on the method of complex variable function. The analytical solution of dynamic stress concentration around circular pipeline is derived under this type of concrete with density variation. This paper discusses the effect of different dimensionless parameters on the distribution of DSCF around the circular pipeline. It provides a theoretical reference and a basis for analyzing the influence of defects on wave propagation in an inhomogeneous concrete with density variation. The specific conclusions are as follows:

(1) When the density inhomogeneous parameters β_1, β_2 and the reference wave number k_0R change, all the maximum values of DSCF always appear on the back wave surface of the circular pipeline. In most cases, the maximum value of DSCF is concentrated within the range of 20~40° and 320~340° at the position of circular pipeline, and the number of extreme points of DSCF in this range also changes significantly.

(2) When the value of β_2 is used to change the incident angle of SH wave, the peak values of DSCF around the circular pipeline when the values of β_2 is positive are much higher than that when β_2 is negative. Meanwhile, the distribution of DSCFs around the circular pipeline is more regular.

(3) At 30° of the circular pipeline, with the dimensionless parameter increasing, the DSCF values at this position will have an obvious oscillation phenomenon. At this position, when β_2 remains unchanged, β_1 and k_0R changes, the maximum and minimum values of different DSCF curves in the same fluctuation cycle are the same. When β_2 is negative, the DSCFs at the position of 30° are higher than that of positive β_2.

The conformal transformation method used in this paper requires a high level of expression for the non-uniformity of concrete density, which is not yet achievable in reality. Although the inhomogeneous concrete model we presented does not exist in reality, we hope that in the future, the changing form and structural model of concrete density we proposed can be applied to concrete materials, and our research method can be applied to the elastic dynamic research of other inhomogeneous concrete forms.

Author Contributions: Conceptualization, Z.Y.; methodology, C.S.; software, G.J.; validation, Y.S. and X.L.; investigation, Y.Y.; data curation, C.S.; writing—original draft preparation, C.S.; writing—review and editing, C.S. All authors have read and agreed to the published version of the manuscript.

Funding: This research was funded by the Scientific Research Fund of Institute of Engineering Mechanics, China Earthquake Administration, Grant No. 2021EEEVL0201, and the National Natural Science Foundation of China, Grant No U2239252 and the Research Team Project of Heilongjiang Natural Science Foundation, Grant No TD2020A001.

Institutional Review Board Statement: Not applicable.

Informed Consent Statement: Not applicable.

Data Availability Statement: Not applicable.

Conflicts of Interest: The authors declare no conflict of interest.

Appendix A. An Example of Choosing Values for Inhomogeneous Density Parameter and Reference Wave Number

Reference [37] presents an inhomogeneous medium with a circular cavity, and its density expression is given as:

$$\rho = \rho_0 \left[4\alpha^2 \left(x^2 + y^2 \right) + 4\alpha\beta x + \beta^2 \right], \tag{A1}$$

where, α and β represent inhomogeneous parameters. When $\beta/\alpha < 5$, the density of the medium is considered to vary continuously in two dimensions.

In the example, in order to analyze the variation of DSCF around a circular cavity under different reference wave numbers, reference wave numbers of 0.1, 0.5, 1.0, and 2.0 were selected. Therefore, in this article, we selected reference wave numbers of 0.5, 1.0, 1.5, and 2.0.

In the process of selecting values for inhomogeneous parameters in this example, it was controlled within the range of $\beta/\alpha < 5$. The selected values for α and β are listed in Table A1.

Although the density variation form we provide is different from the density form in the example, we still give the range of variation for β_1 values from 0.1 to 2.0 and β_2 values from −2.0 to −0.1 and 0.1 to 2.0. However, it should be noted that the selection of values for inhomogeneous parameters can affect the accuracy of the results, and different values may be more suitable for different applications.

To illustrate the impact of inhomogeneous density and reference wave number values on the results, we analyzed the DSCF variation around a circular pipeline using different values for these parameters. We found that the selection of the density function exponent

(β_2 value) has a relatively large impact on the results compared to the values of β_1 and reference wave number. This is because the selection of beta value can not only change the distribution form of density in the material, but also change the direction of density distribution.

In summary, our research results indicate the importance of carefully selecting the parameters used in inhomogeneous density concrete models, and emphasize the need for further research to better understand the impact of inhomogeneous density parameters and reference wave numbers on the DSCF obtained from these models.

Table A1. The values of α and β.

	α	β
Case 1		0.4
Case 2		0.5
Case 3	1.0	0.6
Case 4		1.0
Case 5		1.8
Case 6		2.0
Case 7	0.4	
Case 8	0.5	
Case 9	0.6	1.0
Case 10	1.8	
Case 11	2.0	

References

1. Li, Z.; Liu, D. *Waves in Solids*; Science Press: Beijing, China, 1995. (In Chinese)
2. Yu, T.; Su, X.; Wang, X. The present situation and the tendency of the research on elastoplastic waves. *Adv. Mechan.* **1992**, *22*, 347–357.
3. Gao, M. *A Novel Metamaterial Wave Barrier to Isolate Antiplane Elastic Waves*; Beijing Jiaotong University: Beijing, China, 2020.
4. Li, J.; Li, Y. *Elastic Wave Propagation in Solids: Fundamentals and Applications*; John Wiley & Sons: Hoboken, NJ, USA, 2019.
5. Rose, J.L. *Ultrasonic Waves in Solid Media*; Cambridge University Press: Cambridge, UK, 2014.
6. Wu, T.T.; Achenbach, J.D. *Elastic Wave Propagation and Generation in Seismology*; Elsevier: Amsterdam, The Netherlands, 2016.
7. Kino, G.S.; Maeda, S. *Fundamentals of Acoustic Wave Propagation in Solids*; Springer Science & Business Media: New York, NY, USA, 2012.
8. Zhu, W.; Li, J. *Elastic Wave Propagation in Anisotropic Media*; Springer: Berlin, Germany, 2018.
9. Popovics, J.S.; Subramaniam, K.V.L. Review of Ultrasonic Wave Reflection Applied to Early-Age Concrete and Cementitious Materials. *J. Nondestruct. Evaluat.* **2015**, *34*, 267. [CrossRef]
10. Planes, T.; Larose, E. A review of ultrasonic Coda Wave Interferometry in concrete. *Cem. Concr. Res.* **2013**, *53*, 248–255. [CrossRef]
11. Birgul, R. Hilbert transformation of waveforms to determine shear wave velocity in concrete. *Cem. Concr. Res.* **2009**, *39*, 696–700. [CrossRef]
12. Wu, X.; Yan, Q.; Hedayat, A.; Wang, X. The influence law of concrete aggregate particle size on acoustic emission wave attenuation. *Sci. Rep.* **2021**, *11*, 22685. [CrossRef]
13. Du, Q.; Zeng, Y.; Huang, G.; Yang, H. Elastic metamaterial-based seismic shield for both Lamb and surface waves. *Aip Adv.* **2017**, *7*, 075015. [CrossRef]
14. Delsanto, P.P.; Hirsekorn, S. A unified treatment of nonclassical nonlinear effects in the propagation of ultrasound in heterogeneous media. *Ultrasonics* **2004**, *42*, 1005–1010. [CrossRef]
15. Ziaja, D.; Jurek, M.; Wiater, A. Elastic Wave Application for Damage Detection in Concrete Slab with GFRP Reinforcement. *Materials* **2022**, *15*, 8523. [CrossRef]
16. Yoon, Y.-G.; Lee, J.-Y.; Choi, H.; Oh, T.-K. A Study on the Detection of Internal Defect Types for Duct Depth of Prestressed Concrete Structures Using Electromagnetic and Elastic Waves. *Materials* **2021**, *14*, 3931. [CrossRef]
17. Beniwal, S.; Ghosh, D.; Ganguli, A. Ultrasonic imaging of concrete using scattered elastic wave modes. *NDT E Int.* **2016**, *82*, 26–35. [CrossRef]
18. Guo, Y.C.; Guo, S.H. Propagation Characteristics of Ultrasonic Waves in Concrete Medium with Local Damage. *Adv. Mater. Res.* **2011**, *255–260*, 561–568.

19. Liu, J.; Xie, J.; He, X.Y.; He, Y.S.; Zhong, J.H. Detecting the Defects in Concrete Components with Impact-Echo Method. *Appl. Mechan. Mater.* **2014**, *577*, 1114–1118.
20. Kang, J.M.; Song, S.; Park, D.; Choi, C. Detection of cavities around concrete sewage pipelines using impact-echo method. *Tunnell. Undergr. Space Technol.* **2017**, *65*, 1–11. [CrossRef]
21. Ali, G.; Demarco, F.; Scuro, C. Propagation of Elastic Waves in Homogeneous Media: 2D Numerical Simulation for a Concrete Specimen. *Mathematics* **2022**, *10*, 2673. [CrossRef]
22. Uenishi, K. Elastodynamic Analysis of Underground Structural Failures Induced by Seismic Body Waves. *J. Appl. Mechan.* **2012**, *79*, 1014. [CrossRef]
23. Lu, Z.F.; Liu, M.Y. Shrinkage and Creep Analysis of a Continuous Rigid-Frame Bridge with Density Gradient Concrete. *Adv. Mater. Res.* **2011**, *250–253*, 2506–2509.
24. Zhang, C.; Deng, Z.; Ma, L.; Liu, C.; Chen, Y.; Wang, Z.; Jia, Z.; Wang, X.; Jia, L.; Chen, C.; et al. Research Progress and Application of 3D Printing Concrete. *Bull. Chin. Ceram. Soc.* **2021**, *40*, 6882386.
25. Strieder, E.; Hilber, R.; Stierschneider, E.; Bergmeister, K. FE-Study on the Effect of Gradient Concrete on Early Constraint and Crack Risk. *Appl. Sci.* **2018**, *8*, 246. [CrossRef]
26. Wang, X.; Jia, K.; Liu, Y.; Zhou, H. In-Plane Impact Response of Graded Foam Concrete-Filled Auxetic Honeycombs. *Materials* **2023**, *16*, 745. [CrossRef]
27. Wang, W.; Zhang, Z.; Huo, Q.; Song, X.; Yang, J.; Wang, X.; Wang, J.; Wang, X. Dynamic Compressive Mechanical Properties of UR50 Ultra-Early-Strength Cement-Based Concrete Material under High Strain Rate on SHPB Test. *Materials* **2022**, *15*, 6154. [CrossRef]
28. Wang, W.; Song, X.; Yang, J.; Liu, F.; Gao, W. Experimental and numerical research on the effect of ogive-nose projectile penetrating UR50 ultra-early-strength concrete. *Cem. Concr. Compos.* **2023**, *136*, 104902. [CrossRef]
29. Zhang, Y.; Sun, C.; Wang, S.; Zhu, Y.; Sun, G. Properties and pore structure of foam concrete with different density. *J. Chongqing Univ.* **2020**, *43*, 54–63.
30. Wang, G.; Deng, Z.; Xu, H.; Wang, D.; Lu, Z. Application of Foamed Concrete Backfill in Improving Antiexplosion Performance of Buried Pipelines. *J. Mater. Civ. Eng.* **2021**, *33*, 04021052. [CrossRef]
31. Métais, V.; Chekroun, M.; Marrec, L.L.; Duff, A.L.; Plantier, G.; Abraham, O. Influence of multiple scattering in heterogeneous concrete on results of the surface wave inverse problem. *NDT E Int.* **2016**, *79*, 53–62. [CrossRef]
32. Hei, B.P.; Yang, Z.L.; Wang, Y.; Liu, D.K. Dynamic analysis of elastic waves by an arbitrary cavity in an inhomogeneous medium with density variation. *Math. Mechan. Solids* **2016**, *21*, 931–940. [CrossRef]
33. Yang, Z.L.; Hei, B.P.; Yang, Q.Y. Dynamic analysis on a circular inclusion in a radially inhomogeneous medium. *Chin. J. Theoret. Appl. Mechan.* **2015**, *47*, 539–543.
34. Yang, Z.L.; Hei, B.P.; Wang, Y. Scattering by circular cavity in radially inhomogeneous medium with wave velocity variation. *Appl. Math. Mechan.* **2015**, *36*, 599–608. [CrossRef]
35. Hei, B.P.; Yang, Z.L.; Sun, B.T.; Wang, Y. Modelling and analysis of the dynamic behavior of inhomogeneous continuum containing a circular inclusion. *Appl. Math. Model.* **2015**, *39*, 7364–7374. [CrossRef]
36. Hei, B.P.; Yang, Z.L.; Chen, Z.G. Scattering of shear waves by an elliptical cavity in a radially inhomogeneous isotropic medium. *Earthquake Eng. Eng. Vibrat.* **2016**, *15*, 145–151. [CrossRef]
37. Jiang, G.X.X.; Yang, Z.L.; Sun, C.; Song, Y.Q.; Yang, Y. Analytical study of SH wave scattering by a cylindrical cavity in the two-dimensional and approximately linear inhomogeneous medium. *Waves Random Complex Media* **2020**, *31*, 1799–1817. [CrossRef]

Disclaimer/Publisher's Note: The statements, opinions and data contained in all publications are solely those of the individual author(s) and contributor(s) and not of MDPI and/or the editor(s). MDPI and/or the editor(s) disclaim responsibility for any injury to people or property resulting from any ideas, methods, instructions or products referred to in the content.

Article

Study on Mass Erosion and Surface Temperature during High-Speed Penetration of Concrete by Projectile Considering Heat Conduction and Thermal Softening

Kai Dong, Kun Jiang *, Chunlei Jiang, Hao Wang and Ling Tao

School of Energy and Power Engineering, Nanjing University of Science and Technology, Nanjing 210094, China; dongkai@njust.edu.cn (K.D.)
* Correspondence: jkeddy@163.com

Abstract: The mass erosion of the kinetic energy of projectiles penetrating concrete targets at high speed is an important reason for the reduction in penetration efficiency. The heat generation and heat conduction in the projectile are important parts of the theoretical calculation of mass loss. In this paper, theoretical models are established to calculate the mass erosion and heat conduction of projectile noses, including models of cutting, melting, the heat conduction of flash temperature, and the conversion of plastic work into heat. The friction cutting model is modified considering the heat softening of metal, and a model of non-adiabatic processes for the nose was established based on the heat conduction theory to calculate the surface temperature. The coupling numerical calculation of the erosion and heat conduction of the projectile nose shows that melting erosion is the main factor of mass loss at high-speed penetration, and the mass erosion ratio of melting and cutting is related to the initial velocity. Critical velocity without melting erosion and a constant ratio of melting and cutting erosion exists, and the critical velocities are closely related to the melting temperature. In the process of penetration, the thickness of the heat affected zone (HAZ) gradually increases, and the entire heat conduction zone (EHZ) is about 5~6 times the thickness of the HAZ.

Keywords: concrete; penetration; melting and cutting erosion; temperature; heat conduction; coupling numerical calculation

1. Introduction

Underground protective fortifications are mostly built with high-strength concrete and are mainly used to resist the shock wave generated by the explosion and the penetration of kinetic energy projectiles. The research on kinetic energy projectiles penetrating concrete is mostly concentrated on the depth of penetration (DOP), while the change of projectile kinetic energy, the optimization of nose shape, and ballistic stability are perceived as important research aspects [1,2]. According to the different initial velocities of the projectile, the penetration process can be divided into a rigid region of penetration, a semi-fluid region of penetration, and a hydrodynamic region of penetration by studying the interaction between the projectile and the target [3]. With the increase in the initial velocity of penetration, it is found that the projectile has increasingly obvious mass erosion [4], which causes nose deformation and even disintegration of the projectile during the penetration process; consequently, the penetration efficiency and ballistic stability will be seriously affected [5].

Research on mass erosion began in the 1990s. Forrestal and Frew [6,7] carried out a series of tests on the high-speed penetration of sharp projectiles into concrete and focused on the shape changes of the projectile before and after the penetration. Significant mass loss after the penetration of the projectile was observed, mainly manifested in the abrasion and deformation of the projectile nose, which led to a sharp reduction in the efficiency of improving DOP by increasing the projectile velocity. In the theoretical calculation of the DOP for the projectile penetrating concrete, the projectile is regarded as a rigid body with an unchanged mass in most simulation models. This assumption is reasonable when the initial impact velocity of the projectile is relatively lower [8,9]. However, the calculation models of rigid penetration under the conditions of medium or low initial velocity are not suitable for predicting the penetration with erosion at high initial velocity due to the disregard of mass loss. The linear relationship between the mass loss and the initial kinetic energy of the projectile was established through fitting the test data by Sliding et al. [10] and Chen et al. [11]. The calculation efficiency was greatly improved by the fitting model, but the applicability of the model was limited due to the lack of theoretical connotation.

In order to further understand the erosion mechanism and establish a scientific penetration-erosion theoretical model, Jones [12], He [13], and Guo [14] analyzed the metallography of the projectile after the test and found that there was a sign of metal melting and quenching in the heat affected zone (HAZ) of the nose surface. Adiabatic shear bands were also observed in some areas, indicating that the temperature of the nose was higher than the melting temperature during penetration. The molten metal layer would separate from the surface of the projectile and produce new molten liquid metal on the new layer. Obvious furrow scratches were also detected on the nose and fine aggregate particles were also embedded in the surface after the test, indicating that the surface of the nose was subjected to temperature softening and cut by the aggregate. Zhao [15] found, using the results of Forrestal's test, that with the increase in aggregate hardness, the mass loss of the projectile became more obvious. In summary, the reasons for the mass reduction after penetration could mainly be attributed to the melting of the nose surface under high-speed friction and the cutting by the concrete aggregate at the interface. These findings provide a strong scientific basis for the mechanism and the numerical calculation of mass erosion.

Based on the erosion mechanism, He [16] and Li [17] successively carried out theoretical research on the model of erosion calculation and established a mass erosion calculation method combined with a high-speed friction theory and melting model. A coupling model of erosion calculation taking into account aggregate cutting and thermal melting was established by Ning [18], in which the accumulated friction energy was converted into thermal melting energy. Based on the assumption that the penetration is an adiabatic process, the mass loss is calculated by the receding of the nose surface. Guo [19] calculated the temperature within the thickness of the thin layer on the surface of the nose using the axial one-dimensional heat conduction model and obtained the temperature rise ratio of the typical position generated by friction heat.

There is a coupling relationship between the temperature change of the projectile and mass erosion. It is an important prerequisite for numerical calculation to further understand the energy-force-heat conversion mechanism of the projectile during the process of penetrating concrete. The softening, material flow, and mass loss of the projectile resulting from a high temperature and high-stress state are the mechanisms of material failure during the penetration process [14]. The essence of the projectile temperature rise is the conversion of plastic work to heat, and the heat generated by high-speed friction between the projectile and the target when the projectile impacts the concrete. The rapid rise of temperature can lead to the melting of the projectile surface, and the cutting efficiency was affected by the change of hardness of the HAZ on the projectile surface. On the other hand, when the temperature gradient is generated on the surface of the nose, part of the thermal energy will be propagated from the high-temperature zone to the low (including the interior of the nose and the concrete), and the conduction and dissipation of this part of energy should be considered when calculating the temperature and mass erosion of the nose.

Alloy steel with thermal softening behavior (including AISI 4340 alloy steel, et al.) is selected as the projectile material, and its strength and hardness will be reduced when exposed to higher temperatures [20,21]. The influence of thermal softening on friction cutting has not been discussed in existing theory. With further research, the temperature of the projectile in the process of penetrating concrete has gradually been paid attention to. It is unscientific to regard the nose surface as an adiabatic layer in the existing erosion model when calculating temperature. For the same projectile shape and concrete strength, the penetration duration is mainly determined by the initial velocity and the mass of the projectile. If the surface is taken as an adiabatic layer, the calculation will be less accurate with a longer penetration time; therefore, this consideration is key to scientifically predicting the erosion and temperature evolution law to carry out the coupling model with heat generation, heat conduction, and mass erosion of the projectile.

We conducted coupling calculations on the mass erosion and surface temperature during the process of projectiles penetrating into concrete. Based on the heat transmission theory, high-speed friction theory, and the conversion of plastic work into heat theory, the coupling model of projectile mass erosion and temperature on the nose surface in the process of penetrating concrete are investigated in this paper. In Section 2, the hardness caused by the temperature rise on the surface of the nose is described as dimensionless by the material constitutive model considering the temperature, and the cutting mass loss model during penetration is modified. The projectile temperature rise caused by flash temperature heat conduction and conversion of plastic work into heat is also considered. Models of coupling penetration, mass loss, and heat conduction were established. The erosion calculation of the penetration process was achieved through the assumptions and calculation process provided in Section 3. The numerical calculation of the dynamic mechanical parameters, surface temperature, and mass loss during the projectile penetration into concrete was carried out in Section 4. The effectiveness of the coupling erosion model was verified by comparing the calculation results and tests, while the mass erosion and surface temperature of the projectile were also calculated and analyzed.

2. Thermal and Dynamic Model of Penetration Process

Due to the interaction between the projectile nose and target during the process of penetrating concrete, the plastic deformation caused by extrusion and high-speed friction in the relative sliding process mainly appears at the nose of the projectile. Some researchers have shown that a small number of scratches were observed on the body, but they are negligible compared with the mass loss of the nose. Therefore, the mass erosion calculation always ignores the body. Compared with a medium and low initial velocity, the mechanical essence is more complex when the projectile penetrates concrete at high velocity. In particular, the conversion of work to heat has an apparent influence on the deceleration and temperature of the projectile in the penetration process. Thus, the calculation should couple the thermal and dynamic mechanical processes of the projectile in high-speed penetration.

In this paper, we summarize the research results of scholars, including their calculation models, which mainly consist of the dynamic mechanical model of the penetration, the conversion of plastic work into heat model, the temperature rise model, and the conduction of heat model of the projectile. These models are further researched in the next chapter.

2.1. Dynamic Mechanical Model in Penetration Process

The dynamic mechanical parameters of the penetration process have been studied in detail. The theoretical aspect is mostly based on the cavity expansion theory. The cavity expansion theory used for the calculation of the penetration process was originally proposed by Forrestal and Tzou [22] in 1997. The surface pressure of the projectile nose can be calculated by this theory, based on which the resistance function of the projectile can be described as well. The positive pressure on the surface of the projectile nose can be expressed as:

$$\sigma_n = S f_c + \rho_c v_n^2 \qquad (1)$$

where v_n is the cavity expansion velocity of the target at the interface of the nose and target during penetration, which can be expressed as $v_n = v_p \cos\varphi$; v_p is the projectile velocity; and φ is the included angle between the projectile axis and the normal direction of the nose surface, as shown in Figure 1. Here, ρ_c is the density of concrete target and f_c is the compressive strength without confining pressure of concrete. The parameter $S = 82.6 f_c^{-0.544}$ is fitted by test results [7].

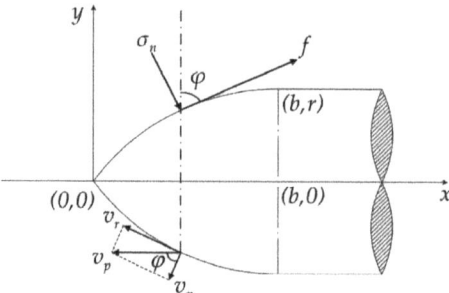

Figure 1. Two-dimensional sectional coordinates of projectile nose.

(a) Crater stage:

It is assumed that there is no mass loss in the crater stage of the mass erosion calculation model [18]. Because the accumulated plastic work during the crater stage is relatively little, the temperature rise of the projectile is not significant as well, so the temperature of the nose surface does not reach the metal melting temperature. In this paper, the dynamic relationship of this stage is coupled with mass erosion in the numerical calculation for more accuracy, and the depth of the crater stage can be expressed as:

$$H_1 = k'd \qquad (2)$$

where k' is a parameter which can be expressed as $k' = 0.707 + h_0/d$, in which h_0 is the length of the nose and d is the diameter of the projectile.

When the penetration depth is $x \leq H_1$, the axial resistance of the projectile in the crater stage can be expressed as:

$$F = cx \qquad (3)$$

where c is a constant. According to the theoretical formula proposed by Forrestal et al. [23], the velocity of the projectile at the end of the crater stage can be expressed as:

$$v_1^2 = \frac{m_0 v_0^2 - (pd^3 k'/4) S f_c}{m_0 + (pd^3 k'/4) N^* r_c} \qquad (4)$$

where m_0 and v_0 are the initial mass and velocity of the projectile impacting the target, respectively. N^* is the shape factor of the projectile nose, which can be expressed as $N^* = (8\psi - 1)/24\psi^2$, where ψ is the initial caliber-radius-head(CRH) of the projectile nose, defined as the ratio of the nose radius to the projectile diameter. According to the dynamic cavity expansion theory, considering the integral effect of penetration resistance on the projectile nose, the shape factor can also be defined as [24]:

$$N^* = -\frac{8}{d^2} \int_0^b \frac{yy'^3}{1+y'^2} dx \qquad (5)$$

where $y = y(x)$ is the boundary function describing the shape of the nose.

(b) Tunnel stage:

When the penetration depth is $x > H_1$, the projectile enters the tunnel stage. The pressure on the nose surface is divided by area and the axial resistance on the projectile is [23]:

$$F = \pi d^2 \left(S f_c + N^* \rho_c v^2 \right)/4 \qquad (6)$$

The basic parameters such as the relationship between deceleration, velocity, penetration depth, the pressure of the nose surface, and the time of projectile in the process of penetration can be calculated based on the above dynamic mechanical models. During high-speed penetration, the mass erosion at the nose of the projectile involves physical and mechanical parameters including projectile friction, melting, temperature rise, and heat conduction, which need to be calculated in combination with the thermodynamic theoretical model.

Through the conclusions obtained from current research results, the results were calculated with fewer errors when the generation of heat on the nose part was considered only and the body part of the projectile was ignored. Therefore, the mass erosion and heat conduction of the projectile nose during penetration can be divided into four areas, as shown in Figure 2, which are, respectively: the mass loss zone, the high-temperature heat affected zone (HAZ), the heat conduction zone, and the undisturbed zone. Each zone has a different calculation method due to its generation mechanism, and adjacent zones are continuously distributed in the surface space of the nose and have a close thermodynamic relationship. These models in different regions based on their generation mechanisms will be analyzed and established in this paper.

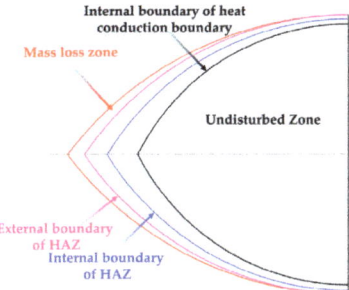

Figure 2. Thermal boundary of the projectile nose during penetration.

2.2. Temperature Rise of Projectile Caused by the Conversion of Plastic Work into Heat at the Boundary

The pressure on the nose of the projectile is extremely high during the penetration, which causes elastic-plastic deformation of the surface. Moreover, the instantaneous dislocation of the lattice will produce an increase in internal energy, resulting in a temperature rise that is distributed in a gradient from the surface of the projectile to its interior part. The stress distribution of the projectile nose can be calculated iteratively using the surface pressure given by Equation (1).

The conversion of plastic work into heat is most evident in the HAZ. According to hardness analysis of the surface of the projectile after penetration by Jerome et al. [25], it has been found that the hardness in the HAZ region is relatively higher. This phenomenon indicates that the hardening process occurs after quenching when the projectile material is heated to the austenite transformation temperature. Therefore, on the basis of scientific evidence, we demarcate the temperature index of the HAZ inner layer as the boundary condition. In addition, the temperature (T_H) range of the normal austenitizing of AISI-4340 steel for direct quenching is 1088K~1118K.

Molinari et al. [26] proposed a formula to calculate the HAZ thickness, which can be expressed as Equation (7), where t_p is the time of heat conduction and the HAZ thickness, $H(t)$, is the heat diffusion length in the characteristic time:

$$H(t) = \frac{1}{\varsigma}\left(\frac{\lambda_p}{\rho_p c_p}\frac{L_p}{v_r}\right)^{1/2} \qquad (7)$$

Here, $\varsigma = 1/\sqrt{2\pi}$ is the characteristic constant, c_p is the specific heat capacity, ρ_p is the density of projectile material, L_p is the length of the projectile, and v_r is the relative sliding velocity of the projectile and target. For 4340 steel, the thermal conductivity is $\lambda_p = 44.5 \, \text{W}/(\text{m} \cdot \text{K})$.

In the process of penetration, severe plastic deformation occurs on the contact surface of the concrete, which consumes a lot of energy and results in a temperature rise. As one of the most widely used and classic thermodynamic constitutive models of metal materials at high strain rates, the Johnson-Cook (J-C) constitutive model composed of three polynomials [27] is applied to the calculation of projectile stress when penetrating the concrete. The strain rate term and temperature term are combined with the traditional stress-strain expression relationship. Thus, this model is applicable for the calculation of high strain rates and high temperatures in the penetration process. The flow stress, σ, can be expressed as a function of the effective plastic strain, ε, the effective strain rate, $\dot{\varepsilon}$, and the temperature, T. The J-C model is expressed as:

$$\sigma = (A + B\varepsilon^n)\left(1 + Cln\frac{\dot{\varepsilon}}{\dot{\varepsilon}_0}\right)\left[1 - \left(\frac{T - T_0}{T_m - T_0}\right)^m\right] \tag{8}$$

where $\dot{\varepsilon}_0$ is the reference strain rate; T_0 is the room temperature; T_m is the melting temperature of the projectile material; and A, B, C, n, and m are material constants.

The numerical method for constitutive equation calculation using the integration rate is introduced as the algorithm of stress update. The stress, $\sigma_{ij}(t + \Delta t)$, at $t + \Delta t$ can be obtained by integrating the stress rate:

$$\sigma_{ij}(t + \Delta t) = \sigma_{ij}(t) + \dot{\sigma}_{ij}dt \tag{9}$$

According to the normal stress of the projectile described in Equation (1) and the J-C constitutive model, the relationship between the temperature gradient and time in the HAZ can be calculated. Furthermore, elastic-plastic work is converted to heat energy in the extrusion process, and the temperature rise of the nose can be expressed as [28]:

$$T_w = \frac{\beta}{\rho_p c_p}\int \sigma d\varepsilon \tag{10}$$

where β is the coefficient of the conversion of plastic work into heat and c_p is the specific heat of the projectile material.

In summary, the temperature rise caused by the conversion of plastic work into heat in the HAZ and inner zone can be calculated by Equations (8)–(10).

2.3. Friction Cutting Considering Thermal Softening of Material

It should be noted that the surface of the nose is a special kind of metal with high temperature and high stress resistance that directly contacts the aggregate during the penetration process. In the present cutting model, it is assumed that the hardness of the projectile and aggregate is constant [17,18]. Because the aggregate particles are constantly updated on the surface of the nose in the penetration process, this assumption mentioned above will cause rare errors. However, cutting and melting are interrelated. In other words, the hardness changes caused by the high temperature of the surface have an important influence on the cutting of the nose.

The relationship between temperature, stress, and strain can be established based on the J-C model. Agreements are obtained that the hardness of the nose surface and the aggregate in the concrete both affect cutting erosion. The hardness of steel is related to heat-treatment. A large number of tests have shown that the hardness and yield strength of steel can be regarded as a linear relationship, approximately. According to the test results shown in Figure 3, the relationship between the tensile strength, σ_b(MPa), of steel and Brinell hardness, H_p, is expressed as follows [21,29,30]:

$$\sigma_b = K \times H_p + b \tag{11}$$

where K and b are the fitting parameters obtained from the tests. For alloy steel, K can be taken as 3.36 and b can be taken as 30.92.

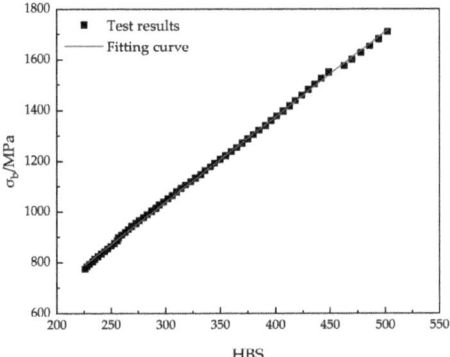

Figure 3. Relationship between Brinell hardness and tensile strength [29].

With a tensile strength of 1080 MPa, the corresponding Brinell hardness is about 320 HBS; AISI-4340 alloy steel is often chosen as a material in the design of penetrating projectiles. The hardness decreases when the strength of the material reduces because of the high temperature. Therefore, the hardness value can be obtained through the strength value. Since the Mohs hardness of quartz aggregate has been adopted in the theoretical formula in literature [18] for calculation, the parameter of relative hardness is introduced in order to adapt to the general calculation of the model, and the dimensionless analysis of the softening behavior of AISI-4340 steel can be described. The softening relative hardness ratio of projectile material is defined as the ratio of the initial hardness under room temperature to the softening hardness under high temperature:

$$\eta(t) = \frac{H_{\text{p-Normal}}}{H_{\text{p-Soft}}(t)} \tag{12}$$

In the process of penetration, the hardness of concrete aggregate, H_c, is defined as a constant. In Equation (12), the hardness of alloy steel, $H_{\text{p-Soft}}(t)$, is a variable parameter related to the penetration time. The initial hardness is the test result at laboratory temperature. The hardness of surface metal will decrease with the rise of temperature.

In the classical Rabinowicz cutting theory [31], using single-point abrasive particles to apply a load of p on a soft metal surface and press it into the depth of h, the volume expression of cutting when the wear sliding unit distance on the metal surface is:

$$V = K\frac{p}{H_m} \tag{13}$$

where H_m is the Mohs hardness of the bearing surface and K is the wear coefficient which depends on the hardness of abrasive particles and the hardness of the metal.

Through microscopic inspection and observation on the surface of the recovered projectile after penetration, it is found that the furrow shape of the nose surface distribution is similar to the particle wear. Equation (13) can be borrowed in cutting calculation. For penetration, the pressure, p, of particles can apply the nose surface pressure described in Equation (1). Since the hardness of steel is proportional to the yield strength, the relative hardness described in Equation (12) can be introduced to modify the classical cutting model. Therefore, the change of hardness on the nose surface caused by melting is considered in the model, and the modified cutting volume of a unit area can be expressed as:

$$dV_c = \eta K_1 \frac{\sigma_n}{Y} v_r dt \tag{14}$$

where K_1 is the wear parameter applicable to the projectile penetrating concrete, which is related to the relative hardness of the contact. The specific calibration method has been studied in the literature [18]. Y is the yield strength of the projectile material and v_r is the relative sliding velocity of the projectile target surface, which can be expressed as $v_r = v_p \sin\varphi$ (see Figure 1). Therefore, for the mass loss caused by cutting, the cutting quality per unit area can be expressed as:

$$\Delta m_c = \rho_p \times \Delta V_c = \rho_p K_1 \int_0^{t_e} \eta \frac{\sigma_n v_r}{Y} dt \qquad (15)$$

The strength, temperature, and hardness of the projectile can be calculated, respectively, and the mass loss models of the projectile considering cutting and thermal softening during the penetration of concrete are established based on these equations above.

2.4. Melting and Heat Conduction

For the calculation of the temperature rise of the projectile nose, it is not scientific to assume the melting of the layer on the surface of the projectile is an adiabatic process. The temperature rise caused by the conversion of plastic work into heat is applied to the whole nose. The friction on the surface of the projectile will produce a "heat supply". There will be heat conduction from the surface of the projectile to the interior when the "heat supply" capacity is higher than the temperature rise of the plastic work.

2.4.1. The Heat Conduction of Flash Temperature

The friction under high-speed sliding is very different from that under low-speed sliding on the surface. During high-speed penetration, the surface temperature will reach and even exceed the melting temperature [32,33]. Meanwhile, the large deformation, phase transformation, and melting at the friction interface make the analysis more complex in penetration. The friction heat generated between the projectile and the target during penetration is also a complex physical problem to solve. Moreover, the heat transfer rate involved in the calculation of heat conduction is a time-dependent parameter. The heat energy is concentrated on the thin-layer surface area of the projectile nose, and the temperature rises in a very short time, in what is known as "flash temperature".

During high-speed penetration, the temperature of the nose surface rises rapidly due to the intense heat flow of the surface melting, resulting in the heat conduction to the internal projectile, to be specific, which is called the heat conduction of flash temperature. In the flash temperature stage, the surface temperature of the nose is much higher than that of the interior. Although the metal melting only takes a very short time and can be considered to be completed instantaneously, the continuous high-temperature liquid melting material has a great influence on the temperature of the internal projectile, especially after the temperature of the exposed surface layer has changed and affected the subsequent penetration parameters.

2.4.2. Melting Temperature and Mass Loss

The molten metal on the surface of the projectile is separate from the projectile. The key to the calculation is the temperature of the liquid metal layer covering the surface of the nose. In this paper, the hypothesis that there is no temperature gradient in the liquid metal is proposed. If the conversion of plastic work into heat in the melting zone is ignored, it can be assumed that all the heat comes from friction. The heat flux density in the melting zone can be calculated according to the following equation:

$$q = f \cdot v_r = q_p + q_c \qquad (16)$$

where q_p is the heat flowing to the projectile nose per area and q_c is the heat flowing to the concrete target per area; q_p and q_c can be expressed as:

$$q_p = -\lambda_p \frac{\partial T}{\partial \vec{n}_p}, q_c = -\lambda_c \frac{\partial T}{\partial \vec{n}_c} \qquad (17)$$

where $\vec{n}_p$ and $\vec{n}_c$ are defined as the normal direction along the relative sliding surface of projectile and concrete, respectively.

The friction is related to the positive pressure, σ_n, on the surface of the projectile nose. The coefficient of dynamic friction is taken as μ and the friction can be expressed as [34,35]:

$$f = \mu \sigma_n + \tau_0 \tag{18}$$

where τ_0 is the shear strength of concrete, and the relationship between the shear strength and the compressive strength without confining pressure is expressed as $\tau_0 = f_c/\sqrt{3}$. The dynamic friction coefficient is the key parameter in Equation (18). Many tests have shown that the coefficient of friction is not a constant value, but varies when the projectile is at different relative sliding velocities. In this paper, the coefficient of friction proposed by Klepaczko is used in the model. The coefficient of friction takes into account adiabatic shear, thermal conductivity, and other factors, and is expressed as:

$$\mu = \frac{\tau_0}{p}\left(\frac{c'}{\Lambda}\right)\left[\left\langle 1-(\Theta_b(v))^2\right\rangle + \frac{\beta(\Theta_b)v}{\tau_0 h}\right] \cdot [1-(1-f_{a0})exp(-D(p-p_0))] \tag{19}$$

The parameters in Equation (19) can be found in the literature [36]. As shown in Figure 4, the relationship between the coefficient of friction and the relative sliding velocity clearly indicates that the friction coefficient decreases with the increase in velocity.

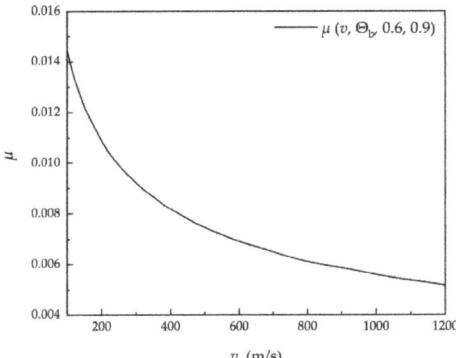

Figure 4. Relationship between coefficient of friction μ and sliding velocity v_r [36].

Conventional projectiles are usually axisymmetric. The nose section area is divided into N_x and N_y points in the horizontal and vertical directions, so the number of grids is $(N_x - 1) \times (N_y - 1)$. Therefore, the temperature at any point can be calculated. The melting area of every time-step under two-dimensional conditions can be calculated based on the temperature. The melting volume can be obtained by integrating along the symmetrical axis. Then, the melting volume per unit time-step (t_{n-1} to t_n) can be expressed as:

$$\Delta V(T,t) = V(T,t_n) - V(T,t_{n-1}) \tag{20}$$

Therefore, the temperature rise of the projectile nose during penetration can be calculated by Equation (10), while the loss mass of melting per unit area of the nose can be calculated by Equation (21) from the initial time to the end time of penetration, t_e:

$$\Delta m_m = \rho_p \times \Delta V_m = \int_0^{t_e} \Delta V(T,t)dt \tag{21}$$

2.4.3. Calculation of Temperature Distribution of Projectile Nose during Penetration

According to Section 2.2, the calculation of the temperature distribution of the projectile nose includes the flash temperature rise caused by friction, the temperature rise caused by the conversion of plastic work into heat, and the heat conduction of the nose. Based on Fourier's law and heat balance theory, the basic theory for the heat conduction of the projectile follows Equation (22):

$$\rho_p c_p \frac{\partial T_t}{\partial t} = \lambda_p \left(\frac{\partial^2 T_t}{\partial x^2} + \frac{\partial^2 T_t}{\partial y^2} \right) \tag{22}$$

The initial condition for solving Equation (22) is $T(x, y, 0) = T_0 = 298$ K; the boundary condition of heat flow flash temperature is $q_p(x, y, t) = -\lambda_p \frac{\partial T}{\partial \vec{n}}$; the boundary temperature of the inner layer in HAZ is $T_H = 1088$ K~1118 K.

It can be seen from Equation (16) that the heat flow boundary is closely related to relative velocity, and the flash temperature period mainly exists in the high-speed stage of penetration. Melting and cutting erosion may have different proportions due to different speed stages, which are introduced in detail below.

3. Calculation Algorithm of The Coupling Model

3.1. Assumptions

Since the penetration process is an extremely complex mechanical process, the numerical calculation needs to be carried out with some assumptions, as follows:

- The projectile penetrates concrete normally, and the projectile is regarded as a standard axially symmetrical structure;
- The mass loss and conduction of heat during the penetration process only occurs at the projectile nose;
- The material of the projectile is isotropic, its density and thermal conductivity remain stable during penetration;
- The influence of phase changes of material on penetration and heat conduction is ignored;
- The aggregate particles in concrete are evenly distributed.

3.2. Discretization

With a thickness scope of the HAZ from microns to millimeters, a multi-scale discretization method is adopted in order to improve the calculation efficiency. As shown in Figure 5, the micro-scale grid division is employed on the surface of the projectile and the independent verification of the size grid is conducted during the calculation.

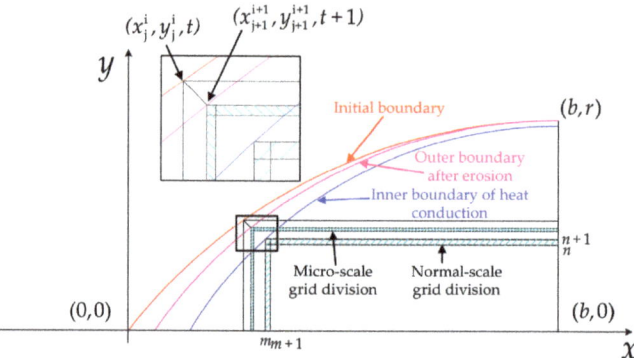

Figure 5. The micro-scale grid division of the receding model of the nose at time t.

3.3. Coupled Algorithm: Simulation of Penetration Process

When the assumptions described in Section 3.1 are determined, the thermal-mechanical-erosion of the projectile penetrating concrete can be calculated based on the model established in Section 2. The calculation process is as follows:

1. Inputting the initial parameters and initial boundary conditions of the projectile and target;
2. Discretization of the projectile nose;
3. Calculating the mechanical parameters in the dynamic process of the projectile in the crater stage and tunnel stage;
4. Calculating the surface pressure and stress of the projectile nose;
5. Calculating the flash temperature boundary, conduction of heat, plastic work, and total temperature of the nose surface.
6. Updating the stress state and the relative hardness of the surface;
7. Calculating the cutting mass loss and the melting mass loss;
8. Updating the nose shape and the mass of the projectile;
9. Increasing the time-step and repeating calculations from step (3) till the projectile velocity drops to zero.

4. Results and Discussion

4.1. Comparison between Theoretical Calculation and Tests

4.1.1. Theoretical Prediction and Experimental Results of Mass Erosion

In the theoretical calculation of the DOP prediction, the most representative formula that calculates the depth of penetration when projectiles penetrate semi-infinite targets was proposed by Forrestal [37], which is a semi-empirical formula based on the test results, and is expressed as follows:

$$H = \frac{2m_0}{\pi d^2 \rho N} ln\left(1 + \frac{\rho N v_1^2}{S f_c}\right) + 2d \tag{23}$$

According to the test results, Silling points out that when the penetration velocity does not exceed 1000 m/s [10], the mass loss of the projectile has a linear relation with the initial kinetic energy. The mass loss remains at a constant level while the velocity is over 1000 m/s, and the mass loss rate of the projectile can be described by Equation (24).

$$\delta = \frac{\Delta m}{m_0} = \begin{cases} c_0 \cdot v_0^2/2, & v_0 \leq 1000 \text{ m/s} \\ c_0/2, & v_0 > 1000 \text{ m/s} \end{cases} \tag{24}$$

Here, Δm is the mass loss after penetration and c_0 is the constant fitted according to the tests.

The molten metal separated from the surface of the projectile and the Mohs hardness of the aggregate in the target are taken into comprehensive consideration for the mass loss calculation of the projectile by He [38]. Ignoring the application limiting of the penetration velocity, the model proposed by Jones et al. [12] is modified as follows:

$$\Delta m = \eta_a \frac{\pi d^2 \tau_0 N_1^* H}{4\kappa Q} \tag{25}$$

where η_a is the parameter related to Mohs hardness of the aggregate, N_1^* is the initial value of the dimensionless longitudinal cross-sectional area of the nose, $\kappa = 4.18$ J/cal is the mechanical equivalent of heat, Q is the melting heat of the unit mass projectile material, and H is the ultimate DOP of the rigid projectile. Those models are representatives for predicting DOP and mass loss, which will be compared with the present model in the numerical calculation in the next section.

4.1.2. Comparison and Analysis of Calculation Results with Test Results

In order to determine the rationality of the model considering the temperature to predict the mass loss of the projectile established in this paper, the test results in the literature [6,7] are used for comparison. The parameters and material properties of the projectile and target required for calculation in the test are shown in Tables 1–4.

Table 1. Parameters of projectiles (4340 Steel).

	Case 1	Case 2	Case 3	Case 4
m_0 (kg)	1.6	0.478	1.62	0.478
d (mm)	30.5	20.3	30.5	20.3
CRH	3	3	3	3
L_n (mm)	50.5	33.7	50.5	33.7
L_p (mm)	304.8	203.2	304.8	203.2

Table 2. Parameters of concrete.

	Case 1	Case 2	Case 3	Case 4
Aggregate Material	Quartz	Quartz	Limestone	Limestone
f_c (MPa)	51	62.8	58.4	58.4
ρ_c (kg/m^3)	2300	2300	2320	2320
Moh's hardness of the aggregate	7	7	3	3
K_1	2×10^{-4}	2×10^{-4}	6×10^{-5}	6×10^{-5}

Table 3. Johnson-Cook constitutive model of 4340 steel material.

ρ_p (kg/m^3)	A	B	C	n	m	T_0 (K)	T_m (K)
7850	792	510	0.014	0.26	1.03	298	1793

Table 4. Parameters required for heat conduction and temperature calculation.

k_p (W/(m·K))	c_p (J/kg·K)	k_c (W/(m·K))	c_c (J/kg·K)	β
44.5	477	1.65	880	0.9

The consistency between test results and the theoretical prediction is the standard for verifying the effectiveness of theoretical predictions. Based on the parameters in Tables 1–4, the DOP in Case 1–Case 4 obtained by using the model and algorithm established in this paper is shown in Figure 6. It can be seen that the calculation results of DOP by the model and calculation method considering the erosion effect established in this paper is in good agreement with the test results. The DOP of the projectile considering mass loss is lower than when mass loss is not considered. Because the parameter S, which affects the DOP, is obtained by result fitting, it is necessary to calibrate this value for the requirements of erosion penetration calculation.

The consistency for deceleration is found in the influence of mass erosion for the calculation of Case 1–Case 4, so we only take Case 1 to analyze in this paper. The deceleration was calculated and the results, whether considering mass erosion or not, are shown in Figure 7. In the process of projectile penetration, the mass of the projectile and the CRH of the nose decreases gradually, resulting in an increase in the deceleration. It can be seen that mass erosion is one of the most important factors affecting the penetration efficiency of the projectile. The mass loss is considered in the numerical calculation in the crater stage. Due to the influence of mass erosion, the deceleration in the crater stage is greater than that without mass erosion consideration, especially for low initial velocity conditions. With the increase in initial velocity, the influence of mass loss on the deceleration gradually decreases in the crater stage. This is mainly because the mass loss in the crater stage is a parameter related to time. The projectile with higher initial velocity has a shorter crater time and less melting mass loss for the same crater depth, so it has less influence on deceleration.

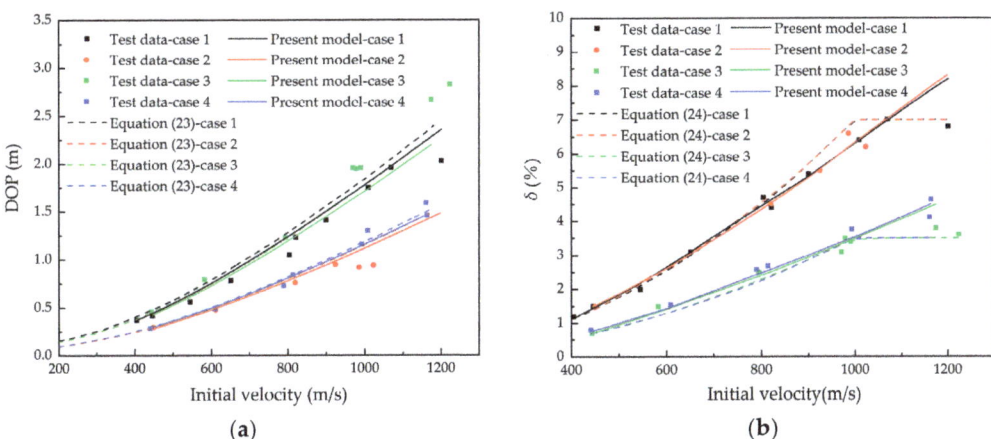

Figure 6. Comparing predicted results with different models and test data in Cases 1–4: (**a**) DOP, (**b**) δ.

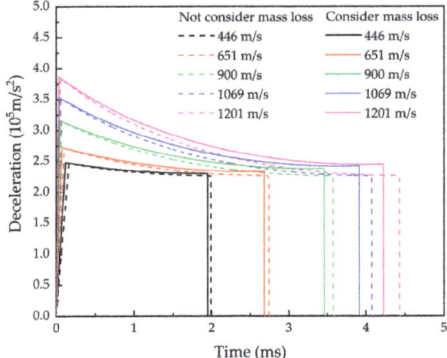

Figure 7. The relationship between deceleration and time considering mass loss.

Due to the influence of mass erosion, the deceleration is higher than in the calculation data which does not consider the mass loss, especially in the middle and final stages of the whole penetration process, so the penetration time is shortened. With the increase in the initial penetration velocity, the cumulative mass erosion has a deeper influence on the deceleration in the tunnel stage.

4.1.3. Analysis of the Proportion of Mass Loss for Two Mechanisms

The proportion of mass loss caused by cutting and melting in total mass erosion is different. In order to further reveal the mechanism of mass loss during high-speed penetration, quantitative analysis of the two mechanisms of mass erosion is convenient for the design and optimization of the projectile. The percentage of cutting erosion and melting erosion in the total mass loss is calculated and shown in Figure 8. It can be clearly seen that the mass loss caused by melting is relatively higher in the velocity scope of the study, indicating that the melting of the nose surface is the main factor of mass erosion. In Case 1 and Case 2, the mass loss caused by melting contributes to about 81% of the total mass loss. In Case 3 and Case 4, the mass loss caused by melting contributes to 91% of the total mass loss due to the lower hardness of the aggregate. It can be seen that the proportion of the mass loss caused by the two mechanisms is closely related to the hardness of the aggregate. When the projectile uses the same material, the higher the hardness of the aggregate, the higher the cutting erosion proportion. In the velocity range from 400 to 1200 m/s, the melting-cutting erosion ratio and velocity have low relevance, and the mass loss of the projectile is approximately linear with the initial velocity. In this view the linear fitting of Equation (24) in this velocity range is reasonable.

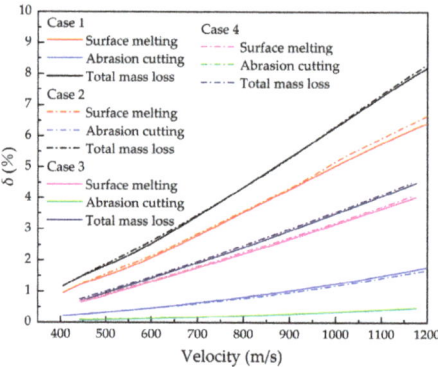

Figure 8. Relationship between mass loss of two mechanisms and initial penetration velocity.

4.1.4. Critical Velocity of Erosion Affected by Melting v_s, v_c

Case 3 had the condition that the projectile penetrates concrete at an initial velocity of 445 m/s. When the velocity decreases to about v_s = 370 m/s, the melting mass decreases rapidly and the inflection point appears. The surface of the nose will not melt when the velocity decreases to v_c = 106 m/s, and only cutting erosion occurs in the subsequent penetration process, as shown in Figure 9. When the projectile penetrates at an initial velocity of 1069 m/s, there is no disappearance of melting erosion even as the velocity decreases to the critical velocity. This phenomenon is related to the time accumulation of heat conduction of flash temperature. Because the velocity of heat conduction is higher than the receding velocity of erosion, the HAZ layer is a zone with a temperature gradient. When the projectile penetrates at low initial velocity, the influence of the initial melting flash temperature on the heat conduction is negligible, resulting in a small thickness of the HAZ layer. The surface internal energy is not enough to melt the material when the velocity is reduced to the critical velocity, v_c, and only cutting erosion occurs after that. In the case of high-speed penetration, the HAZ surface temperature of the projectile surface is still high even though the projectile has reduced to the critical velocity, and the energy generated by friction is enough to heat the surface to the melting temperature.

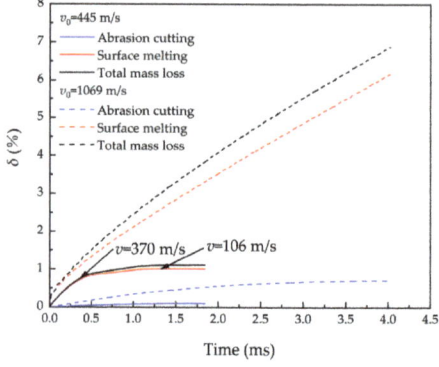

Figure 9. Relationship between mass erosion and time at two different initial velocities (445 m/s; 1069 m/s).

4.1.5. The Influence of the Melting of Mass Erosion

It can be inferred from Section 4.1.3 that when metal materials with high melting temperatures are used on the surface of the nose, mass erosion can be largely reduced. For high melting temperatures such as wolfram (melting temperature: 3708 K) and molybdenum (melting temperature: 2918 K), the final mass erosion at different initial penetration velocities is calculated and compared with 4340 steel, assuming that the hardness of the two materials is equal with that of steel. As shown in Figure 10, the material with a high melting temperature as the projectile surface can significantly reduce the mass loss in the process of penetration at the same initial penetration velocity.

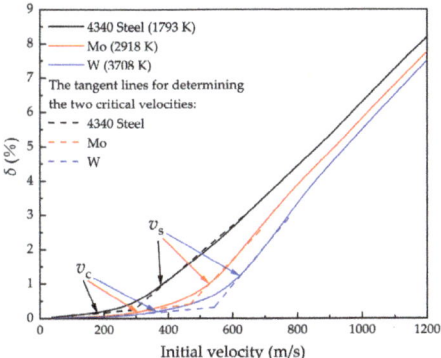

Figure 10. Effect of metals with different melting temperatures on mass erosion.

It can be seen from the previous section that there is a critical velocity, v_c, for losing the melting erosion. In order to explore the law of melting at different initial velocities, the mass erosion during penetration at different initial velocities is calculated. It is found that v_c increases with the increase in the melting temperature of materials.

When the velocity is lower than v_c, the mass loss is caused by cutting completely. The surface temperature of the projectile nose is relatively low, especially at low speed, and does not reach the melting temperature, so the mass loss of the projectile is less due to the thermal softening behavior. When the initial velocity exceeds the v_c, melting erosion emerges. With the increase in initial velocity, the proportion of melting erosion increases gradually. At critical velocity, v_s, the ratio of mass loss caused by melting and cutting tends to be stable, and melting erosion is the main factor of mass loss. For Case 1 in this paper, the v_c of 4340 steel is 175 m/s and the v_s is 379 m/s; the v_c of molybdenum under the same hardness is 311 m/s and the v_s is 530 m/s; and the v_c of wolfram under the same hardness is 364 m/s and the v_s is 616 m/s.

The core reason for the reduction or disappearance of melting erosion is that the melting efficiency is reduced or even insufficient due to the reduction in heat on the projectile surface. The critical velocity, v_c and v_s, is affected by multiple factors such as velocity, melting temperature of the material, nose shape, properties of concrete, and the temperature of the HAZ caused by heat conduction during penetration. Further quantitative analysis is not conducted in this paper, and detailed numerical calculation is required for specific conditions.

4.2. Temperature Field of the Nose Surface

Analyze the condition in which the projectile penetrates concrete at an initial velocity of 1069 m/s in Case 1. The temperature fields of the nose at different times during the penetration process are shown in Figure 11. The molten liquid metal with flash temperature (over 3000 K) is eliminated. Due to the extremely short time for the conduction of heat, the HAZ is only on the surface of the projectile and the surface temperature is between 1500 K to 1793 K, which is consistent with the law obtained by Guo [19]. However, the dynamic boundary effect of the erosion process is also considered in this paper, and the results are more scientific and accurate.

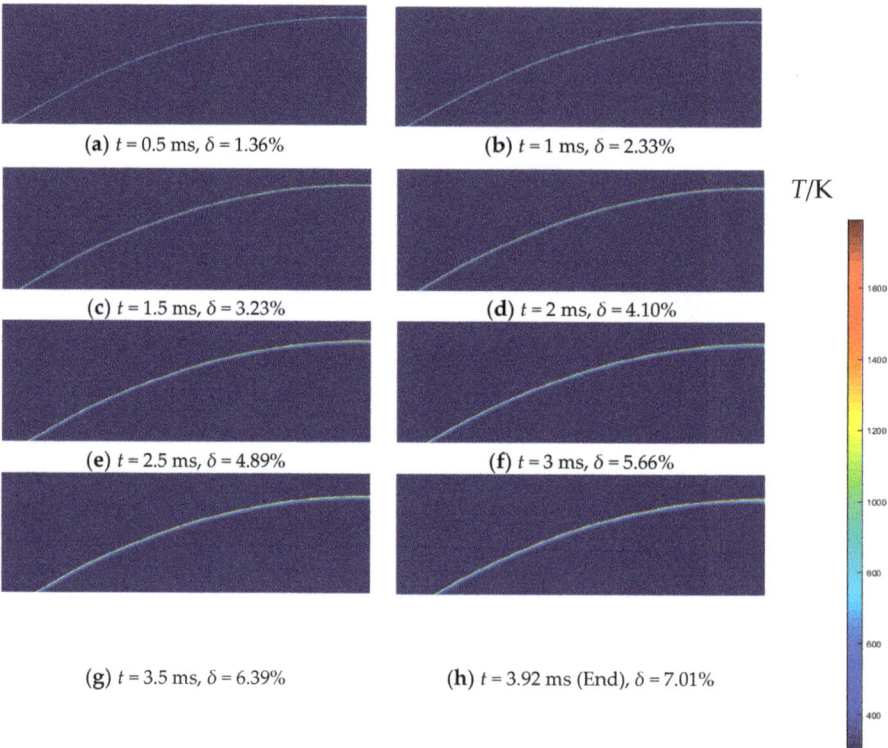

Figure 11. Temperature field of the nose surface (Case 1, v_0 = 1069 m/s). (**a**–**h**) each picture shows the outer contour of the projectile nose after erosion at current time, and displays the temperature gradient of the nose cross-section. Where δ represents the percentage of mass loss at the current time. Due to the axisymmetric structure of the projectile, only half of the projectile nose is displayed.

The austenitizing temperature for direct quenching of AISI-4340 steel is about T_H = 1088 K. It is defined as the HAZ where the surface temperature is higher than T_H, and the heat conduction zone where the surface temperature is higher than T_0 = 298 K. The thickness changes of the HAZ and the entire heat conduction zone (EHZ) on the nose surface at the nose tip (Point A), the nose middle (Point B), and the nose end (Point C) during the penetration process are shown in Figure 12. The result shows that the thickness of the HAZ increases with the increase in the penetration time, and the thickness of the EHZ is about 5~6 times of the thickness of HAZ. The thickness of the HAZ and EHZ at the tip is significantly higher than those at the middle and end of the nose and the thicknesses of the HAZ at the middle and end of the nose are approximately equal. The thickness of the heat conduction zone at the end of the nose is slightly higher than that at the middle. This distinction is mainly caused by the difference in erosion thickness. The erosion thickness gradually decreases from the tip to the end of the nose, and the receding displacement at the tip is significantly higher than that at the middle and end. The difference in HAZ thickness between the middle and the end of the nose is quite slight, but the difference in erosion thickness has an influence on the EHZ.

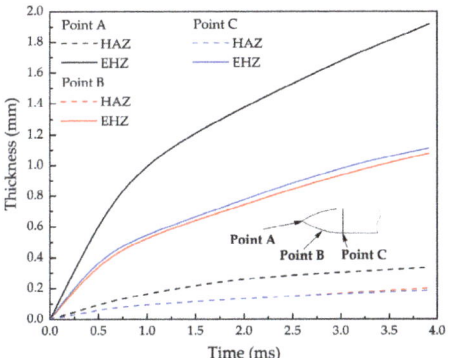

Figure 12. Thickness of the HAZ and EHZ on nose surface (Case 1, v_0 = 1069 m/s).

5. Conclusions

In this paper, a coupling calculation model of penetration, mass erosion, and heat conduction in the process of the projectile penetrating concrete at high speed is established, and the numerical calculation of dynamic mechanical parameters, surface temperature, and mass loss in the process of penetration is carried out. The effectiveness of the coupling erosion model is verified by comparison with experimental data, and the internal mechanism of the erosion process is analyzed. The maximal velocity of projectile for the erosion study is 1201 m/s in this paper and the projectile may be destroyed once it exceeds this limited velocity. The conclusions are as follows:

(1) When the projectile penetrates concrete at high speed, the influence of mass erosion on the deceleration of the projectile increases with the increase in initial velocity. The proportion of melting erosion is higher than that of cutting erosion. The proportion of cutting and melting is closely related to the hardness of the aggregate in concrete. The higher the hardness of the aggregate, the higher proportion of cutting mass loss.

(2) There are critical initial velocities of the projectile without melting erosion and critical initial velocities with stable proportional for melting and cutting erosion. The critical velocity is mainly related to the melting temperature of the material on the nose surface. When the velocity is lower than critical velocity v_c, only cutting erosion occurs; when it is higher than v_c but lower than v_s, the proportion of melting increases with the increase in velocity. When the velocity is higher than critical velocity v_s, the ratio of melting and cutting erosion does not change with the different initial velocities.

(3) The thickness of the high-temperature heat affected zone and the entire heat conduction zone on the surface of the projectile nose are closely related to the time of penetration, and both increase with the increase in penetration time. The thickness of the entire heat conduction zone is about 5~6 times of the high-temperature heat affected zone.

Author Contributions: Conceptualization, K.D.; methodology, K.D., K.J. and L.T.; software, K.D. and C.J.; validation, K.D. and K.J.; formal analysis, K.D. and C.J.; investigation, K.D.; resources, K.J. and H.W.; data curation, K.D. and L.T.; writing—original draft preparation, K.D.; writing—review and editing, K.J.; project administration, K.J.; funding acquisition, H.W. All authors have read and agreed to the published version of the manuscript.

Funding: This research received no external funding.

Institutional Review Board Statement: Not applicable.

Informed Consent Statement: Not applicable.

Data Availability Statement: Not applicable.

Conflicts of Interest: The authors declare no conflict of interest.

Abbreviations

The following abbreviations are used in this manuscript:
DOP Depth of penetration
CRH Caliber-radius-head
HAZ Heat affected zone
EHZ Entire heat conduction zone

References

1. Chen, X.; Lu, F.; Zhang, D. Penetration trajectory of concrete targets by ogived steel projectiles—Experiments and simulations. *Int. J. Impact Eng.* **2018**, *120*, 202–213. [CrossRef]
2. Wang, W.; Song, X.; Yang, J.; Liu, F.; Gao, W. Experimental and numerical research on the effect of ogive-nose projectile penetrating UR50 ultra-early-strength concrete. *Cem. Concr. Compos.* **2023**, *136*, 104902. [CrossRef]
3. Chen, X.; Li, Q. Transition from Nondeformable Projectile Penetration to Semihydrodynamic Penetration. *J. Eng. Mech.* **2004**, *130*, 123–127. [CrossRef]
4. Kong, X.; Wu, H.; Fang, Q.; Zhang, W.; Xiao, Y. Projectile penetration into mortar targets with a broad range of striking velocities: Test and analyses. *Int. J. Impact Eng.* **2017**, *106*, 18–29. [CrossRef]
5. Wu, H.; Huang, F.; Wang, Y.; Duan, Z.; Shan, Y. Mass Loss and Nose Shape Change on Ogive-nose Steel Projectiles During Concrete Penetration. *Int. J. Nonlinear Sci. Numer. Simul.* **2012**, *13*, 273–280. [CrossRef]
6. Forrestal, M.J.; Frew, D.J.; Hanchak, S.J.; Brar, N.S. Penetration of grout and concrete targets with ogive-nose steel projectiles. *Int. J. Impact Eng.* **1996**, *18*, 465–476. [CrossRef]
7. Frew, D.J.; Hanchak, S.J.; Green, M.L.; Forrestal, M.J. Penetration of concrete targets with ogive-nose steel rods. *Int. J. Impact Eng.* **1998**, *21*, 489–497. [CrossRef]
8. Bian, H.; Jia, Y.; Shao, J.; Pontiroli, C. Numerical study of a concrete target under the penetration of rigid projectile using an elastoplastic damage model. *Eng. Struct.* **2016**, *111*, 525–537. [CrossRef]
9. Kong, X.; Wu, H.; Fang, Q.; Peng, Y. Rigid and eroding projectile penetration into concrete targets based on an extended dynamic cavity expansion model. *Int. J. Impact Eng.* **2017**, *100*, 13–22. [CrossRef]
10. Silling, S.A.; Forrestal, M.J. Mass loss from abrasion on ogive-nose steel projectiles that penetrate concrete targets. *Int. J. Impact Eng.* **2007**, *34*, 1814–1820. [CrossRef]
11. Chen, X.; He, L.; Yang, S.Q. Modeling on mass abrasion of kinetic energy penetrator. *Eur. J. Mech. A Solids* **2010**, *29*, 7–17. [CrossRef]
12. Jones, S.E.; Foster, J.C.; Toness, O.A.; DeAngelis, R.J. An Estimate for Mass Loss From High Velocity Steel Penetrators. In Proceedings of the ASME PVP-435 Conference on Thermal-Hydraulic Problems, Sloshing Phenomena, and Extreme Loads on Structures, Vancouver, BC, Canada, 5–9 August 2002; Volume 422, pp. 227–237.
13. He, L.; Chen, X.; Fan, Y. Metallographic observation of reduced scale advanced EPW after high-speed penetration. *Explos. Shock Waves* **2012**, *32*, 515–522. (In Chinese)
14. Guo, L.; He, Y.; Zhang, X.; Pang, C.; Qiao, L.; Guan, Z. Study mass loss at microscopic scale for a projectile penetration into concrete. *Int. J. Impact Eng.* **2014**, *72*, 17–25. [CrossRef]
15. Zhao, J.; Chen, X.; Jin, F.; Xu, Y. Depth of penetration of high-speed penetrator with including the effect of mass abrasion. *Int. J. Impact Eng.* **2010**, *37*, 971–979. [CrossRef]
16. He, L.; Chen, X.; Xia, Y. Representation of nose blunting of projectile into concrete target and two reduction suggestions. *Int. J. Impact Eng.* **2014**, *74*, 132–144. [CrossRef]
17. Li, Z.; Xu, X. Theoretical Investigation on Failure Behavior of Ogive-Nose Projectile Subjected to Impact Loading. *Materials* **2020**, *13*, 5372. [CrossRef]
18. Ning, J.; Li, Z.; Ma, T.; Xu, X. Failure behavior of projectile abrasion during high-speed penetration into concrete. *Eng. Fail. Anal.* **2020**, *115*, 104634. [CrossRef]
19. Guo, L.; He, Y.; Zhang, X.; He, Y.; Deng, J.; Guan, Z. Thermal-mechanical analysis on the mass loss of high-speed projectiles penetrating concrete targets. *Eur. J. Mech. A Solids* **2017**, *65*, 159–177. [CrossRef]
20. Marques, A.; Souza, R.A.; Pinto, G.A.M.; Galdino, A.G.S.; Machado, M.L.P. Evaluation of the softening mechanisms of AISI 4340 structural steel using hot torsion test. *J. Mater. Res. Technol.* **2020**, *9*, 10886–10900. [CrossRef]
21. Wu, S.; Qu, R.; Liu, Z.; Li, H.; Wang, X.; Tan, G.; Zhang, P.; Zhang, Z. Locating the optimal microstructural state against dynamic perforation by evaluating the strain-rate dependences of strength and hardness. *Int. J. Impact Eng.* **2021**, *152*, 103856. [CrossRef]
22. Forrestal, M.J.; Tzou, D.Y. A spherical cavity-expansion penetration model for concrete targets. *Int. J. Solids Struct.* **1997**, *34*, 4127–4146. [CrossRef]
23. Forrestal, M.J.; Altman, B.S.; Cargile, J.D.; Hanchak, S.J. An empirical equation for penetration depth of ogive-nose projectiles into concrete targets. *Int. J. Impact Eng.* **1994**, *15*, 395–405. [CrossRef]
24. Li, Q.; Chen, X. Dimensionless formulae for penetration depth of concrete target impacted by a non-deformable projectile. *Int. J. Impact Eng.* **2003**, *28*, 93–116. [CrossRef]

25. Jerome, D.M.; Tynon, R.T.; Wilson, L.L. Experimental observations of the stability and survivability of ogive-nosed, high-strength steel alloy projectiles in cementious materials at striking velocities from 800–1800 m/s. In Proceedings of the 3rd Joint Classified Ballistics Symposium, San Diego, CA, USA, 1 May 2000; Professional Engineering Publishing: San Diego, CA, USA, 2000; pp. 1–4.
26. Molinari, A.; Estrin, Y.; Mercier, S. Dependence of the coefficient of friction on the sliding conditions in the high velocity range. *J. Tribol.* **1999**, *121*, 35–41. [CrossRef]
27. Johnson, G.R.; Cook, W.H. A constitutive model and data for metals subjected to large strains, high strain rates, and high temperatures. In Proceedings of the Seventh International Symposium on Ballistics, International Ballistics Committee, The Hague, The Netherlands, 19–21 April 1983; pp. 541–547.
28. Wang, X. Effects of constitutive parameters on adiabatic shear localization for ductile metal based on JOHNSON-COOK and gradient plasticity models. *Trans. Nonferrous Met. Soc. China* **2006**, *16*, 1362–1369. [CrossRef]
29. *GB/T 1172-1974*; Conversion of Hardness and Strength for Ferrous Metal. PTCA (Part A: Physical Testing). Standards Press of China: Beijing, China, 2001; Volume 37, pp. 406–409.
30. Tabor, D. The hardness and strength of metals. *J. Inst. Met.* **1951**, *79*, 1–18.
31. Rabinowicz, E.; Dunn, L.A.; Russell, P.G. A study of abrasive wear under three-body conditions. *Wear* **1961**, *4*, 345–355. [CrossRef]
32. Archard, J.F. The temperature of rubbing surfaces. *Wear* **1959**, *2*, 438–455. [CrossRef]
33. Blok, H. The flash temperature concept. *Wear* **1963**, *6*, 483–494. [CrossRef]
34. Forrestal, M.J.; Longcope, D.B.; Norwood, F.R. A Model to Estimate Forces on Conical Penetrators Into Dry Porous Rock. *J. Appl. Mech.* **1981**, *48*, 25–29. [CrossRef]
35. Forrestal, M.J.; Norwood, F.R.; Longcope, D.B. Penetration into targets described by locked hydrostats and shear strength. *Int. J. Solids Struct.* **1981**, *17*, 915–924. [CrossRef]
36. Klepaczko, J. *Surface Layer Thermodynamics of Steel Penetrators at High and Very High Sliding Velocities*; Report AFRL-MN-EG-TR-2001-7076; Air Force Research Laboratory, Eglin AFB: Greene County, OH, USA, 2001.
37. Forrestal, M.J.; Frew, D.J.; Hickerson, J.P.; Rohwer, T.A. Penetration of concrete targets with deceleration-time measurements. *Int. J. Impact Eng.* **2003**, *28*, 479–497. [CrossRef]
38. He, L.; Chen, X.; He, X. Parametric study on mass loss of penetrators. *Acta Mech. Sin.* **2010**, *26*, 585–597. [CrossRef]

Disclaimer/Publisher's Note: The statements, opinions and data contained in all publications are solely those of the individual author(s) and contributor(s) and not of MDPI and/or the editor(s). MDPI and/or the editor(s) disclaim responsibility for any injury to people or property resulting from any ideas, methods, instructions or products referred to in the content.

Article

Corrosion-Effected Bond Behavior between PVA-Fiber-Reinforced Concrete and Steel Rebar under Chloride Environment

Xuhui Zhang, Xun Wu * and Yang Wang

College of Civil Engineering, Xiangtan University, Xiangtan 411105, China
* Correspondence: xunwu.ada@outlook.com; Tel.: +86-183-7324-0951

Abstract: Corrosion-effected bond behavior between polyvinyl-alcohol-fiber-reinforced concrete and steel rebar under a chloride environment is the experimental subject studied in the present work. Twenty-four pull-out specimens are designed and subjected firstly to an accelerated corrosion test. The effects of polyvinyl alcohol fibers on the cracking behavior, chloride penetration of concrete members and the corrosion loss of steel rebars during the corrosion test are discussed. After this, these corroded specimens are subjected to a pull-out test. The failure mode, the bond-slip curves and the typical bond-stress values are measured during the test. The effects of polyvinyl alcohol fibers and corrosion loss on bond behavior between polyvinyl-alcohol-fiber-reinforced concrete and steel rebar are clarified. Results show that the polyvinyl-alcohol-fiber-reinforced concrete exhibits worse resistance to corrosion damage than plain concrete. The cracking width, chloride penetration depth in concrete and the corrosion loss of steel rebar are more serious for the specimens with more polyvinyl alcohol fibers. The polyvinyl alcohol fibers also negatively affect bonding in ascending branches for both the specimens, but improve the bonding in descending branches after peak stress in the case of splitting. In the present test, the bond strength of corrosive specimens is increased slightly and then decreases gradually with the deepening of corrosion loss. The failures of specimens change from pull-out to splitting-pull-out as the corrosion time exceeds 30 days. Compared with uncorroded specimens, the maximum degradation of bond strength is about 50.1% when the corrosion is increased from 0% to 15%.

Keywords: polyvinyl alcohol fiber; bond degradation; corrosive damage; experimental study

1. Introduction

Polyvinyl alcohol (PVA) fiber is a kind of organic synthetic material, which has many advantages compared with other fiber materials. PVA fiber has high strength and ductility with a relatively high elastic modulus [1], outstanding resistance to corrosion without toxicity, high hydrophilicity and tolerance of the alkaline environment in concrete, strong bonding with the cement matrix [2], low cost, and effective restriction of cracking of concrete over long term [2]. PVA fiber can substantially improve the post-cracking behavior of concrete, enhancing the ductility [3] and toughness [4]. It also increases the splitting tensile strength [5] and flexure strength [6]. Besides that, the frost resistance [7] and fatigue life of concrete are enhanced with the addition of PVA fiber [8,9].

At present, PVA fibers have been widely used in civil engineering such as in fiber-cement as asbestos replacement, Engineered Cementitious Composites (ECC), strengthening of enlarged section for concrete structures, and various kinds of shotcrete, which play a promising role in toughening and reducing cracking in concrete [10]. Meanwhile, some studies have been reported that the content of PVA fibers significantly affect the fluidity of concrete mixtures [11]. The flowability of concrete decreases obviously with the increase in PVA fibers, which will lead to increased porosity and micro-damage in the matrix [12]. In this case, PVA-fiber-reinforced concrete would be susceptible to environmental erosion, resulting in a serious durability problem. Some studies have been carried out on the durability [13,14] of PVA concrete, such as tests of its frost and permeability resistance [15]. Nowadays, however, very few works have been performed to clarify the corrosion-effected bond behavior between PVA-fiber-reinforced concrete and steel rebar subjected to a chloride environment.

Extensive works have been carried out to investigate the effect of corrosion on the bonding between concrete and rebar [16–18]. Goksu and Inci [19] found that the better mechanical interlock between slightly corroded rebar and surrounding concrete increased the bond strength between the rebar and surrounding concrete, as was also reported in the studies by Choi [20] and Kim [21]. However, for a higher degree of corrosion, the breaking down of the mechanically weak layers of the corrosion products with increasing corrosion, and the ductility of the structure, more rapidly decrease in this case [21,22]. Rakesh et al. [23] presented that the bond strength was enhanced prior to corrosion cracking and then rapidly reduced with increase in corrosion level. Khaled and Ted [24] studied the effect of corrosion on the bond between rebar and concrete; they found that the 15% corrosion loss in steel bar mass can decrease approximately 35.6% of the bond strength. The material properties and the mechanism of PVA-fiber-reinforced concrete are significantly different from those of plain concrete. It has been reported that PVA fiber increases the splitting tensile strength [25] and flexure strength [26,27] of concrete as compared with plain concrete. Besides that, the bridging effect of fibers in concrete result in a difference in the bonding mechanism. The bond behavior between the PVA-fiber-reinforced concrete and steel rebar corroded under a chloride environment is still not well understood.

In this paper, an experimental test is proposed to study the corrosion-effected bond behavior between PVA-fiber-reinforced concrete and steel rebar under a chloride environment. In the subsequent sections, the paper carries out work on the following parts. In Section 2, four specimens with different PVA fiber contents were designed and subjected to accelerated corrosion in chloride solution. The bond stress and slip of the specimens were tested by pull-out tests, and corrosion loss and chloride penetration depth were measured. In Section 3, chloride penetration depth, corrosion loss of steel rebars and corrosion-induced concrete cracking of specimens with different contents of PVA fibers are described. In Section 4, the effects of PVA fiber and corrosion loss on bond behavior between concrete and steel rebar are clarified. Based on the research, several conclusions are drawn in Section 5.

2. Experimental Program

2.1. Detail of Specimens

Twenty-four pull-out specimens were prepared in the present test, which were designed based on the Standard for Test Methods for Concrete Structures (GB 50152-1992) [28]. All the specimens were designed with the same size, which consisted of a 150 mm × 150 mm × 150 mm concrete cube in which a steel rebar was embedded, as shown in Figure 1. The diameter and the length of the steel rebar were 12 mm and 360 mm, respectively. The embedment length between the steel rebar and concrete was set as 60 mm for all the pull-out specimens, i.e., five times the diameter of the rebar, which was obtained by installing PVC tubes on both ends of the specimen to avoid the contact between concrete and rebar. The embedded length in the present work were designed based on the standard GB 50152-1992 [28] and similar experimental studies in the published papers [29–31].

The factors considered in the tests included four fiber volume contents (0%, 0.2%, 0.4% and 0.6%) and six salt solution immersive times (0 days, 10 days, 20 days, 30 days, 40 days and 50 days). Designation of specimens was coded with two parts. The first part represents the content of PVA fibers. The second part indicates the immersive times where 0 d, 10 d, 20 d, 30 d, 40 d and 50 d stand for immersive times of 0 days, 10 days, 20 days, 30 days, 40 days and 50 days, respectively. For example, PVA0.2-20d means that the fiber volume content was 0.2% and the immersive time was 20 days.

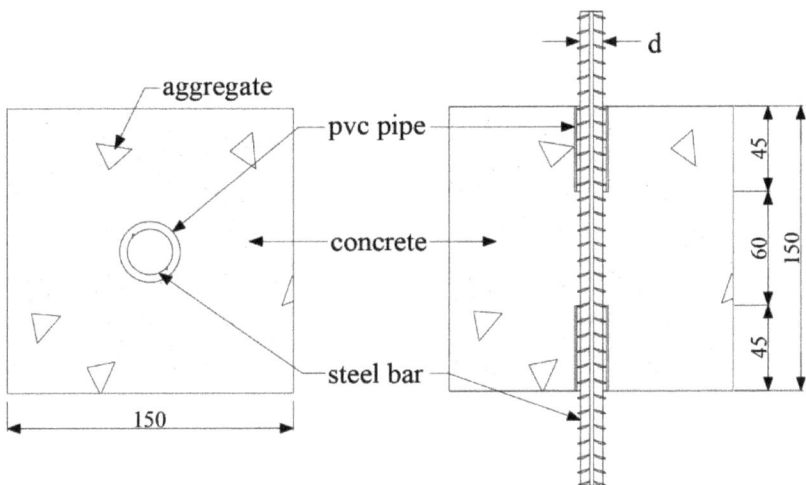

Figure 1. Detail of the specimen (Units: mm).

2.2. Materials and Mixture Design

The concrete was mixed using ordinary Portland cement, river sand, coarse aggregate, water, polycarboxylate superplasticizer and PVA fibers. The cement utilized in the test was ordinary Portland cement of 42.5 grade. The maximum coarse aggregate size was 20 mm with continuous graded crushed stone. The fine aggregate is natural river sand, and its fineness modulus is 1.82 by sieving test. PVA fibers produced by Kuraray in Tokyo, Japan were employed in the test. The diameter and length were 39 μm and 12 mm, respectively. The elastic modulus and the tensile strength of the fibers were 42 GPa and 1.6 GPa. The tested yield strength and ultimate tensile strength of the steel rebar were 454 and 521 MPa, respectively.

All specimens had the same mix proportions except the content of PVA fiber. The content of the cement was 485 kg/m^3. The water–cement ratio was 0.42. The contents of coarse aggregate and fine aggregate were 1092 kg/m^3 and 619 kg/m^3, respectively. The content of the water-reducing agent was 0.8% by weight of cement. The content of PVA fiber was the percentage of concrete volume.

In the present test, the mixture was mixed with a forced mixer for 6 min. The fine and coarse aggregate were first dry mixed for 2 min. After that, water and water-reducing agent were pre-mixed and added into the mixture by 70% and mixed for 1 min. The PVA fibers were then dispersed gradually into the mixture and mixed thoroughly for 2 min. Finally, the remaining water and water-reducing agent were added the mixture, then mixed for another 1 min.

After the mixing procedure was completed, tests were conducted on the fresh mixture to determine the fluidity of the concrete. The slumps were 180 mm, 102 mm, 45 mm and 12 mm for concrete mixtures with 0%, 0.2%, 0.4% and 0.6% PVA fibers, respectively. It was found that the slump decreased by about 43%, 75% and 93% when the fibers were increased from 0% to 0.6%. This indicated that the PVA fiber obviously affected the fluidity of the concrete. This is consistent with the results reported in previous research studies [11]. This could be attributed to the roughness of the fiber surface, and the increase in the interface friction between cement and fiber.

Pull-out specimens and some reserved samples with the same content of PVA fibers were cast using the same mixing batches, which were demolded after 24 h and cured in the room at 20 ± 2 °C and 95% relative humidity. At the age of 28 days, the reserved samples were subjected to a mechanical property test and the pull-out specimens were subjected to the accelerated corrosion (see detail in the following section). The compressive strength, splitting tensile strength and flexural strength were tested according to GB/T 50081-2019 [32] and direct tensile strength tests were undertaken based on the method suggested by Dashti [33]. The mechanical property strengths of the samples with different PVA fiber contents and the coefficient of variation (Cv) are shown in Table 1. The coefficient of variation of the data obtained is less than 10%, which is weak variability, and thus the mechanical properties obtained below are accurate.

Table 1. Mechanical property strength of specimens.

Specimen Series	Fiber Content (%)	Compressive Strength/Cv (MPa/%)	Splitting Tensile Strength/Cv (MPa/%)	Flexural Strength/Cv (MPa/%)	Direct Tensile Strength/Cv (MPa/%)
PVA0	0	39.98/3	2.76/3	3.32/3	2.85/3
PVA0.2	0.2	38.21/3	3.35/3	3.48/2	3.00/2
PVA0.4	0.4	35.06/3	3.47/2	3.74/5	3.10/2
PVA0.6	0.6	37.23/5	3.69/2	3.81/2	3.25/4

2.3. Accelerated Corrosion

The electrochemical method was employed in the present study to accelerate the corrosion of specimens. Specimens were laid flat along the rebar on the corrosion tank and partially submersed in 5% sodium chloride solution. The stainless bar was connected with the negative terminal of the supply, and the steel rebar was connected with the positive terminal. The direct current flowed from the positive terminal to the rebar, and then through the saturated concrete and saline solution to the stainless bar, and finally back to the negative terminal. Figure 2 shows the setup of the accelerated corrosion system.

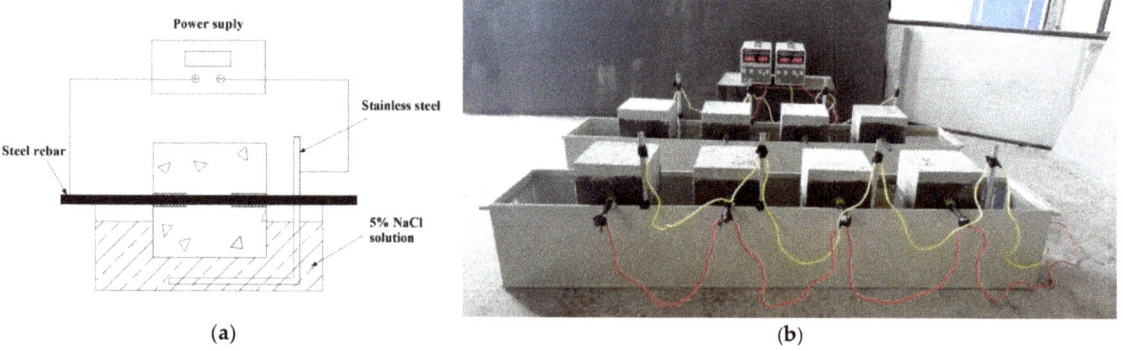

Figure 2. Setup of accelerated corrosion. (**a**) Setup device; (**b**) scene for the corrosion.

Before the accelerated corrosion test, specimens were partially submersed in sodium chloride solution for 3 days to saturate the concrete. The solution was kept approximately 10 mm below the rebar and the ends of the specimens were coated with anti-rust oil to prevent the permeability of chlorine salts along the rebar–concrete interface. A constant current was applied with the corrosion current density of 300 μA/cm^2.

In this study, the corrosion current and pH of the solution were monitored and adjusted twice a day to ensure that they were in a stable state. To ensure the effectiveness of the liquid at a constant concentration, the solution was renewed at weekly intervals. Moreover, corrosion of rebars could induce the cracking of concrete covers. Therefore, the cracking on concrete surfaces was observed and measured to investigate the restraint of PVA fibers on corrosion-induced concrete cracking.

2.4. Pull-Out Test

After accelerated corrosion, the specimens were subjected to a pull-out test, the test instrument is MTS-809 in the laboratory of Xiangtan University, the manufacturer of the instrument is the American MTS company (Monroe, NC, USA), and the maximum axial force capacity is 100 kN. A special loading frame was designed in the present test, as shown in Figure 3. The loading end of the specimens and the loading frame were fixed by the upper and lower clamps of the MTS, respectively. The position of the upper plate of the loading frame can be adjusted to fully contact with the specimen surface, which was used to minimize the effect of uneven stress and to prevent unexpected lateral movement of specimens during the loading process. The pull-out force was automatically measured by the test machine. The relative slip between the steel bar and concrete was measured by an extensometer clamped at the free end of the specimens. Additionally, a dial indicator was set transversely on the specimen surface to measure the development of the crack during the loading test.

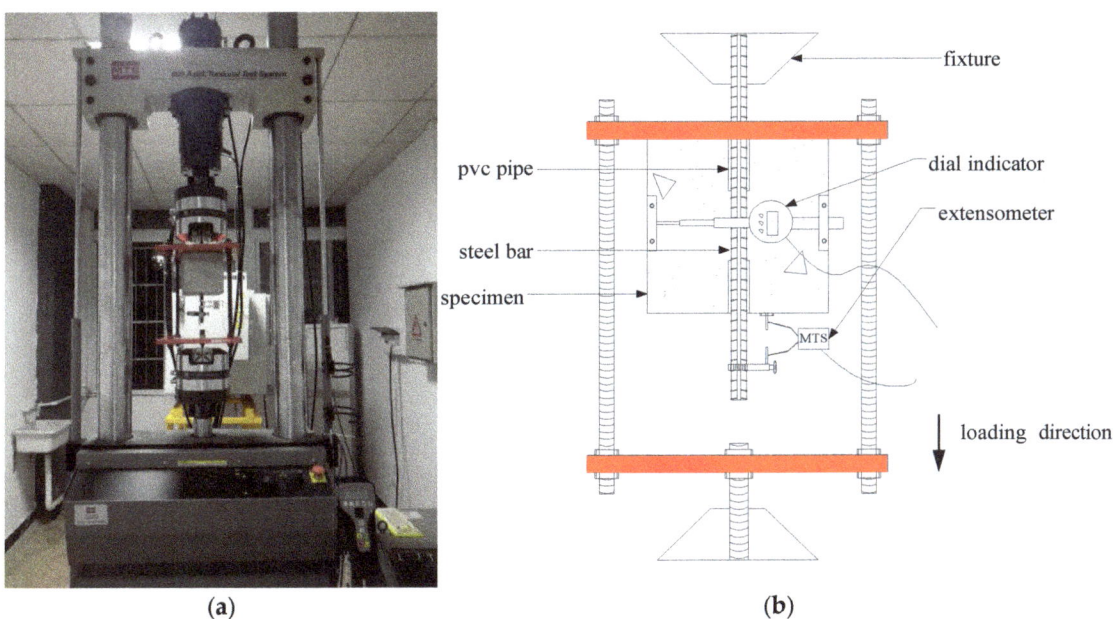

Figure 3. Loading instrument and detail device. (**a**) Loading instrument; (**b**) detail of device.

The pull-out tests were monotonically loaded using a displacement-control mode at a speed of 1 mm/min. The experimental data, including pull-out force, relative slip and crack width were synchronously collected at the rate of 1 datum/s. The test was stopped as the splitting of specimens or the residual pull-out force trended to be constant. The measured slip between concrete and rebar at the free end was employed to represent the slip characteristic of the specimens. Additionally, the average bond stress along the embedded length of the rebar was used to represent the stress characteristic, which can be expressed as

$$\tau = \frac{F}{\pi d L} \quad (1)$$

where F is the measured pull-out force; d is the nominal diameter of steel bar; and L is the embedded length of the steel bar in concrete. Bond stress and slip discussed in the following sections were obtained based on the above method.

2.5. Measurement of Corrosion Loss and Chloride Penetration Depth

After the pull-out tests, the specimen was split along the direction of rebar and the rebar was taken out to measure the actual corrosion loss. The rebars were cleaned with oxalic acid solution and then neutralized with alkali. After drying, the corrosion region, i.e., the bond region, of the steel rebar was cut off. The mass of the cut rebar was measured by using an electronic scale with the precision of 0.01 g. The actual mass loss of the corroded rebar was calculated as follows:

$$\rho_c = \frac{m_1 - m_0}{m_1} \quad (2)$$

where ρ_c is the actual corrosion loss of rebar; m_1 is the mass of the original rebar; m_0 is the mass of the corroded rebar.

Besides that, the split specimen was employed to measure the penetration depth of chloride in PVA-reinforced concrete based on the $AgNO_3$ colorimetric method [34]. The split specimen was cut transversely in the middle, and $AgNO_3$ solution with the concentration of 0.1 mol/L was evenly sprayed on the cutting faces of the concrete. The area penetrated by the chloride turned white, whereas the inner area turned brown, as shown in Figure 4, and the depth of chloride penetration was determined by the color difference.

Figure 4. Color comparison.

According the recommendations of the NT Build 492 [35], the measurements of the depths of the white area were employed to represent the performance of chloride penetration, whereas the penetration depths was measured by using a slide caliper. From the center to both edges, tests were undertaken at intervals of 10 mm. Seven depths were measured and the average of the value was the chloride penetration depth, as shown in Figure 5b. The depth in the edge area of the specimen was not measured to avoid the edge effect caused by uneven saturation.

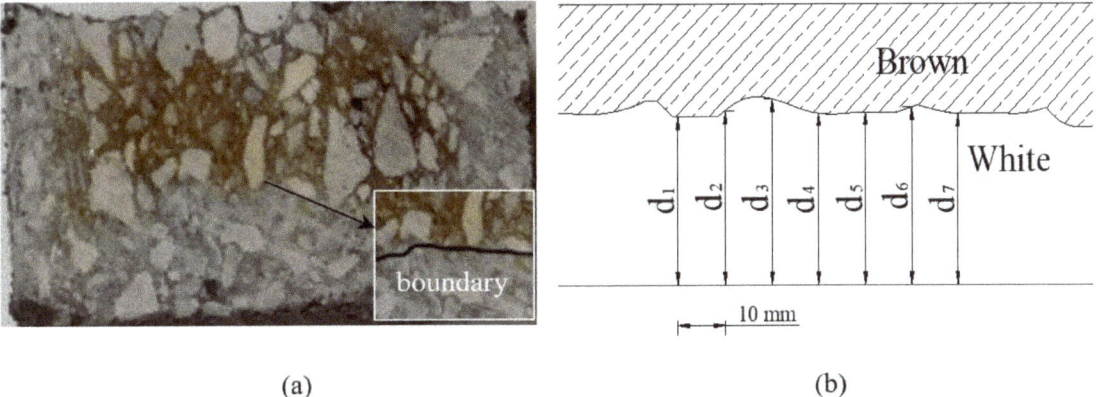

(a)   (b)

Figure 5. Discoloration border: (**a**) discoloration boundary; (**b**) illustration of chloride discoloration depth measurement.

3. Corrosion Damages

3.1. Crack Behavior

In the corrosion process, the electrical potential applied to the positively charged steel bars attracted negatively charged chloride ions from the salt solution into the concrete. When the chloride ions permeated into the surface of the steel rebar, the bar began to corrode. However, corrosion deteriorated the steel ribs and filled the concrete–steel interface with rust products [36]. The expansive nature of the corrosion products could result in the initiation of a corrosive crack [37]. In these cases, the specimens were monitored with the method of periodic observation to determine the formation and propagation of corrosive cracks. The typical crack observation for specimens is shown in Figure 6.

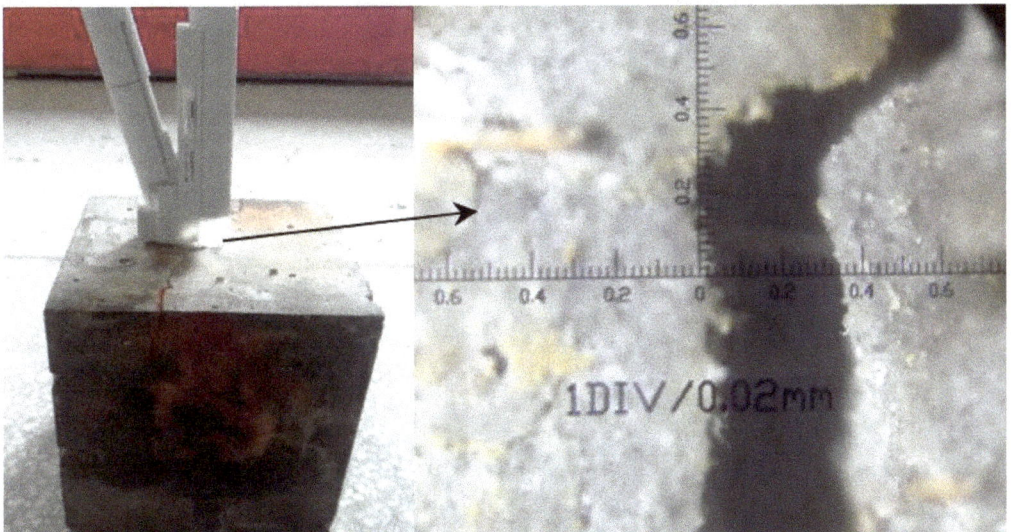

Figure 6. Crack observation.

Figure 7 shows the crack behavior of the PVA-12 series in different corrosive days. In particular, for the specimen with 0% PVA fibers, no sign of cracking was observed during the corrosion. The initial longitudinal crack along the direction of the rebar was observed on the bottom of the other specimens after about 25 days. The width of the initial crack increased obviously with the increment of PVA fibers, and after the corrosion time reached 30 days, the rising slope of the crack width growth curve also increased with the increase in PVA fiber. The slope of the ascending branch reflects the speed of rebar corrosion, which depends mainly on the resistance against chloride penetration for concrete. This indicates that the porous structure caused by addition of PVA fibers degrades the density of concrete. On the other hand, the growth rate of the crack width decreases gradually with the increase in the corrosion times. This could be attributed to the fact that the corrosion products flow out along the crack, which results in a reduction in the cycle of extrusion stress caused by corrosive products.

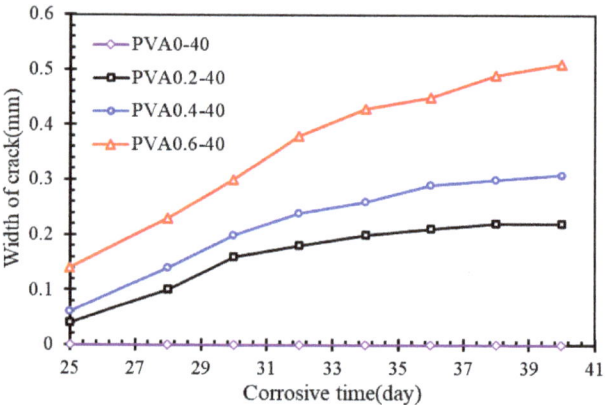

Figure 7. Crack width for different days.

Details of the maximum crack widths for corrosive specimens are shown in Figure 8. All the specimens show an incremental trend in crack width with the increase in corrosion times and PVA fibers. For instance, the mean maximum crack of specimens with 40 d and 50 d corrosion times was 2.24 and 5.19 times that of the specimen with a 30 d corrosion time, respectively. In this test, after adding PVA fiber, when the corrosion time reached 40 days and the PVA fiber content increased from 0.4% to 0.6%, the corrosion crack increment reached a maximum of 66.7%. Therefore, there is no positive influence of PVA fibers on corrosion cracking. As the corrosion level increases, the risk of corrosion of reinforcing steel within the concrete rises, which will harm the durability of the concrete. After the corrosion, the specimens were removed from the setup for visual inspection and pull-out tests.

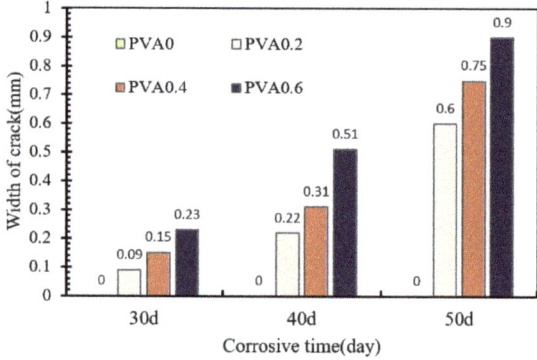

Figure 8. Maximum crack width.

3.2. Corrosion Loss

The corroded steel rebars were measured and the percentage of mass was computed using Equation (2), and the actual corrosion level is shown in Table 2. It can be seen that after the same number of corrosion days, there is significant difference in the corrosion degree of specimens with different fiber doping. The difference was mostly derived from the following three aspects. First, the permeability of the concrete with different PVA fiber contents was not included in the calculation of the theoretical level of corrosion. Although the specimens were immersed in the solution for three days prior to the accelerated corrosion, it would have taken a longer period for the chloride permeate to reach the surface of the steel rebar. Secondly, the incorporation of fibers increases the pore space of concrete, making the epoxy coating less protective than expected, which means that the corrosion range was larger than expected and resulted in a current density under 300 μA/cm². Further, manual measuring errors during the weighting process of the rebar could also have caused the variation in corrosion loss.

Table 2. Experimental parameters and results of pull-out tests.

Test	Content of PVA Fiber (%)	ρ_c (%)	f_{cu} (MPa)	f_t (MPa)	f_{ts} (MPa)	f_{tf} (MPa)	Failure Mode	τ_{ons} (MPa)	τ_{max} (MPa)	S_0 (mm)
PVA0-0	0	0.0	39.98	2.85	2.76	3.32	P	4.95	21.08	1.53
PVA0.2-0	0.2	0.0	38.21	3.00	3.35	3.48	P	5.86	19.57	1.42
PVA0.4-0	0.4	0.0	35.06	3.10	3.47	3.74	P	6.10	19.10	1.34
PVA0.6-0	0.6	0.0	37.23	3.25	3.69	3.81	P	5.75	18.18	1.69
PVA0-10	0	0.6	39.98	2.85	2.76	3.32	P	6.36	22.20	1.94
PVA0.2-10	0.2	0.7	38.21	3.00	3.35	3.48	P	10.14	21.30	1.54
PVA0.4-10	0.4	0.9	35.06	3.10	3.47	3.74	P	10.40	19.41	1.58
PVA0.6-10	0.6	1.0	37.23	3.25	3.69	3.81	P	13.27	18.90	1.11
PVA0-20	0	1.1	39.98	2.85	2.76	3.32	P	10.64	19.38	1.00
PVA0.2-20	0.2	3.9	38.21	3.00	3.35	3.48	P	10.71	18.26	1.12
PVA0.4-20	0.4	4.4	35.06	3.10	3.47	3.74	P	8.65	16.20	1.04
PVA0.6-20	0.6	5.5	37.23	3.25	3.69	3.81	P	5.69	14.05	1.24
PVA0-30	0	1.9	39.98	2.85	2.76	3.32	S-P	17.19	18.60	0.02
PVA0.2-30	0.2	4.8	38.21	3.00	3.35	3.48	S-P	14.24	15.55	0.26
PVA0.4-30	0.4	7.3	35.06	3.10	3.47	3.74	S-P	13.13	13.70	0.21
PVA0.6-30	0.6	9.9	37.23	3.25	3.69	3.81	S-P	11.83	12.45	0.19
PVA0-40	0	2.3	39.98	2.85	2.76	3.32	S-P	18.00	18.17	0.02
PVA0.2-40	0.2	8.6	38.21	3.00	3.35	3.48	S-P	12.14	14.56	0.09
PVA0.4-40	0.4	11.3	35.06	3.10	3.47	3.74	S-P	9.55	12.59	0.08
PVA0.6-40	0.6	13.2	37.23	3.25	3.69	3.81	S-P	9.48	9.97	0.05
PVA0-50	0	3.1	39.98	2.85	2.76	3.32	S	16.98	17.88	0.07
PVA0.2-50	0.2	12.7	38.21	3.00	3.35	3.48	S-P	13.54	13.72	0.03
PVA0.4-50	0.4	14.6	35.06	3.10	3.47	3.74	S-P	9.94	10.13	0.04
PVA0.6-50	0.6	16.9	37.23	3.25	3.69	3.81	S-P	8.91	9.38	0.19

Notes: P means pull-out failure of specimens; S means splitting failure of specimens; S-P means splitting-pull-out failure of specimens.

Figure 9 shows the typical appearance of rebar with different corrosion levels. From top to bottom, the rebars were sequentially removed from the concrete with PVA fiber contents of 0%, 0.2%, 0.4% and 0.6%. The actual corrosion loss of the rebar was marked correspondingly. Compared with the appearance of the uncorroded rebar shown in Figure 9a, the rebar with lower corrosion loss, shown in Figure 9b,c, presented slightly concave corrosion pits, and the ribs coming out of the corrosion pits showed slight damage. At 9% and 12% corrosion levels, the area of the corrosion pits increased, accompanied by heavily damaged ribs, as shown in Figure 9d,e. At 15% corrosion level, the area of the corrosion pits increased further, as shown in Figure 9f, and the ribs had almost disappeared, resulting in a relatively smooth surface. In this case, the corrosive damage gradually deepened on the surface of the rebar with the increase in corrosion loss.

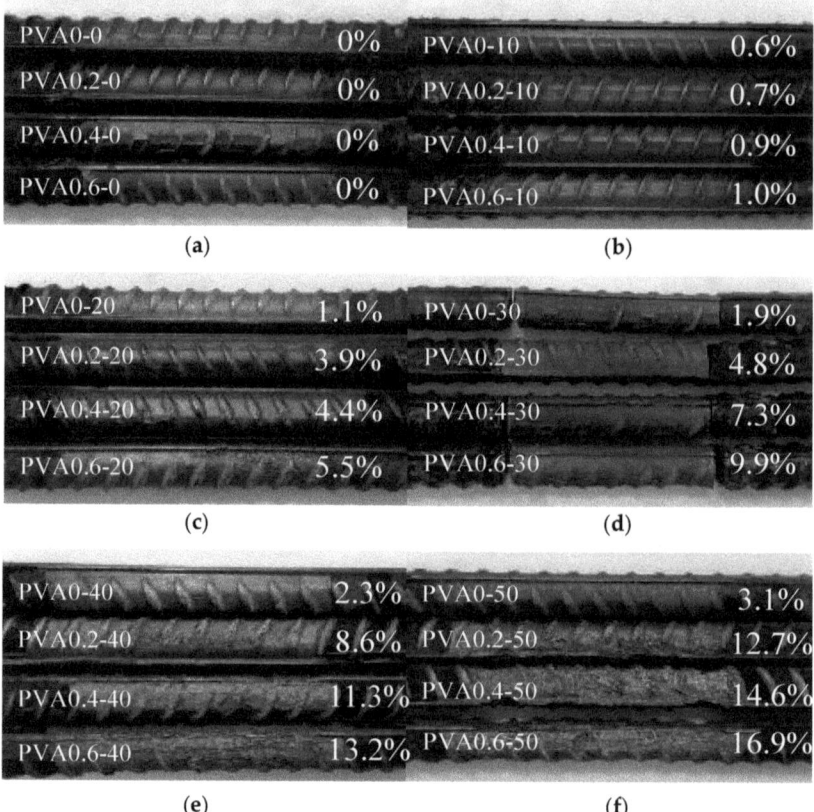

Figure 9. Appearance of corroded rebar: (**a**) 0%; (**b**) 3%; (**c**) 6%; (**d**) 9%; (**e**) 12% and (**f**) 15%.

Based on the test results, the PVA-fiber-reinforced concrete specimens exhibited relatively worse corrosion resistance in the chlorine salt environment compared with the specimen with 0% fiber content. The more PVA fiber, the higher the corrosion loss of rebar with the same corrosive time. For example, the increase in PVA fiber volume content from 0% to 0.2%, 0.4% and 0.6% in the PVA-15 series specimens increased the corrosion loss by about 9.6%, 1.9% and 2.3%, respectively. The corrosion loss of the specimens without the addition of PVA fibers was usually lower in the present test, which may be the reason why corrosion cracking did not occur. Therefore, the addition of the PVA fibers led to smaller and more closely spaced cracks in the concrete, resulting in reduced permeability resistance of the concrete.

3.3. Chloride Penetration

Figure 10 shows the colorimetric pictures of specimens with different contents of PVA fibers and salt solution immersive times. From left to right, the concrete sections were taken from the specimens with PVA fiber contents of 0%, 0.2%, 0.4% and 0.6%. As can be seen from Figure 10, the precipitated reaction products and color variation were clear on the concrete surface. The color change boundary was obviously different with the increase in immersive time. During the time of chloride penetration, all directions of the concrete were affected, which caused the white area to appear around the concrete cross-section. The color of part of the concrete surface became darker, which can be attributed to products of the oxidation reaction.

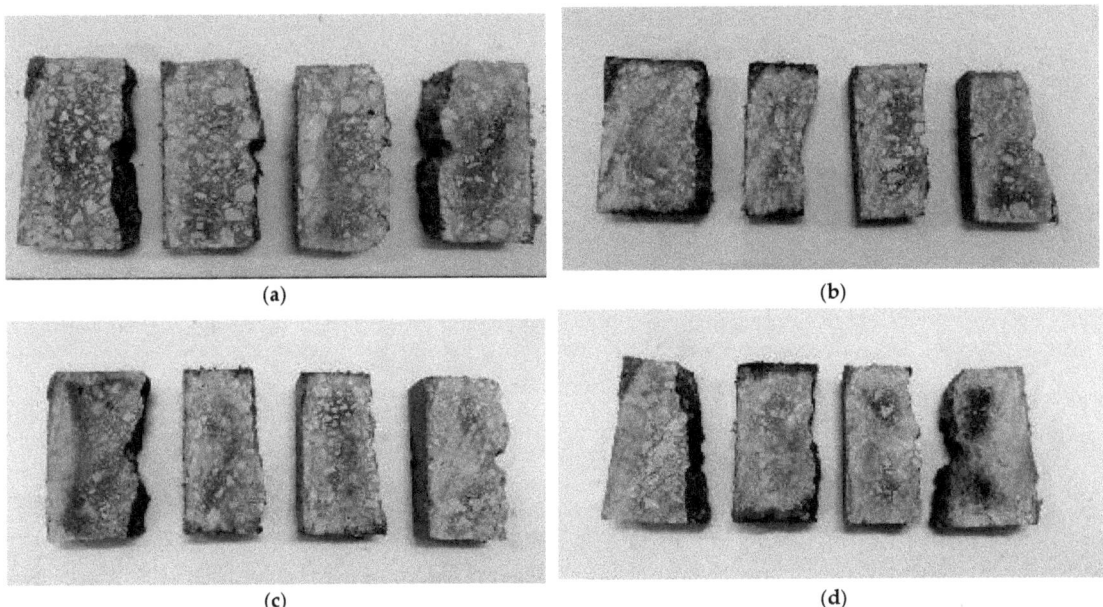

Figure 10. AgNO$_3$ colorimetric results: (**a**) 20 d; (**b**) 30 d; (**c**) 40 d; (**d**) 50 d.

In this test, the chloride penetration depths were measured from the side of the concrete. The color rendering within the boundary range of the concrete surface was not measured, to eliminate the influence of the boundary effect. The results of the depth measurement are summarized in Figure 11. Almost all of the chloride penetration depths are raised with the increase in immersive time. For example, the increase in corrosive time from 20 d to 30 d, 40 d and 50 d in the PVA0 series specimens increased the penetration depth by about 12.1%, 20.3% and 28.7%, respectively. The more the PVA fiber contents, the more obvious the penetration depth. For instance, the penetration depth of PVA0.2 series specimens increased by about 15.6%, 24.9% and 35.1% under the same corrosive conditions compared with the PVA0 series. It can also be noticed that when the same immersive times were used, incorporation of PVA fibers provided an additional penetrative path and increased the chloride penetration rate into the concrete, as compared with the concrete of the PVA0 series. The increase in depth was pronounced in the PVA0.6 series. In general, the addition of PVA fibers showed a negative effect on resistance to the chloride ion penetration.

After that, the concrete was cut into flat pieces, for which the area of 50 × 50 mm was selected for image processing to observe the tiny holes on the concrete surface through Image-J software. Firstly, the picture of the concrete pieces was gray processed as shown in Figure 12a. It can be seen that the concrete aggregate and mortar are lighter in color than the holes. Then arbitrarily determine a line segment that passes through a region with significantly higher grayscale values, such as the a-b line in Figure 12a. Then, the gray value along the a-b line segment, shown in Figure 12b, was determined. The curve shows that the gray value at the hole was significantly reduced. The gray level threshold of 130 was preliminarily selected in the present test and areas with a gray level below this threshold were identified as concrete pores.

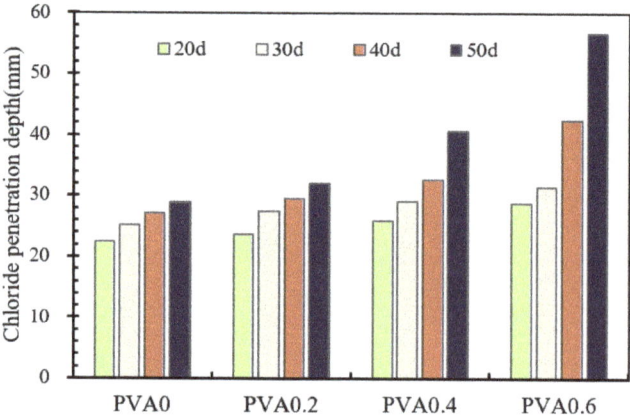

Figure 11. Chloride penetration depth.

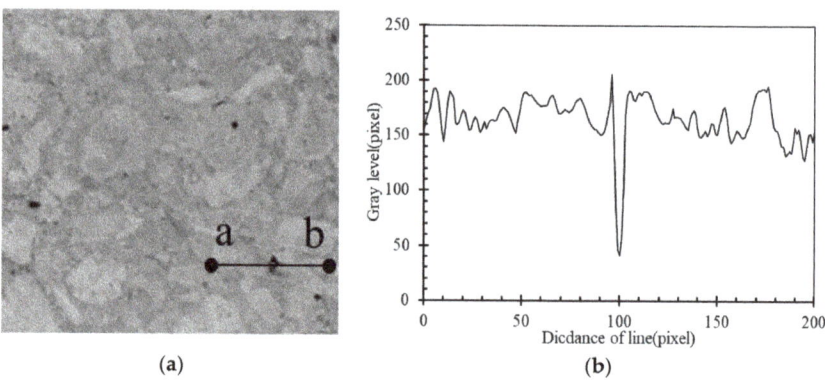

(a) (b)

Figure 12. Detail of the specimen pieces and gray level. (**a**) Specimen pieces; (**b**) gray level.

Figure 13 shows the binary processing image of concrete pieces in different PVA fiber volume contents. The number of pore structures on the surface of PVA0 series concrete was small, and the diameter of the holes was much smaller than that of the fiber concrete. The diameter and number of harmful holes were increased obviously with the increase in fiber contents. In this case, the incorporation of PVA fiber reduces the overall density of the concrete, which is consistent with the decrease in the compressive strength of fiber-reinforced concrete.

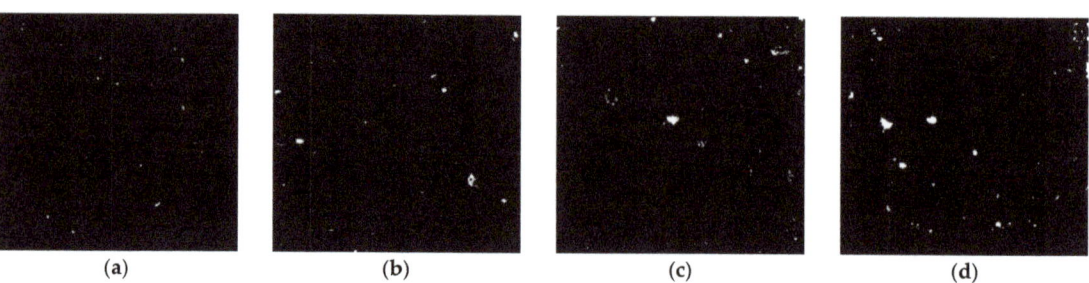

Figure 13. Binary processing image. (**a**) PVA0; (**b**) PVA0.2; (**c**) PVA0.4; (**d**) PVA0.6.

4. Bond Behaviors

Test results of the pull-out specimens, including the failure mode, the actual corrosion level ρ_c, the bond stress at onset slip of free end τ_{ons}, the maximum bond stress τ_{max} and the slip corresponding to the maximum stress S_0 are summarized in Table 2. Because there is only one pull-out specimen per group, there is no coefficient of variation (Cv) here.

4.1. Failure of Pull-Out Specimens

The specimens failed in three different modes: pull-out failure, splitting-pull-out failure and splitting failure. For the first one, the specimens failed gradually with the pull-out of the steel rebar without concrete cracking. For the second one, the specimens retained their integrity after the concrete splitting and the bond stress was not completely lost. For the third one, the specimens failed suddenly with the splitting of concrete accompanied by a loud crash. After that, the concrete members were completely broken apart. These three failure modes are shown in Figure 14.

Figure 14. Typical failure of specimens: (**a**) pull-out failure; (**b**) splitting-pull-out failure and (**c**) splitting failure.

Failure modes of all specimens are summarized in Table 2. It can be seen that the failure modes were affected by the corrosion level and PVA fiber volume contents. The specimens with 0 d, 10 d and 20 d corrosive times failed in pull-out mode due to the lower corrosion level and larger relative concrete cover depth. For the specimens of 30 d, 40 d and 50 d corrosive times, there was greater expansibility of corrosion products, with the result that the failure was splitting-pull-out mode. In particular, the specimen PVA0-50 failed in splitting mode. It could be attributed to the damage caused to the concrete and rebar by prolonged immersion in chloride solution. This indicates that the failure mode of the specimen tends towards splitting-pull-out damage as the corrosion time exceeds 30 days.

4.2. Effects of PVA Fibers

Figure 15 shows the bond stress–slip curves for specimens with different PVA fiber volume contents in the same rebar corrosion level. Since the corrosion degree of the steel bars in the PVA0 series specimens was generally low, this curve has been omitted from the comparison of the curves with a high corrosion degree, as shown in Figure 15d–f, below. As mentioned before, the curves are significantly affected by the failure modes. In addition, the additional PVA fibers also show an evident effect on bond stress–slip curves. The slope of the ascending branch of the bond stress–slip curve and the bond strength decreased with increasing PVA fibers for all specimens except for the 1% series specimen PVA0.6. The slope of the ascending branch reflects the bond stiffness of specimens, which depends mainly on the mechanical action between concrete and rebars. This indicates that the porous structure caused by the addition of PVA fibers degrades the mechanism between concrete and rebar.

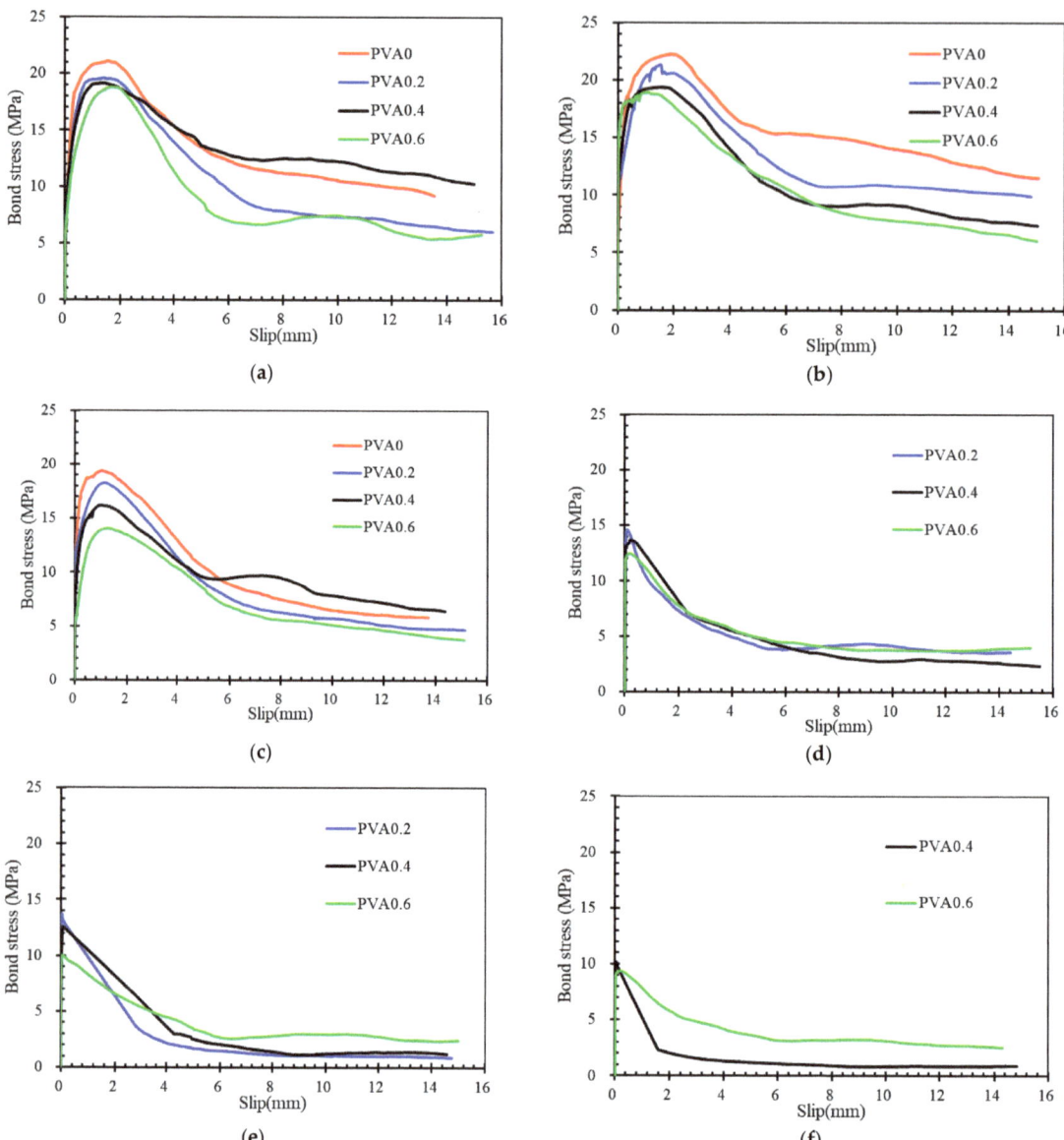

Figure 15. Bond stress–slip curves for the specimens with different PVA volume contents: (**a**) 0% series; (**b**) 1% series; (**c**) 5% series; (**d**) 8% series; (**e**) 12% series; (**f**) 15% series.

The addition of PVA fibers shows a positive effect on the descending branches for specimens which failed in splitting-pull-out mode. After the curve has passed the peak load stress, the decline in bond stress can be effectively relieved with the increase in fiber content. Besides that, the residual stress was obvious for specimens with the addition of PVA fibers. The reasons should be attributed to the bridging effects of PVA fibers after the peak load stress of specimens, which effectively restrains the cracking and slows down the decrease in bond stress.

Details about the comparison of bond strength for specimens with different fiber volume contents are summarized in Figure 16. All the specimens show a descending trend in bond strength with the increase in PVA fiber. For example, the increase in PVA fiber volume content from 0% to 0.2%, 0.4% and 0.6% in specimens from the 1% corrosion level series decreased the bond strength by about 4.05%, 8.89% and 2.61%, respectively. The maximum decrement is about 20.85% in the present test.

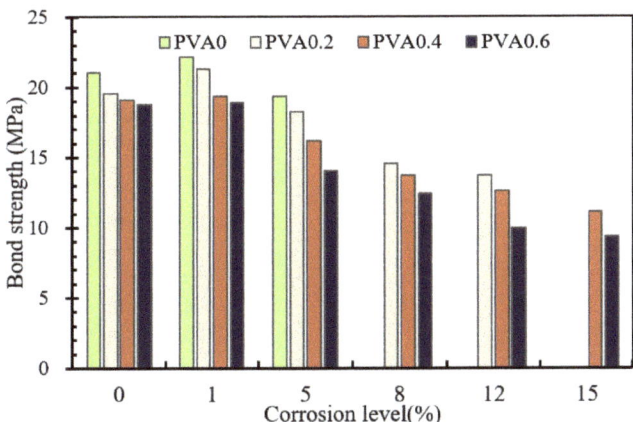

Figure 16. Bond strength of specimens with different contents of PVA fibers.

Figure 17 shows the comparison of crack width before and after the pull-out test for PVA-40 series specimens with different fiber volume contents. As mentioned before, the larger the volume content of PVA fibers, the larger the corrosion crack width of the concrete. However, in the pull-out test, it was found that the anti-drawing cracking property of the specimens gradually increased with the increase in PVA fiber content. Here, 0%, 0.2%, 0.4% and 0.6% of the PVA fibers caused the propagation of crack widths of about 3.5 mm, 2.6 mm, 1.1 mm and 0.7 mm, respectively. In this case, PVA fibers play a positive role in restricting the split-induced cracking. This indicates that PVA fibers restrict both the micro-cracking and macro-cracking of specimens which failed by splitting. The restriction effects on macro-cracking seem much more significant than that on micro-cracking.

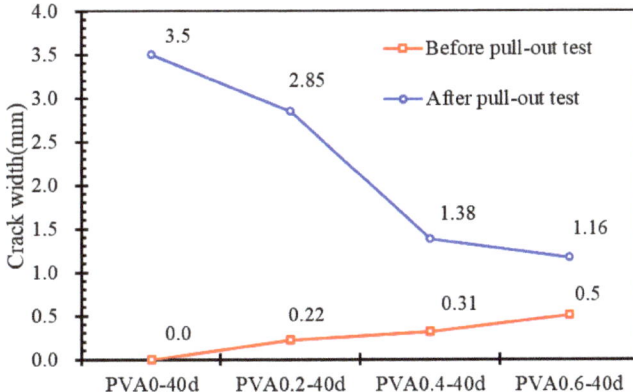

Figure 17. Crack width before and after pull-out test.

4.3. Effects of Corrosion Loss

Figure 18 shows the bond stress–slip curves for the specimens with different corrosion loss. The specimens in each subgraph have similar PVA fiber volume contents, but different corrosion levels. The curves match well with the failure modes for specimens with different corrosion levels. The curves show a gradual descent for specimens with low corrosion levels after the maximum bond stress, whereas there is an obvious rapid segment for specimens with high levels. Besides that, corrosion level also affects the initial bonding stiffness and bond strength of specimens. The initial bonding stiffness of each specimen usually increases gradually and then decreases with an augment in the corrosion of the steel rebar, except for individual specimens. The more the PVA fibers, the greater the difference of initial bond stiffness caused by different corrosion levels. In addition, the bond stress increases slightly at 1.0% corrosion loss. Specimens with high corrosion level have low bond stress.

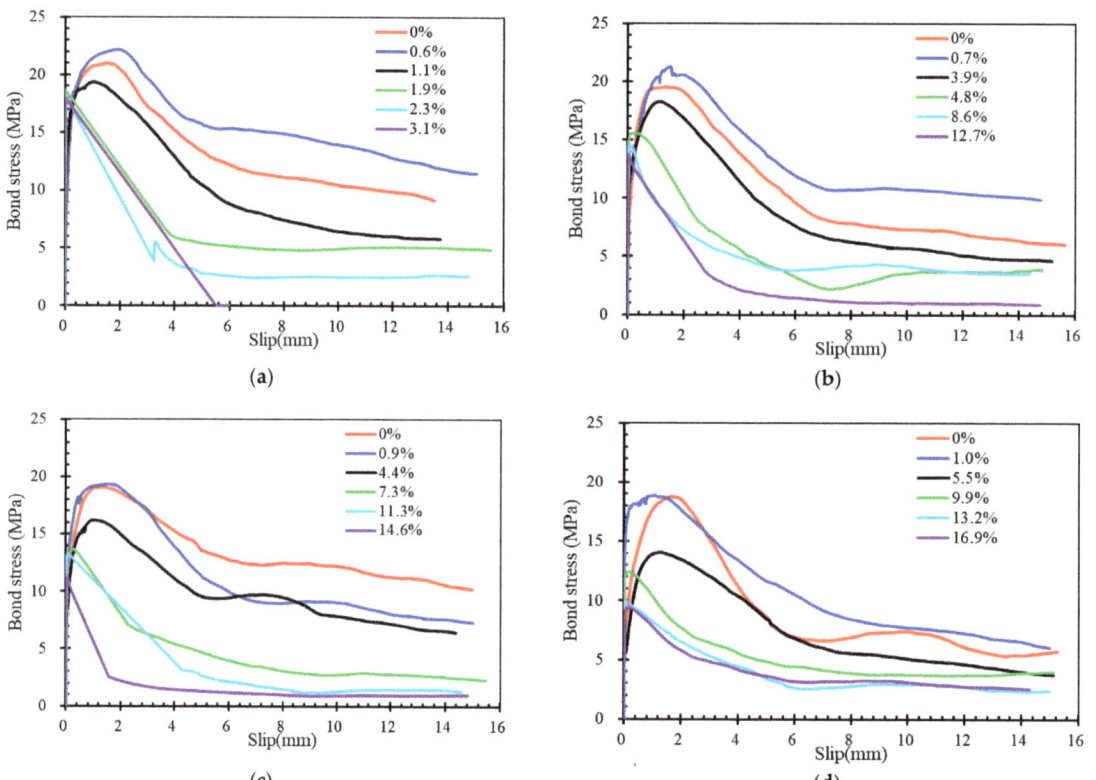

Figure 18. Bond stress–slip curves for the specimens with different corrosion loss: (**a**) PVA0 series; (**b**) PVA0.2 series; (**c**) PVA0.4 series; (**d**) PVA0.6 series.

The reason could be mainly attributed to the difference of the mechanical interlocking force between rebar and concrete. For cases with low corrosion levels, the positive effect was generated on bond properties because the slight corrosion of the steel bar improves the frictional resistance on the contact surface. However, with further increases in the corrosion level, the relative height of the ribs of the steel rebar gradually decreased. This caused the mechanical interlocking force between the concrete and the ribs of the rebar to significantly diminish. Besides that, cracking due to the volume expansion of the corrosion products reduced the confinement effect of the concrete. For the initial bonding stiffness of most specimens, however, the inflection point occurs at a higher corrosion level. The uneven corrosion of the rebar results in the corrosion damage on the upper surface being less than that on the lower surface. In this case, it could provide a certain friction at the beginning of the pull-out process. As the load increased, the friction was consumed and had no effect on the maximum bond stress. Therefore, the bond stress of all specimens initially increased slightly and then gradually decreased with further corrosion of the rebar.

Details about the comparison of bond strength and slip corresponding to bond strength for specimens with different corrosive times are summarized in Figure 19. As mentioned before, almost all of the bond strength and slip corresponding to bond strength increase with the increase in corrosive times. In particular, the bond strength and slip corresponding to bond strength increases slightly at low corrosive levels. For instance, the increase in corrosive time from 0 d to 10 d in PVA0.2 series specimens increased the bond strength by about 8.8% and the corresponding slip by about 8.4%. However, the increase in corrosive time from 10 d to 20 d, 30 d, 40 d and 50 d in specimens from the same series decreased the bond strength by about 14.3%, 14.8%, 6.3% and 5.8%, and decreased the corresponding slip by about 27.0%, 77.2%, 65.0% and 60.9%. Compared with uncorroded specimens, the maximum degradation of bond stress is about 50.1%, of which the corrosive time is 50 d. The more the PVA fibers, however, the greater the difference in the data, which is caused by corrosion loss. In general, the corrosion level of the rebar shows an obvious effect on bond strength and slip corresponding to bond strength for all specimens.

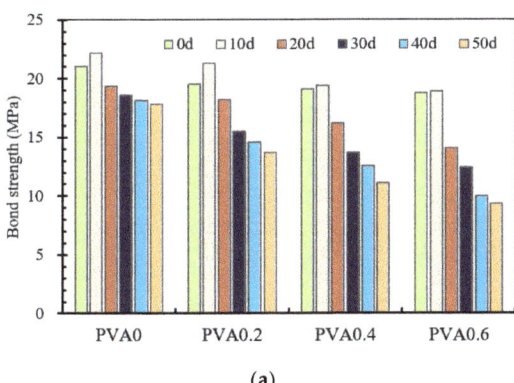
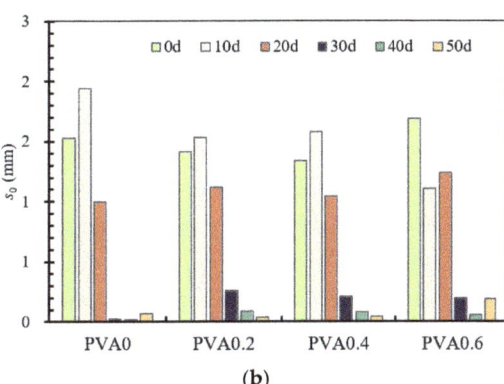

Figure 19. Bond stress–slip curves for the specimens with different corrosion loss. (**a**) Bond strength; (**b**) slip corresponding to bond strength.

5. Conclusions

This study experimentally investigated the bond behavior between PVA-fiber-reinforced concrete and steel rebar corroded under a chloride environment. The effects of PVA fibers and corrosion loss on bond behavior were clarified. The following conclusions may be drawn based on the present study:

- The PVA-fiber-reinforced specimens exhibited worse resistance to corrosion damage than plain specimens; the harmful fine pores in fiber concrete provide channels for chloride penetration. The maximum increment of crack width is about 66.7% in the present test for PVA-fiber-reinforced specimens. With the increase in the fibers, the corrosive cracking become more obvious.
- PVA fiber generally showed a negative effect on bond behavior, but a positive effect on the descending branches for the case with splitting failure. PVA fibers decreased both the initial bond stiffness and bond strength in the present test. The maximum decrement of bond strength was about 31.49%, for samples with PVA fiber contents of less than 0.6%. The lowest extension of crack width was about 0.7 mm with the addition of PVA fibers

- in the pull-out test, in which the PVA fibers can restrict the split-induced cracking and protect against the failure of specimen in a more ductile way.
- With the deepening of corrosion loss, the bond strength of corrosion specimens first slightly increased, and then gradually decreased. Compared with plain specimens, the maximum degradation of bond stress was about 50.1%, for which the corrosion level was 15%. Specimens with a greater corrosion level usually had a greater initial bonding stiffness, but lower bond strength than specimens with a high level after uneven corrosion.
- There are no forces between the PVA-fiber-reinforced concrete and the reinforcement in the chloride environment explored in this paper. In engineering practice, PVA-fiber-reinforced concrete is often in the load-bearing state received under the influence of the external environment; the rust characteristics in the load-holding state remain to be further analyzed and studied.

Author Contributions: Writing—original draft, X.Z., X.W. and Y.W.; Writing—review & editing, X.W. All authors have read and agreed to the published version of the manuscript.

Funding: This research was funded by "Education Department Research Foundation of Hunan Province (Grant No. 21B0170)", "the Natural Science Foundation of Hunan Province of China (Grant No. 2021JJ40540)", "Guangdong Basic and Applied Basic Research Foundation (grant No. 2019A1515111032)", "National Natural Science Foundation (grant No. 52008107)" and "the Special Funds for the Construction of Innovative Provinces in Hunan, China (Nos. 2019RS1059)".

Informed Consent Statement: Informed consent was obtained from all subjects involved in the study.

Data Availability Statement: All data generated or analyzed during this study are included in this article.

Conflicts of Interest: The authors declare no conflict of interest.

References

1. Pakravan, H.R.; Jamshidi, M.; Latifi, M. Study on fiber hybridization effect of engineered cementitious composites with low- and high-modulus polymeric fibers. *Constr. Build. Mater.* **2016**, *112*, 739–746. [CrossRef]
2. Juarez, C.; Fajardo, G.; Monroy, S.; Duran-Herrera, A.; Valdez, P.; Magniont, C. Comparative study between natural and PVA fibers to reduce plastic shrinkage cracking in cement-based composite. *Constr. Build. Mater.* **2015**, *91*, 164–170. [CrossRef]
3. Wang, J.; Dai, Q.; Si, R.; Guo, S. Investigation of properties and performances of Polyvinyl Alcohol (PVA) fiber-reinforced rubber concrete. *Constr. Build. Mater.* **2018**, *193*, 631–642. [CrossRef]
4. Liu, F.; Ding, W.; Qiao, Y. Experimental investigation on the flexural behavior of hybrid steel-PVA fiber reinforced concrete containing fly ash and slag powder. *Constr. Build. Mater.* **2019**, *228*, 116706. [CrossRef]
5. Nuruddin, M.F.; Khan, S.U.; Shafiq, N.; Ayub, T. Strength Prediction Models for PVA Fiber-Reinforced High-Strength Concrete. *J. Mater. Civ. Eng.* **2015**, *27*, 04015034. [CrossRef]
6. Abushawashi, N.; Vimonsatit, V. Material Classification and Composite Elastic Modulus of Hybrid PVA Fiber Ferrocement. *J. Mater. Civ. Eng.* **2016**, *28*, 04016073. [CrossRef]
7. Kim, G.; Lee, B.; Hasegawa, R.; Hama, Y. Frost resistance of polyvinyl alcohol fiber and polypropylene fiber reinforced cementitious composites under freeze thaw cycling. *Compos. Part B Eng.* **2016**, *90*, 241–250.
8. Jang, J.; Kim, H.; Kim, T.; Min, B.; Lee, H. Improved flexural fatigue resistance of PVA fiber-reinforced concrete subjected to freezing and thawing cycles. *Constr. Build. Mater.* **2014**, *59*, 129–135. [CrossRef]
9. Şahmaran, M.; Özbay, E.; Yücel, H.E.; Lachemi, M.; Li, V.C. Frost resistance and microstructure of Engineered Cementitious Composites: Influence of fly ash and micro poly-vinyl-alcohol fiber. *Cem. Concr. Compos.* **2012**, *34*, 156–165. [CrossRef]
10. Said, M.; El-Azim, A.A.A.; Ali, M.M.; El-Ghazaly, H.; Shaaban, I. Effect of elevated temperature on axially and eccentrically loaded columns containing Polyvinyl Alcohol (PVA) fibers. *Eng. Struct.* **2020**, *204*, 110065. [CrossRef]
11. Celik, K.; Meral, C.; Mancio, M.; Mehta, P.K.; Monteiro, P.J.M. A comparative study of self-consolidating concretes incorpo-rating high-volume natural pozzolan or high-volume fly ash. *Constr. Build. Mater.* **2014**, *67*, 14–19. [CrossRef]
12. Tabatabaeian, M.; Khaloo, A.; Joshaghani, A.; Hajibandeh, E. Experimental investigation on effects of hybrid fibers on rhe-ological, mechanical, and durability properties of high-strength SCC. *Constr. Build. Mater.* **2017**, *147*, 497–509. [CrossRef]
13. Raj, B.; Sathyan, D.; Madhavan, M.K.; Raj, A. Mechanical and durability properties of hybrid fiber reinforced foam concrete. *Constr. Build. Mater.* **2020**, *245*, 118373. [CrossRef]
14. Teng, S.; Afroughsabet, V.; Ostertag, C.P. Flexural behavior and durability properties of high performance hybrid-fiber-reinforced concrete. *Constr. Build. Mater.* **2018**, *182*, 504–515. [CrossRef]

15. Wang, L.; He, T.; Zhou, Y.; Tang, S.; Tan, J.; Liu, Z.; Su, J. The influence of fiber type and length on the cracking resistance, durability and pore structure of face slab concrete. *Constr. Build. Mater.* **2021**, *282*, 122706. [CrossRef]
16. Yang, O.; Zhang, B.; Yan, G.; Chen, J. Bond Performance between Slightly Corroded Steel Bar and Concrete after Exposure to High Temperature. *J. Struct. Eng.* **2018**, *144*, 04018209. [CrossRef]
17. Coronelli, D.; François, R.; Dang, H.; Zhu, W. Strength of Corroded RC Beams with Bond Deterioration. *J. Struct. Eng.* **2019**, *145*, 04019097. [CrossRef]
18. Farhan, N.A.; Sheikh, M.N.; Hadi, M.N.S. Experimental Investigation on the Effect of Corrosion on the Bond Between Reinforcing Steel Bars and Fibre Reinforced Geopolymer Concrete. *Structures* **2018**, *14*, 251–261. [CrossRef]
19. Goksu, C.; Inci, P.; Ilki, A. Effect of Corrosion on Bond Mechanism between Extremely Low-Strength Concrete and Plain Reinforcing Bars. *J. Perform. Constr. Facil.* **2016**, *30*, 04015055. [CrossRef]
20. Choi, Y.S.; Yi, S.-T.; Kim, M.Y.; Jung, W.Y.; Yang, E.I. Effect of corrosion method of the reinforcing bar on bond characteristics in reinforced concrete specimens. *Constr. Build. Mater.* **2014**, *54*, 180–189. [CrossRef]
21. Yin, S.; Jing, L.; Lv, H. Experimental Analysis of Bond between Corroded Steel Bar and Concrete Confined with Textile-Reinforced Concrete. *J. Mater. Civ. Eng.* **2019**, *31*, 04019208. [CrossRef]
22. Huang, C.-H. Effects of Rust and Scale of Reinforcing Bars on the Bond Performance of Reinforcement Concrete. *J. Mater. Civ. Eng.* **2014**, *26*, 576–581. [CrossRef]
23. Paswan, R.; Rahman, R.; Singh, S.K.; Singh, B. Bond Behavior of Reinforcing Steel Bar and Geopolymer Concrete. *J. Mater. Civ. Eng.* **2020**, *32*, 04020167. [CrossRef]
24. Soudki, K.; Sherwood, T. Bond Behavior of Corroded Steel Reinforcement in Concrete Wrapped with Carbon Fiber Rein-forced Polymer Sheets. *J. Mater. Civ. Eng.* **2020**, *15*, 358–370. [CrossRef]
25. Sun, L.; Hao, Q.; Zhao, J.; Wu, D.; Yang, F. Stress strain behavior of hybrid steel-PVA fiber reinforced cementitious composites under uniaxial compression. *Constr. Build. Mater.* **2018**, *188*, 349–360. [CrossRef]
26. Pan, J.; Cai, J.; Hui, L.; Christopher, K.Y. Development of Multiscale Fiber-Reinforced Engineered Cementitious Composites with PVA Fiber and $CaCO_3$ Whisker. *J. Mater. Civ. Eng.* **2018**, *30*, 04018106. [CrossRef]
27. Shafiq, N.; Ayub, T.; Khan, S.U. Investigating the performance of PVA and basalt fibre reinforced beams subjected to flexural action. *Compos. Struct.* **2016**, *153*, 30–41. [CrossRef]
28. GB 50152-1992; Standard for Test Methods for Concrete Structures. Domestic-National Standards-State Administration of Market Supervision and Administration CN-GB: Beijing, China, 2012.
29. Zhang, P.; Gao, Z.; Wang, J.; Wang, K. Numerical modeling of rebar-matrix bond behaviors of nano-SiO_2 and PVA fiber reinforced geopolymer composites. *Ceram. Int.* **2021**, *47*, 11727–11737. [CrossRef]
30. Zhang, X.; Zhang, W.; Cao, C.; Xu, F.; Yang, C. Positive effects of aligned steel fiber on bond behavior between steel rebar and concrete. *Cem. Concr. Compos.* **2020**, *114*, 103828. [CrossRef]
31. Zhang, X.; He, F.; Chen, J.; Yang, C.; Xu, F. Orientation of steel fibers in concrete attracted by magnetized rebar and its effects on bond behavior. *Cem. Concr. Compos.* **2023**, *138*, 104977. [CrossRef]
32. GB/T 50081-2019; Standard Test Method for Physical and Mechanical Properties of Concrete. The State Administration for Market Regulation: Beijing, China, 2019.
33. Dashti, J.; Nematzadeh, M. Compressive and direct tensile behavior of concrete containing Forta-Ferro fiber and calcium aluminate cement subjected to sulfuric acid attack with optimized design. *Constr. Build. Mater.* **2020**, *253*, 118999. [CrossRef]
34. Filho, J.H.; de Medeiros, M.H.F.; Pereira, E.; Helene, P.; Isaia, G. High-Volume Fly Ash Concrete with and without Hydrated Lime: Chloride Diffusion Coefficient from Accelerated Test. *J. Mater. Civ. Eng.* **2013**, *25*, 411–418. [CrossRef]
35. NT Build 492; Concrete, Mortar and Cement-Based Repair Materials: Chloride Migration Coefficient from Non-Steady-State Migration Experiments. Nordtest Method: Espoo, Finland, 1999.
36. Zhang, X.; Wang, L.; Zhang, J.; Liu, Y. Bond Degradation–Induced Incompatible Strain between Steel Bars and Concrete in Corroded RC Beams. *J. Perform. Constr. Facil.* **2016**, *30*, 04016058. [CrossRef]
37. Cao, C.; Cheung, M.M.; Chan, B.Y. Modelling of interaction between corrosion-induced concrete cover crack and steel corrosion rate. *Corros. Sci.* **2013**, *69*, 97–109. [CrossRef]

Disclaimer/Publisher's Note: The statements, opinions and data contained in all publications are solely those of the individual author(s) and contributor(s) and not of MDPI and/or the editor(s). MDPI and/or the editor(s) disclaim responsibility for any injury to people or property resulting from any ideas, methods, instructions or products referred to in the content.

Article

Mechanical Behaviour Evaluation of Full Iron Tailings Concrete Columns under Large Eccentric Short-Term Loading

Xinxin Ma [1], Jianheng Sun [1,*], Fengshuang Zhang [1,*], Jing Yuan [1], Mingjing Yang [1], Zhiliang Meng [1], Yongbing Bai [1] and Yunpeng Liu [2]

[1] Civil Engineering Department, Hebei Agricultural University, Baoding 071001, China; maxinxin1987@126.com (X.M.)
[2] School of Materials Science and Engineering, University of Science and Technology Beijing, Beijing 100083, China
* Correspondence: sjh@hebau.edu.cn (J.S.); lgzhfsh@hebau.edu.cn (F.Z.)

Abstract: In this study, full iron tailings concrete (FITC) was created using iron tailings from a tailings pond in Qian'an, China. Iron tailings account for 86.8% of the total mass of solid raw materials in the FITC. To enable large-scale use of FITC, a comprehensive investigation of the structural behaviour of full-iron tailing-reinforced concrete (FITRC) specimens is warranted. Therefore, eight rectangular reinforced concrete (RC) columns with conventional reinforced concrete (CRC) as a control were tested to investigate the effects of section dimensions, initial eccentricities, and concrete strengths, on the structural behaviour of FITRC columns under large eccentric short-term loading. The experimental and analytical results indicated that the sectional strain of the FITRC columns satisfied the plane-section assumption under short-term loading, and the lateral deflection curve agreed well with the half-sinusoidal curve. In addition, the FITRC columns exhibited a slightly lower cracking load and lower ultimate load capacity than the CRC columns, and the crack widths were larger than those of the CRC columns. The reduction in the load capacity observed in the FITRC was within the permissible range stated in the design code, thereby satisfying the code requirements. The deformation coefficients of the FITRC and CRC columns were identical, and the cracking and ultimate loads calculated according to the current code and theories were in good agreement with the measured results.

Keywords: iron tailings; RC column; large eccentric; short-term load; bearing capacity; ductility

1. Introduction

Tailings are a major source of solid waste generated by the separation of valuable fractions from ore during mining operations [1,2]. Improper disposal of tailings occupies land resources, pollutes the environment, and releases hazardous substances [3]. In China, tailings are widely distributed [4] with a low utilisation rate [5]. Figure 1 shows photographs of piles of iron tailings. In 2018, 475 million tonnes of iron tailings were produced in China, accounting for 39.22% of the total tailings [6] (Figure 2). In 2019, 536 million tonnes of tailings were produced in China, of which 116 million tonnes were comprehensively utilised [7]. This represents a utilisation rate of 21.6%, which is significantly lower than the average utilisation rate in developed countries [8]. In addition, tailings dams pose a debris-flow hazard. Sudden failures of tailings dams occurred in Xiangfen County, Shanxi Province, China, in 2008 [9], and Jiaokou County, Shanxi Province, China, in 2022 [10]. These disasters injured hundreds of people and caused significant losses. Because of these problems, tailings have received considerable attention from the scientific community [11,12]. The optimal approach for dealing with the problems associated with iron tailings is to develop technologies for their large-scale use [13].

Figure 1. Photos of iron tailings stacking dam in Hebei, China: (**a**) mining and tailings site; (**b**) tailings site.

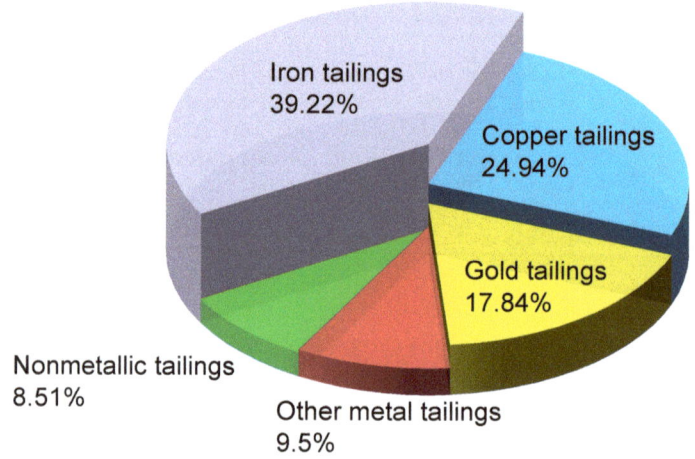

Figure 2. Breakdown of tailings production in 2018.

Concrete is a leading construction material, with an annual global consumption of approximately 25 billion tonnes [14]. The construction sector is responsible for over 30% of global greenhouse gas emissions [15]. In China, cement production is expected to reach approximately 2.38 billion tonnes in 2021 [16]. Several studies have recommended the use of tailings in the cement and concrete industry [17–24] to significantly reduce carbon dioxide emissions and consumption of natural resources [25].

Iron tailings are inert at room temperature [26,27]. Previous studies have used iron tailings as coarse [28,29] and fine aggregates [30–35], as well as silica materials for the production of aerated concrete at high temperatures [36–38]. Some studies have also used iron tailings as an admixture [39–44] and examined the activation of iron tailings powder [45,46]. The use of iron tailings in concrete has the potential to reduce the storage of iron tailings, reduce land use and reduce environmental pollution. At the same time, the replacement of fly ash with iron tailings powder as an admixture could reduce the cost of concrete and promote the sustainable development of concrete. In addition, iron tailings can improve the carbonation resistance, frost resistance, and sulphate erosion resistance of concrete [47,48].

To enable the extensive application of iron tailings in reinforced concrete, it is imperative to validate the structural behaviour of reinforced concrete samples containing iron tailings compared with plain concrete. A study was conducted to compare the flexural behaviour of iron tailings sand concrete beams and conventional concrete (CC) beams [49]. The results showed that the two types of concrete beams exhibited similar flexural behaviours and that the calculation of the crack width of iron tailings sand concrete beams should be corrected when the replacement rate of iron tailings sand is greater than 40%. Another study [50] investigated the axial compressive behaviour of short iron tailings sand concrete columns. The results showed that the iron tailings sand concrete short columns were comparable to CC short columns in terms of axial compressive strength, deformation, and ductility. The seismic behaviour of iron tailings sand concrete columns was investigated [51]. The results showed that the failure patterns of iron tailings sand concrete columns and CC columns were almost identical, while the flexural capacity of the iron tailings sand concrete columns was slightly different from that of the CC columns, and iron tailings sand could completely replace conventional sand. Compared with CC columns, the axial compressive properties of full iron tailing concrete columns satisfied the requirements of the current code [52]. Further, the calculation model that considered the hoop restraint effect could more accurately predict the axial compressive bearing capacity of full iron tailings concrete columns.

These studies show that the mechanical behaviour of reinforced concrete (RC) specimens made from iron tailings sand replacing fine aggregates is marginally different from that of conventional reinforced concrete (CRC) specimens. However, very few studies have been conducted on the mechanical behaviour of full iron tailings reinforced concrete (FITRC) specimens. The novelty of this study is the use of full iron tailings concrete (FITC), which was produced using iron tailings powder as an admixture, iron tailings gravel as a coarse aggregate, and iron tailings sand as a fine aggregate. To the best of our knowledge, this is the first investigation of the structural behaviour of FITRC columns under large eccentric loads. We aimed to comprehensively evaluate the characteristics and limitations of FITRC columns to provide a theoretical basis for the large-scale use of iron tailings. We believe that this study will be of practical significance to structural engineers in China. Previous dam failures in tailings ponds have caused significant casualties and property losses and have adversely affected the environment. Therefore, large-scale use of iron tailings in RC can effectively dispose of iron tailings, reduce the consumption of resources needed for cement production and sand mining, and ensure sustainable development.

2. Materials and Experimental Design

2.1. Materials

Table 1 shows the chemical compositions of the iron tailings, fly ash, river sand, and conventional gravel. Iron tailings are siliceous, with a silica content exceeding 60%.

Table 1. Main chemical composition of concrete raw materials.

Specimens	SiO_2	Fe_2O_3	MgO	Al_2O_3	CaO	$CaCO_3$
Iron tailings	68.2%	12.5%	6.8%	5.1%	4.8%	0.0%
Fly ash	52.2%	5.6%	1.3%	26.5%	4.2%	0.0%
River sand	44.0%	3.3%	16.0%	6.4%	27.5%	0.0%
Conventional gravel	15.0%	0.3%	16.2%	0.2%	0.0%	68.1%

All the samples were in powder form (<50 μm). The mineral composition of the raw materials was determined using X-ray diffraction (XRD). Quartz is the primary mineral in iron tailings, with trace amounts of anorthite, haematite, and microcline (Figure 3), all of which are inert materials that do not undergo hydration. This indicates that iron tailings have low cementitious reactivity in their original state. The particle sizes of the iron tailings were reduced, the original microstructure of the iron tailings was altered, and the reactivity of the iron tailings was improved by mechanical grinding. Cement consists of primary active cementitious materials, such as dicalcium silicate (C2S), tricalcium silicate (C3S), tricalcium aluminate (C3A), and tetracalcium aluminoferrite (C4AF). The main mineral constituents of fly ash are mullite, sillimanite, and quartz. The primary mineral constituents of river sand are cordierite, dolomite, cossyrite, and quartz. The primary mineral constituents of the conventional gravel are calcite, dolomite, and quartz.

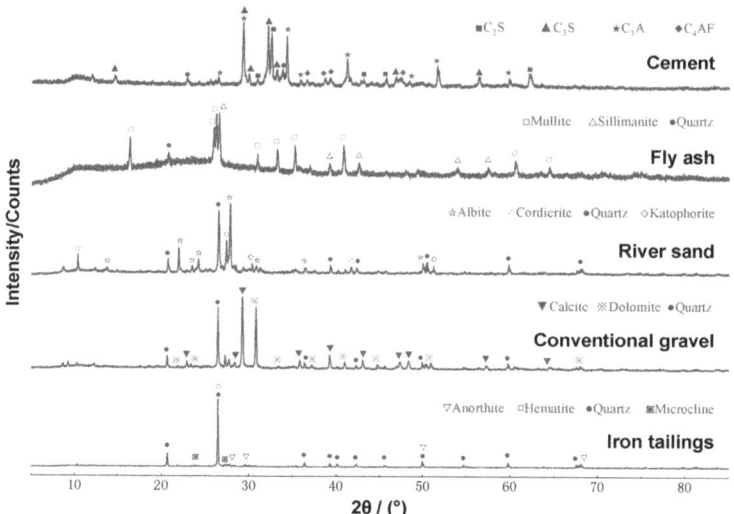

Figure 3. XRD patterns of concrete raw materials.

Figure 4 shows the microscopic morphologies of the cementitious materials. The cement (Figure 4a) and iron tailings powder (Figure 4b) had varying particle sizes and many angularities. The fly ash particles were mostly spherical in shape (Figure 4c) with heterogeneous surfaces and showed some micro-porosities. Compared to iron tailings, fly ash can morphologically reduce the fluidity of concrete mixtures.

(a)

(b)

Figure 4. Cont.

(c)

Figure 4. SEM micrograph at 4000× magnification of particles: (**a**) cement; (**b**) iron tailings; (**c**) fly ash.

Ordinary Portland cement (P.O 42.5) was selected, and its primary indices are listed in Table 2. Iron tailings powder was sourced from Qian'an, Hebei, China, and mechanically activated by grinding in a single-shaft horizontal ball mill. Grade II fly ash was used. Table 3 lists the physical indices of the two mineral admixtures, and Figure 5 shows the particle size distributions of the two admixtures.

Table 2. Primary technical indices of Portland cement.

Apparent Density (kg·m^{-3})	Flexural Strength (MPa)		Compressive Strength (MPa)	
	3 d	28 d	3 d	28 d
3090	6.9	10.8	31.2	52.7

Table 3. Primary technical indices of mineral admixtures.

Specimens	D_{10} (μm)	D_{50} (μm)	D_{90} (μm)	SSA (m^2·kg^{-1})	AD (m^2·kg^{-1})
Iron tailings powder	1.51	15.51	74.84	480.6	2770
Conventional gravel	3.02	26.51	98.18	415.1	2180

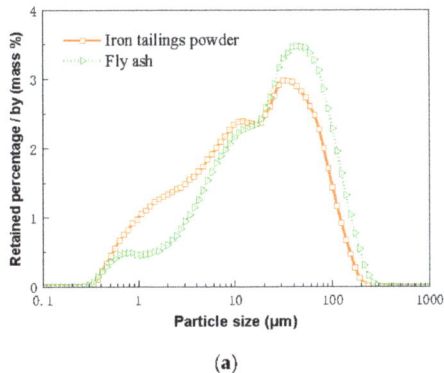

(a)

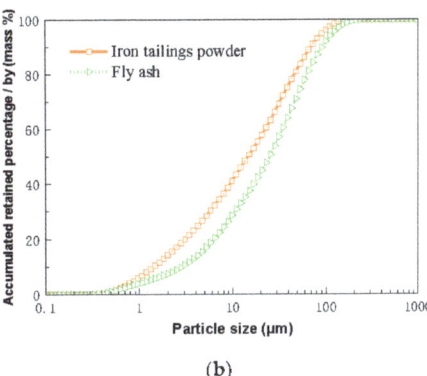

(b)

Figure 5. Particle size distribution curves of two mineral admixtures: (**a**) particle size distribution; (**b**) cumulative particle size distribution.

Table 4 summarises the primary indices of the iron tailings gravel and conventional gravel, showing that iron tailings gravel is superior to conventional gravel in terms of compressive strength and rock crushing index; however, the rate of expansion caused by the alkali–aggregate reaction of iron tailings gravel is inferior to that of conventional gravel. It is also superior to conventional gravel in terms of soundness. Iron tailings sand has a fineness modulus of 2.8, apparent density of 2780 kg·m^{-3} and bulk density of 1666 kg·m^{-3}. River sand has a fineness modulus of 2.6, apparent density of 2650 kg·m^{-3} and bulk density of 1623 kg·m^{-3}.

Table 4. Main technical indices of coarse aggregate.

Specimens	Compressive Strength of Rock (MPa)	Crushing Index (%)	Soundness (%)	Alkali–Aggregate Reaction (%)	AD (m^2·kg^{-1})	BD (kg·m^{-3})
Iron tailings gravel	62.2	6.05	3.5	0.051	2730	1570
Conventional gravel	60.0	6.56	3.7	0.042	2700	1550

2.2. Experiment Design

The concrete mix proportions were designed according to the JGJ55-2011 specification in China [53]. In total, four concrete mix proportions and two strength grades were tested (grades C35 and C45). The specific proportions are listed in Table 5. Considering that fly ash can reduce the fluidity of concrete mix, FITC is superior to CC in terms of workability at the same water/binder ratio and water-reducing agent dosage. The mass of the iron tailings was 87.0% of the total mass of the FITC.

Table 5. Design of mix proportions.

Specimens	Water Binder Ratio	Water (kg)	Cement (kg)	Iron Tailings Powder (kg)	Iron Tailings Sand (kg)	Iron Tailings Gravel (kg)	Fly Ash (kg)	River Sand (kg)	Conventional Gravel (kg)	Water Reducing Agent (kg)	Slump (mm)	Dispersion (mm)
FITC45	0.32	170	369	158	739	1017	-	-	-	3.1	210	520 × 530
CC45	0.32	170	369	-	-	-	158	706	976	3.1	200	470 × 490
FITC35	0.40	170	301	129	775	1071	-	-	-	2.5	215	515 × 535
CC35	0.40	170	301	-	-	-	129	746	1031	2.5	205	475 × 490

The mechanical behaviour of the concrete, including cubic and prismatic compressive strength, splitting tensile strength, modulus of elasticity, and Poisson's ratio, was tested according to GB/T 50081-2019 [54]. The test results are presented in Table 6. The cubic, prismatic compressive, and tensile strengths of FITC were slightly lower than those of CC, and the modulus of elasticity of FITC decreased by 19.1% and 18.4%, respectively, compared with CC.

Table 6. Mechanical behaviour of concrete.

Specimens	Cubic Compressive Strength f_{cu} (MPa)	Prismatic Compressive Strength f_c (MPa)	Splitting Tensile Strength f_t (MPa)	Static Modulus of Elasticity E_c (MPa)	Poisson's Ratio ν
FITC45	53.5	38.1	3.5	3.22×10^4	0.27
CC45	55.9	41.4	3.7	3.98×10^4	0.24
FITC35	44.4	34.8	3.2	3.05×10^4	0.25
CC35	47.8	36.3	3.4	3.74×10^4	0.23

An HRB400 grade rebar was selected, consisting of 14 mm and 16 mm diameter longitudinal bars and 8 mm diameter stirrups. Figure 6 shows the stress–strain curves of the rebars, and Table 7 lists the primary engineering indices of the rebars. Symmetrical and asymmetrical reinforcements were used for the RC columns. Figure 7a shows the details of the reinforcement. Two axial linear variable differential transformers (LVDTs) and five lateral LVDTs were placed on the RC columns, as shown in Figure 7.

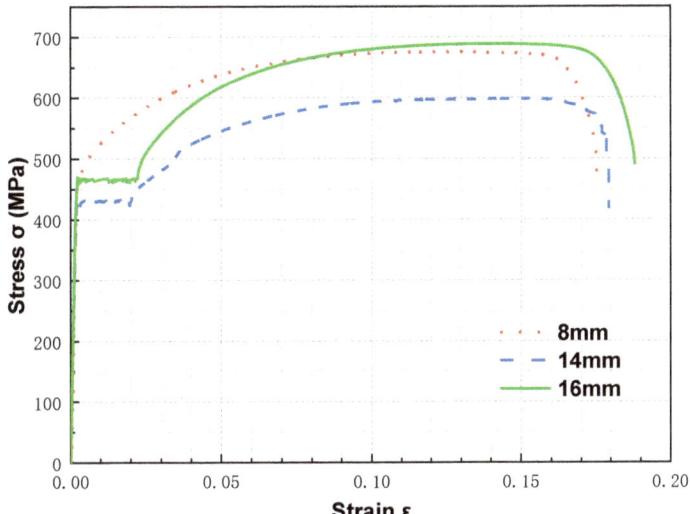

Figure 6. Stress–strain curve of rebars.

Table 7. Mechanical behaviour of rebars.

Reinforcement Diameter (mm)	Yield Strength f_y (MPa)	Ultimate Strength f_u (MPa)	Elastic Modulus E_s (MPa)	Percentage Elongation after Fracture (%)
8	460	676	2.06×10^4	17.2
14	430	598	2.09×10^4	19.0
16	466	689	2.07×10^4	20.3

To mitigate the effect of an additional bending moment on an eccentrically pressurised column specimen, the span-depth ratio (L/h) should not exceed 5. The 1200 mm high specimen had 150 mm × 250 mm sectional dimensions and symmetrical reinforcement of two 14 mm rebars in tension and two 14 mm rebars in compression made of C45 strength concrete. The 1500 mm-high specimen had a 200 mm × 300 mm section and asymmetric reinforcement with three 16 mm rebars in tension and two 16 mm rebars in compression made from C35 concrete. To ensure that the specimens could withstand a large eccentric load, cow legs were placed at the top and bottom of the specimen. The concrete was poured into three FITRC columns and one CRC column.

Before the start of the experiment, the RC columns were placed at the location specified by the initial eccentricity e_0 and preloaded. The preload did not exceed 20% of the ultimate load, and the column stability was evaluated. The columns were formally loaded in several stages of 5%–15% of the calculated ultimate bearing capacity N_u. Once the specified load was reached after each loading stage, the load was maintained for 3 min to fully release the strain in the RC, and the strain and displacement were measured at the corresponding time points.

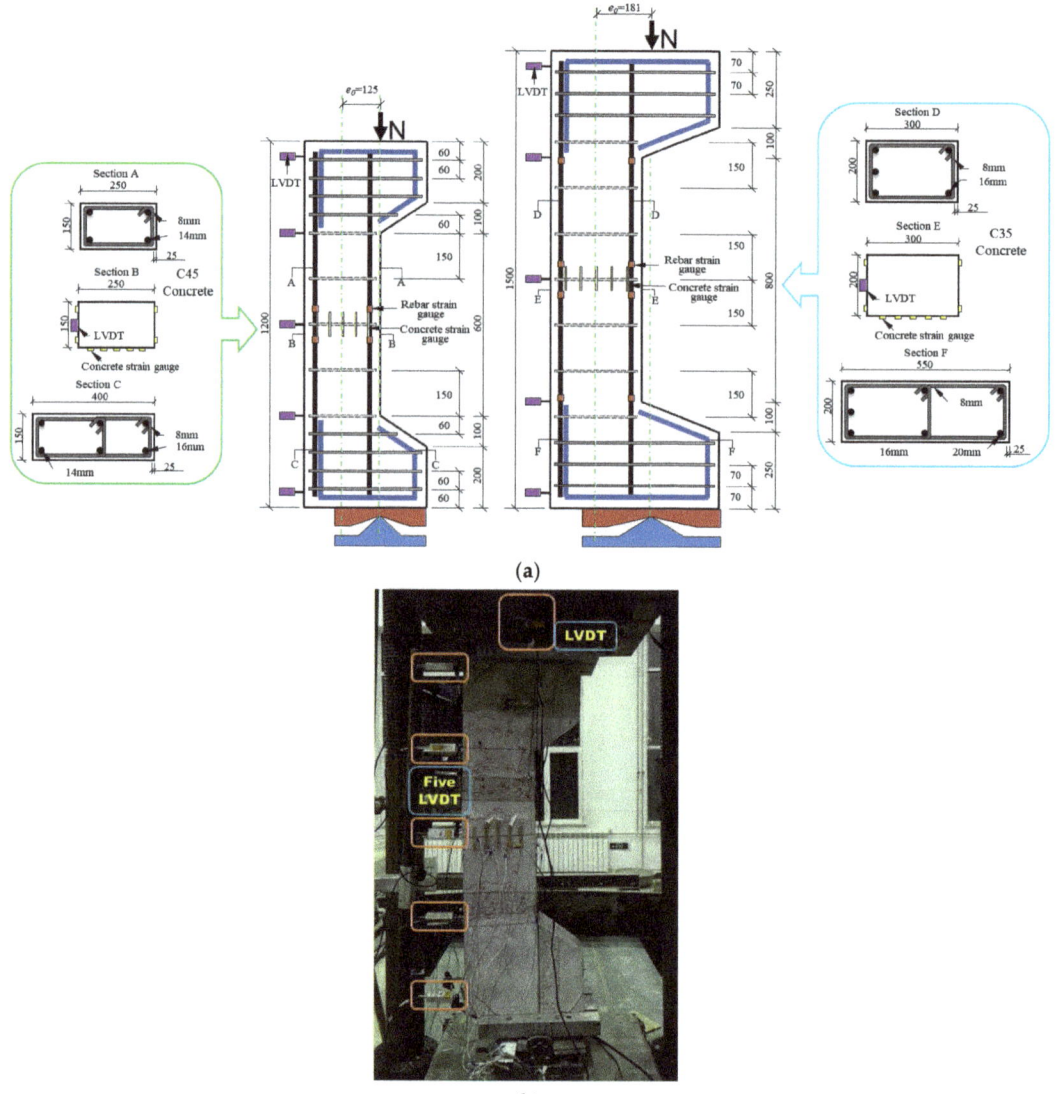

Figure 7. Compression test on RC columns under large eccentric loading: (**a**) geometry and reinforcement details of RC columns (unit: mm); (**b**) test setup.

3. Experimental Results and Discussion

3.1. Failure Modes and Crack Propagation

The failure modes of the FITRC and CRC columns under large eccentric loads were similar. The failure of all specimens was manifested by longitudinal bar yielding in the tension zone, followed by concrete crushing in the compression zone (Figure 8). The eccentricity of the FITRC45 and CRC45 columns was 0.60 and that of the FITRC35 and CRC35 columns was 0.70. In a previous study [55], the eccentric compressive behaviour of iron tailings sand RC columns was investigated, and similar results were obtained.

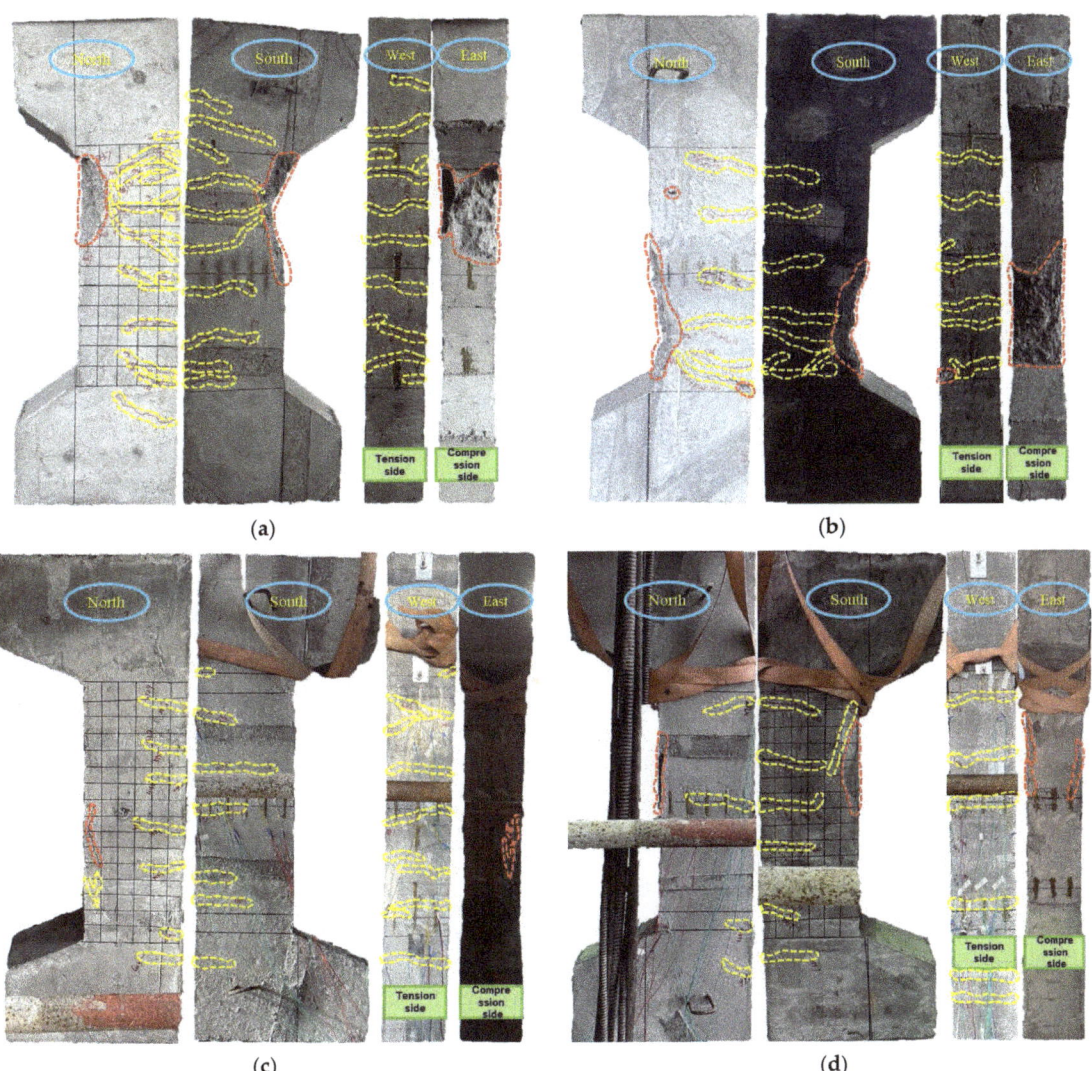

Figure 8. Typical failure modes of RC columns: (**a**) FITRC45-2 column specimen; (**b**) CRC45 column specimen; (**c**) FITRC35-2 column specimen; (**d**) CRC35 column specimen.

Cracks occurred in the tension zone at the midspan point along the lateral depth of the FITRC45 and CRC45 columns when the load reached approximately 0.2 N_u and propagated with increasing load, resulting in new microcracks. When the load reached approximately 0.9 N_u, vertical cracks were observed in the lateral compression zone of the column specimens and began to propagate. When the ultimate load N_u was reached, the concrete cracks in the compression zone elongated and were accompanied by the appearance of new microcracks. Some of the concrete was crushed as the test specimen continued to be loaded, and the ultimate compressive strain of the concrete in the compression zone was reached; thus, the axial load on the column specimens decreased. Similar failure processes were observed for FITRC35 and CRC35 columns.

Regardless of whether the eccentricity was 0.60 or 0.70, the final failure of all the RC columns was demonstrated by the crushed concrete in the outer layer of the stirrups and the intact concrete in the inner layer of the stirrups. This can be attributed to the fact that the stirrups significantly restrained the concrete in the inner layer and inhibited its deformation. In addition, Figure 8 shows that the area of crushed concrete was related to the eccentricity of the specimen section; that is, a smaller eccentricity indicated a larger area of crushed concrete.

Figure 9 shows the load–maximum crack width curves of the RC columns. The crack width propagation trends of FITRC and CRC columns remained similar for eccentricities of 0.60 and 0.70, and the curves of the four RC columns overlapped. Considering that the splitting tensile strength of FITC was lower than that of CC, the crack width of FITRC columns was larger than that of CRC columns at the same load level.

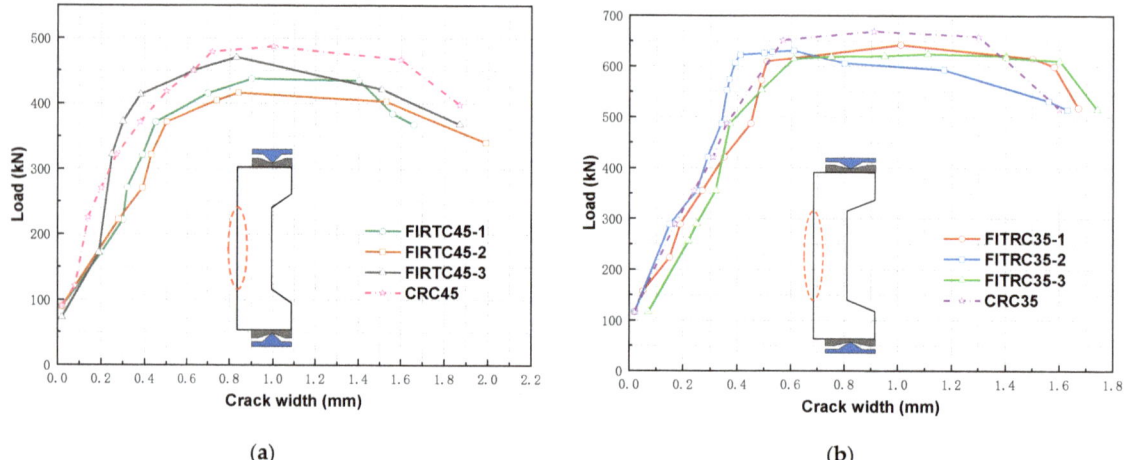

Figure 9. Load–maximum crack width curves of RC columns: (**a**) $e_0/h_0 = 0.60$; (**b**) $e_0/h_0 = 0.70$.

3.2. Load–Deflection Relationships

Figure 10 shows the axial load–deflection relationship (axial and lateral displacements) of RC columns under large eccentric loads. Axial and lateral displacements were measured using a vertical displacement gauge and a displacement gauge at the midspan point along the column depth, respectively. The negative and positive values in the figure indicate the axial and lateral displacements, respectively, at the midspan point along the column depth. Prior to longitudinal bar yielding, there was an approximately linear relationship between the loads and displacements of all the RC columns. After tensile yielding of the longitudinal bars, the stiffness of the RC columns decreased and the load-deflection curve increased nonlinearly until the peak load was reached when the concrete cover was crushed and spalled. After the peak load, the curve shows a decreasing trend; the load of the RC columns showed a rapid reduction of 5–10% followed by a more gradual reduction, and ductile failures occurred in the specimens. The vertical and lateral displacements of the CRC columns were clearly smaller than those of the FITRC columns. In particular, there were smaller lateral displacements compared to vertical displacements, considering that the modulus of elasticity of CC was approximately 19% greater than that of FITC (Table 6). Similarly, the peak loads of the CRC columns were slightly higher than those of the FITRC columns (Table 8).

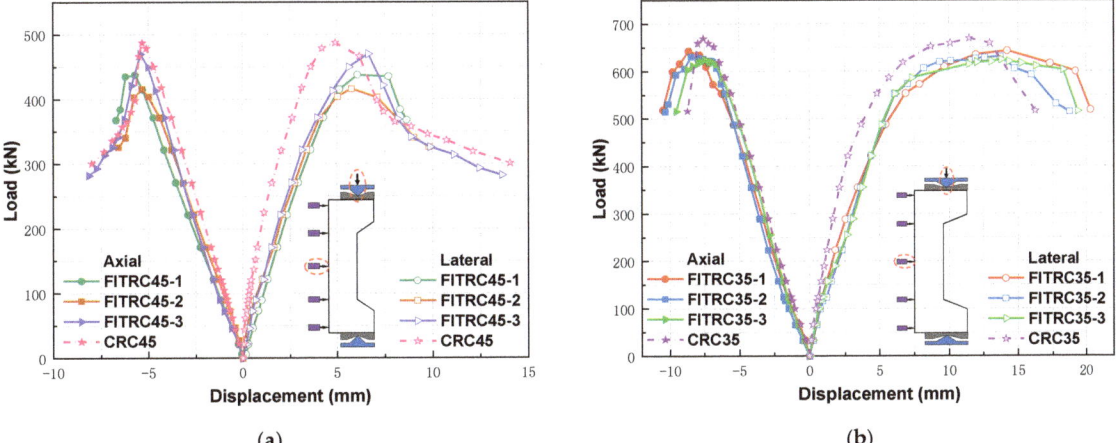

(a)   (b)

Figure 10. Load–deflection curves of RC columns: (**a**) $e_0/h_0 = 0.60$; (**b**) $e_0/h_0 = 0.70$.

Table 8. Main performance indices of RC columns.

Specimens	N_u (kN)	Δp (mm)	Δy (mm)	$\Delta_{0.85}$ (mm)	μ	λ
FITRC45-1	437.7	6.06	5.23	8.04	1.54	1.33
FITRC45-2	416.2	5.75	4.00	8.58	2.15	1.49
FITRC45-3	470.1	6.63	4.71	7.75	1.65	1.17
CRC45	487.4	4.91	2.46	6.89	2.80	1.40
FITRC35-1	642.5	14.23	6.62	19.80	2.99	1.39
FITRC35-2	631.0	13.73	6.72	17.32	2.58	1.26
FITRC35-3	624.0	13.80	6.60	19.17	2.90	1.39
CRC35	669.1	11.46	4.28	15.02	3.51	1.31

Figure 11 shows the lateral displacements of four typical RC columns distributed along the column depth (measured using five displacement gauges). The RC columns exhibited similar lateral displacements at the different loading stages. Furthermore, the columns showed slow and rapid increases in lateral displacement in the early loading stages and as the peak load approached, respectively. The lateral deformations of the RC columns were caused by first- and second-order moments. In accordance with the literature [56], the lateral deformations and sinusoidal waveforms of columns hinged at both ends remained similar. Therefore, the lateral deformation of the RC columns can be expressed as follows:

$$D_L = \Delta_p \sin\left(\frac{\pi \cdot l}{L}\right) \quad (1)$$

where Δ_p is the maximum displacement at the midspan point along the column depth, l is the location along the column depth, L is the column depth, and D_L is the lateral displacement of the column at l. Figure 11 compares the lateral deformations of the RC columns and the sinusoidal model, which agree well for eccentricities of both 0.60 and 0.70, thereby indicating that the sinusoidal model can effectively predict the lateral deformations of FITRC columns at different loading stages.

3.3. Deformation and Ductility

Table 8 lists the primary test results of the RC columns, including the peak load N_u, displacement at the corresponding peak load Δ_p, and yield displacement Δ_y. The behaviour parameters of the RC columns were quantified to determine the deformation capacity of the RC columns. In seismic design, the inelastic deformation capacity of specimens is generally quantified using the displacement ductility factor [57] and deformation coefficient [58]. The displacement ductility factor indicates the ductility behaviour of the specimens, which can be calculated as follows.

$$\mu = \frac{\Delta_{0.85}}{\Delta_y} \quad (2)$$

The deformation coefficient indicates the deformation capacity of the specimens after reaching the peak load and can be calculated as follows.

$$\lambda = \frac{\Delta_{0.85}}{\Delta_p} \quad (3)$$

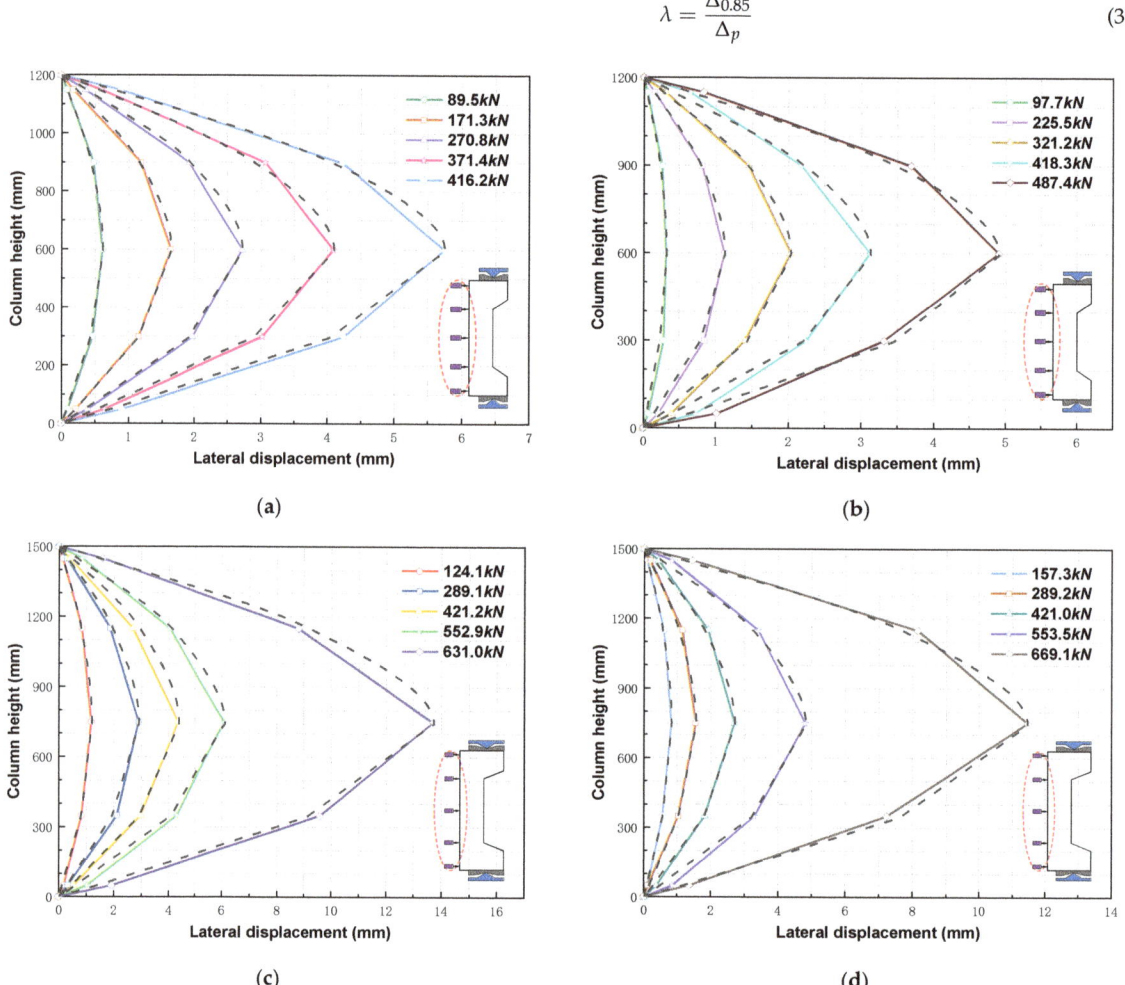

Figure 11. Lateral deformation along the section depth of typical RC columns: (**a**) FITRC45-2 column specimen; (**b**) CRC45 column specimen; (**c**) FITRC35-2 column specimen; (**d**) CRC35 column specimen.

Here, $\Delta_{0.85}$ is the corresponding displacement at N_u of 0.85 in the load decreasing stage [57], Δ_p is the peak lateral deflection at the peak load, and Δ_y is the corresponding displacement at the yield load when the limit is reached in the elastic stage. The yield displacement Δ_y can be obtained using a graphical method [57] (Figure 12).

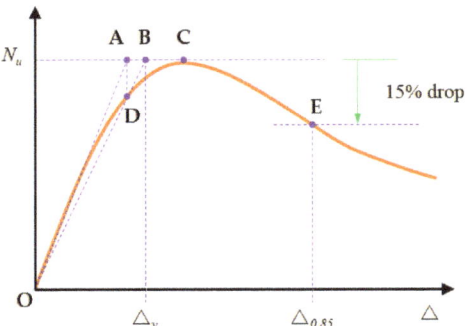

Figure 12. Definitions of yield displacement Δ_y and ultimate displacement $\Delta_{0.85}$.

Table 8 shows that the mean ductility factors of the FITRC45 and FITRC35 columns were 1.78 and 2.82, respectively, and the ductility factors relative to the CRC45 and CRC35 columns were reduced by 36.6% and 19.5%, respectively. Therefore, an increase in section size and eccentricity reduced the difference in ductility between FITRC35 and CRC35 columns. A previous study [52] examined the axial compressive properties of full iron tailings RC columns and also found that the ductility coefficients of FITRC columns were lower than those of CRC columns. The mean deformation coefficients of the FITRC45 and FITRC35 columns were 1.33 and 1.35, respectively, and the deformation coefficients relative to CRC45 and CRC35 columns were reduced and increased by 5.27% and 2.80%, respectively. In the load-decreasing phase, the deformation capacities of FITRC and CRC columns were identical.

As can be seen from Table 8, the lateral deflection Δ_p corresponding to the peak load of the FITRC columns was greater than that corresponding to the peak load of the CRC columns, and the mean values of Δ_p of the FITRC45 and FITRC35 columns increased by 25% and 21%, respectively, compared to the CRC45 and CRC35 columns.

3.4. Load–Strain Relationships

Figure 13 shows the load–rebar strain curves of typical RC columns, where c is A_{sc} of the compressive rebars, which were close to the axial compressive force with a negative strain value, and t is A_{st} of the tensile rebars, which were away from the axial compressive force (Figure 7) with a positive strain value. ε_{y14} and ε_{y16} are the tensile yield strains of rebars with diameters of 14 and 16 mm, respectively. Figure 13 shows that the strains of A_{st} in the RC columns were greater than ε_{y14} and ε_{y16}, and the yield strain of rebars in the compression zone was reached before the peak load was reached. The stress state is a typical feature of compression failure at large eccentricities. In addition, the strain of rebars in the FITRC columns was significantly higher than that in the CRC columns for both A_{sc} and A_{st}, because FITC had a lower modulus of elasticity than CC. In addition, the prismatic compressive and cracking strengths of FITC were slightly lower than those of CC. As a result, the ultimate load of the FITC columns was lower than that of the CRC columns, while the lateral deflection and axial displacement of the FITC columns are greater than those of the CRC columns.

Concrete strain gauges were placed along the section depth of the RC column, as shown in Figure 7a, to study the strain distribution of the concrete during loading. Figure 14 shows the strain distribution of typical concrete. Under large eccentric loads, the concrete in the tension zone of the RC columns cracked, resulting in failure of the concrete strain gauges near the tension zone. In particular, the strain distributions in the concrete before and after cracking were recorded. Figure 13 shows that the concrete strain was linearly distributed along the depth of the RC column section. Therefore, FITRC columns satisfied the planar section assumption, and the flexural capacity of FITRC columns can be theoretically calculated according to the planar section assumption.

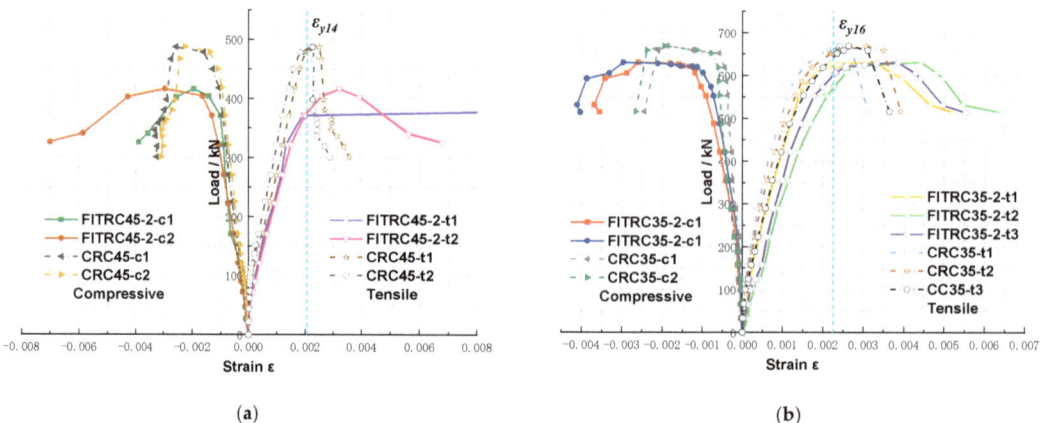

Figure 13. Load–rebar strain curves of typical RC columns: (**a**) $e_0/h_0 = 0.60$; (**b**) $e_0/h_0 = 0.70$.

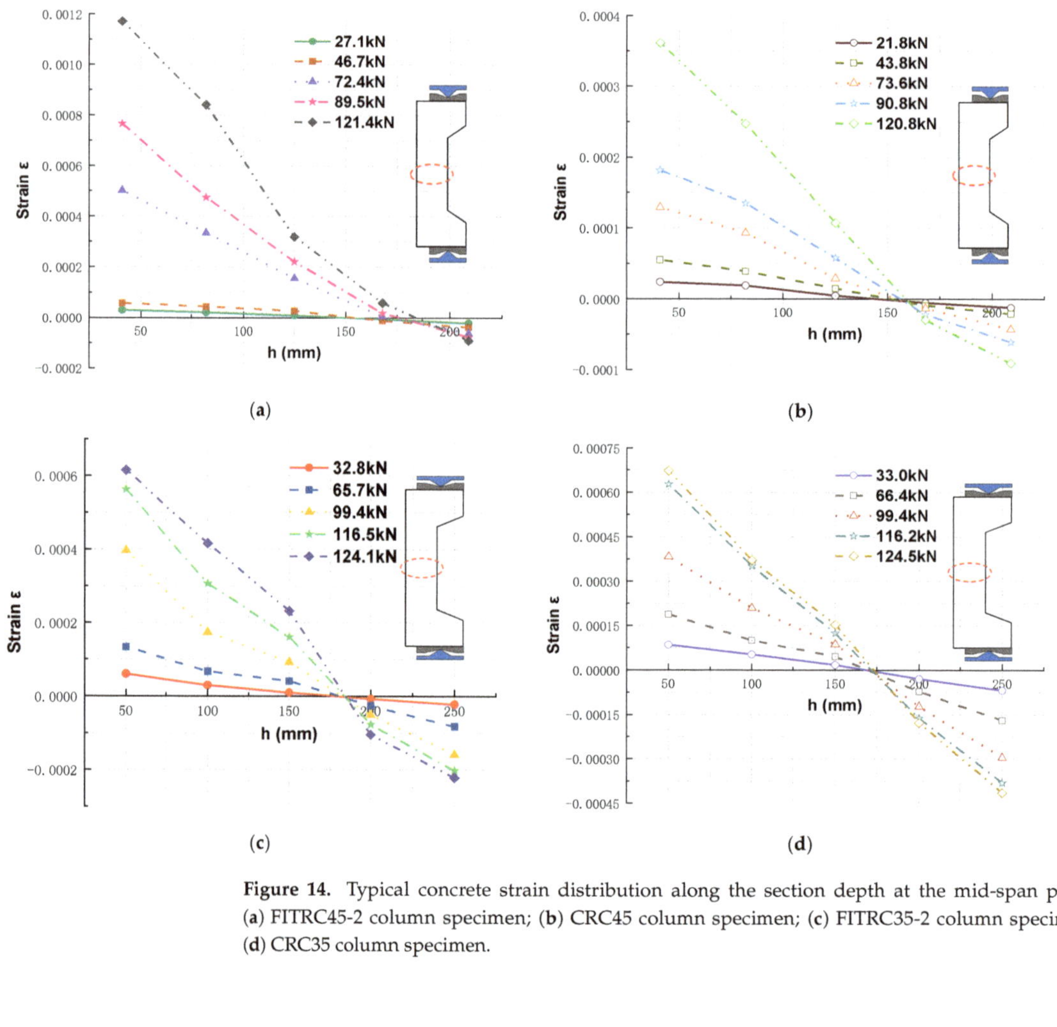

Figure 14. Typical concrete strain distribution along the section depth at the mid-span point: (**a**) FITRC45-2 column specimen; (**b**) CRC45 column specimen; (**c**) FITRC35-2 column specimen; (**d**) CRC35 column specimen.

4. Analysis of Sectional Capacities

4.1. Moment Magnification Factor

Considering the axial and flexural deformations that occurred in the RC columns under large eccentric loads during the experiment, the axial capacity, moment, and crack resistance should be calculated across the section at the mid-span of the column depth. Figure 15 shows a schematic of the second-order effects of the RC columns. Under the eccentric axial load N with initial eccentricity e_0, the lateral displacement Δ_p of the RC columns across the column depth at the midspan allowed the eccentricity of the axial load relative to the centre of mass of the section to reach $\Delta_p + e_0$. Meanwhile, the moment of the RC columns increased from $M_1 = Ne_0$ to $M_{max} = M_1 + M_2 = N(e_0 + \Delta_p)$, which is known as the second-order effect of eccentrically loaded columns [59,60]. During design, the second-order effects are accounted for using the moment augmentation factor [61].

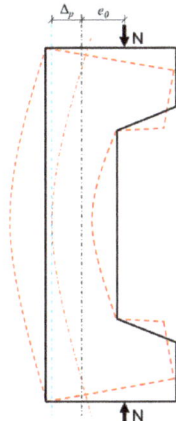

Figure 15. Schematic diagram of the second-order effect.

According to [62], the relationship between the ultimate sectional curvature φ_p and Δ_p of RC columns can be expressed as:

$$\varphi_p = \Delta_p \left(\frac{\pi}{L}\right)^2 \qquad (4)$$

This equation is valid only if the first- and second-order deformations of a column can be expressed as sinusoidal shapes.

Therefore, the moment augmentation factor η can be expressed as:

$$\eta = \frac{e_0 + \Delta_p}{e_0} = 1 + \frac{\varphi_p L^2}{e_0 \pi^2} \qquad (5)$$

In the design code, an additional eccentricity e_a is introduced because of uncertain load locations, uneven concrete quality, and construction variations, and a value of 20 mm is considered with an eccentricity of $e_i = e_0 + e_a$.

4.2. Bearing Capacity

In general, the bearing capacity under eccentric compression is calculated using the equivalent rectangular stress diagram [61], assuming that the FITRC columns satisfy the plane section assumption, and the theoretical calculation can be performed according to GB 50010-2010 [61]. Furthermore, to simplify the calculation process, the tensile strength of the concrete was ignored. Figure 16 shows a simplified diagram for calculating the section of an RC column subjected to compression failure under a large eccentric load.

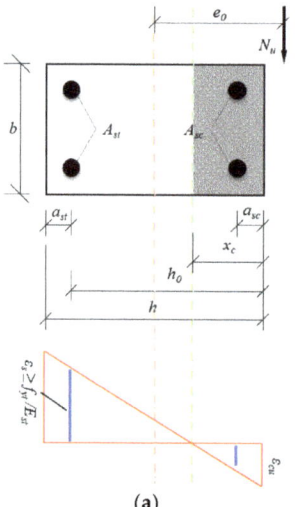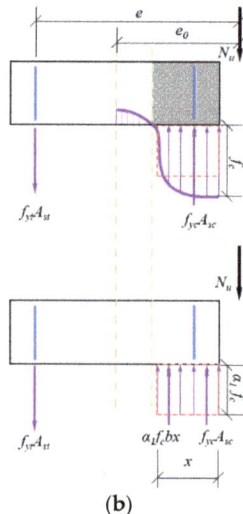

Figure 16. Simplified diagram for calculation of the section subjected to compression failure under large eccentric loading: (**a**) strain diagram of the specimens; (**b**) equivalent calculation diagram.

The compressive stress curve of the concrete in the compression zone was replaced with an equivalent rectangular diagram. According to the equilibrium conditions of the forces, the following formula was obtained.

$$N_u = \alpha_1 f_c b x + f_{yc} A_{sc} - f_{yt} A_{st} \tag{6}$$

Here, α_1 is the equivalent rectangular stress block coefficient of concrete in the compression zone, defined as 1.0; b and h are the sectional dimensions of the RC columns; f_{yt} and f_{yc} are the yield strengths of the tensile and compressive longitudinal bars, respectively; A_{st} and A_{sc} are the sectional areas of the tensile and compressive longitudinal bars, respectively. According to the resultant point of various forces on the tensile specimens in Figure 16b, the moment equilibrium conditions were determined and expressed as:

$$N_u e = \alpha_1 f_c b x (h_0 - 0.5x) + f_{yc} A_{sc} (h_0 - a_{sc}) \tag{7}$$

$$e = \eta e_i + 0.5h - a_{st} \tag{8}$$

where e is the distance from the point of axial force to the resultant point of A_{st} of tensile longitudinal bars, h_0 is the distance from the resultant point of A_{st} of tensile longitudinal bars to the edge of the compressive concrete, x is the depth of concrete in the compression zone, a_{st} is the distance from the resultant point of A_{st} of tensile longitudinal bars to the edge of the tensile concrete, and a_{sc} is the distance from the resultant point of A_{sc} of compressive longitudinal bars to the edge of the compressive concrete.

4.3. Crack-Resistant Load

According to the SL191-2008 standard in China [63], the crack resistance of RC columns under eccentric compression should be calculated as follows:

$$N_{cr} \leq \frac{\gamma_m \alpha_{ct} f_t A_0 W_0}{e_i A_0 - W_0} \tag{9}$$

where γ_m is the plastic section modulus, with the rectangular section set to 1.55; α_{ct} is the tensile stress limit coefficient of concrete, offset to 0.85; A_0 is the area of the transformed section; and W_0 is the elastic section modulus of the tensile edge of the transformed section, calculated as:

$$A_0 = bh + \alpha_E A_{st} + \alpha_E A_{sc} \tag{10}$$

$$W_0 = I_0/(h - y_0) \tag{11}$$

$$I_0 = (0.0833 + 0.19\alpha_E\rho)bh^3 \tag{12}$$

$$y_0 = (0.5 + 0.425\alpha_E\rho)h \tag{13}$$

where I_0 is the moment of inertia of the column section, y_0 is the distance from the axis of gravity of the transformed section to the compression edge, and ρ is the reinforcement ratio of the tensile longitudinal bars.

4.4. Experimental Verification of Theoretical Predictions

According to the aforementioned formulae, the moment augmentation factor and bearing capacity of the RC columns were calculated using the measured mechanical parameters of the rebar and concrete, respectively, and compared with the experimental results. As shown in Table 9, the theoretically calculated cracking load and ultimate bearing capacity were in good agreement with the measured results, with a maximum error of only 16%. In addition, there was a strength safety margin, which verifies that the calculation formulae are effective for FITRC columns under large eccentric loads.

Table 9. Bearing capacity of RC columns.

Specimens	N_{cr} (kN)	N_u (kN)	$N_{cr\text{-}t}$ (kN)	$N_{u\text{-}t}$ (kN)	$\frac{N_{cr}}{N_{cr\text{-}t}}$	$\frac{N_u}{N_{u\text{-}t}}$
FITRC45-1	89.5	437.7	85.3	405.2	1.05	1.08
FITRC45-2	89.5	416.2	85.3	405.2	1.05	1.03
FITRC45-3	90.3	470.1	85.3	405.2	1.06	1.16
CRC45	90.8	487.4	86.8	418.6	1.05	1.16
FITRC35-1	116.5	642.5	110.8	598.9	1.05	1.07
FITRC35-2	116.5	631.0	110.8	598.9	1.05	1.05
FITRC35-3	116.3	624.0	110.8	598.9	1.05	1.04
CRC35	116.2	669.1	111.9	608.4	1.04	1.10

Table 10 lists the moments of the RC columns and the calculated moment augmentation factor. According to the structural design code [61], the moment augmentation factor η should be 1.0 when L/h ratios do not exceed 5.0. Furthermore, the moment augmentation factor of the FITRC columns is greater than that of the CRC columns in the control group. The measured moment augmentation factors of the six FITRC columns are close to each other, with a maximum value of 1.079, whereas the η value of the FITRC columns increases by a maximum of 1.5% compared to the CRC columns. This is because the L/h ratios of the RC columns in this study do not exceed 5.0 and the impact of the second-order effects is marginal.

Table 10. Moments of RC columns.

Specimens	N_u (kN)	Δ_p (mm)	η	M_1 (kN·m)	M_2 (kN·m)	M_{max} (kN·m)	$\frac{M_2}{M_2(CRC)}$
FITRC45-1	437.7	6.06	1.048	54.7	2.7	57.4	1.11
FITRC45-2	416.2	5.75	1.046	52.0	2.4	54.4	1.00
FITRC45-3	470.1	6.63	1.053	58.8	3.1	61.9	1.30
CRC45	487.4	4.91	1.039	60.9	2.4	63.3	1.00
FITRC35-1	642.5	14.23	1.079	116.3	9.1	125.4	1.19
FITRC35-2	631.0	13.73	1.076	114.2	8.7	122.9	1.13
FITRC35-3	624.0	13.80	1.076	112.9	8.6	121.6	1.12
CRC35	669.1	11.46	1.063	121.1	7.7	128.8	1.00

In addition, Table 10 shows that the second-order moments of the FITRC columns are larger than those of the CRC columns. Furthermore, the average ratio of the second-order moments of the three FITRC45 columns to the CRC45 columns is 1.14, and that of the three FITRC35 columns to the CRC35 columns is 1.15. Therefore, if the second-order effect is to be considered in the structural design, the moment augmentation factor η of the FITRC columns should be 1.15 for safety reasons.

5. Conclusions

This study investigated the structural behaviour of FITRC columns under large eccentric loading. Six FITRC columns and two CRC column specimens were examined to investigate the effect of different raw materials, section dimensions, and eccentricities, on the mechanical behaviour of RC columns under large eccentric compression. The following conclusions were drawn from this study:

1. Under large eccentric short-term loads, the failure modes of the FITRC and CRC columns were found to be similar, and the failures were manifested by the yielding of the tensile and compressive rebars and concrete crushing in the compression zone. The long-term behaviour of FITRC (creep and shrinkage) requires further investigation.
2. As the prismatic compressive and tensile strengths of the FITRC columns were slightly lower than those of the CRC columns, the ultimate load capacity of the FITRC columns was slightly lower than that of the CRC columns, and the crack widths of the FITRC columns were greater than those of the CRC columns.
3. The sectional strain of the FITRC columns, which was similar to that of the CRC columns, satisfied the planar section assumption, and the lateral deflection curve agreed well with the half-sinusoidal assumption, which is consistent with the CRC column assumption in the current specifications.
4. Compared with the CRC45 and CRC35 columns, the ductility factors of the FITRC45 and FITRC35 columns were 36.6% and 19.5% lower, respectively. The underlying cause of this phenomenon was the comparatively low modulus of elasticity of FITC, which resulted in a more pronounced lateral deformation of the FITRC columns when subjected to eccentric loading than the CRC columns under equivalent conditions.
5. Based on the current Chinese standards, the theoretical calculations for the cracking load and ultimate load capacity of FITRC columns are relatively accurate. The calculation results indicate that FITRC columns have a certain safety reserve, and that FITC has the potential for practical application in the construction sector.
6. Because the lateral deflection of the FITC columns was greater than that of the CRC columns, the second-order moments of the FITC columns were greater than those of the CRC columns. If the second-order effect is considered in the structural design, the moment augmentation factor of the FITRC columns should be 1.15 for safety reasons. Therefore, FITRC columns with high L/h ratios should be investigated further.

The ductility factor of FITRC columns is much lower than that of CRC columns, so the ductility of FITRC columns can theoretically be improved by stirrup reinforcement. However, this hypothesis requires further investigation.

Author Contributions: X.M.: Conceptualisation, methodology, validation, investigation, data curation, and writing of the original draft. J.S.: Conceptualisation, methodology, writing—review and editing, and supervision. F.Z.: Validation, data curation, resources, and supervision. J.Y.: Investigation and resources. M.Y.: Investigation and resources. Z.M.: Validation and visualisation. Y.B.: Investigation and resources. Y.L.: Investigation and resources. All authors have read and agreed to the published version of the manuscript.

Funding: This work was supported by the Hebei Province Key Research and Development Project, China [grant number 19211502D], and the Hebei Province Graduate Innovation Funding Project, China [grant number CXZZBS2018109].

Institutional Review Board Statement: Not applicable.

Informed Consent Statement: Not applicable.

Data Availability Statement: Not applicable.

Conflicts of Interest: The authors declare no conflict of interest.

References

1. Maruthupandian, S.; Chaliasou, A.; Kanellopoulos, A. Recycling mine tailings as precursors for cementitious binders—Methods, challenges and future outlook. *Constr. Build. Mater.* **2021**, *312*, 125333. [CrossRef]
2. Zuccheratte, A.C.V.; Freire, C.B.; Lameiras, F.S. Synthetic gravel for concrete obtained from sandy iron ore tailing and recycled polyethyltherephtalate. *Constr. Build. Mater.* **2017**, *151*, 859–865. [CrossRef]
3. Arbili, M.M.; Alqurashi, M.; Majdi, A.; Ahmad, J.; Deifalla, A.F. Concrete made with iron ore tailings as a fine aggregate: A step towards sustainable concrete. *Materials* **2022**, *15*, 6236. [CrossRef] [PubMed]
4. Lv, X.; Lin, Y.; Chen, X.; Shi, Y.; Liang, R.; Wang, R.; Peng, Z. Environmental impact, durability performance, and interfacial transition zone of iron ore tailings utilized as dam concrete aggregates. *J. Clean. Prod.* **2021**, *292*, 126068. [CrossRef]
5. Lv, X.; Shen, W.; Wang, L.; Dong, Y.; Zhang, J.; Xie, Z. A comparative study on the practical utilization of iron tailings as a complete replacement of normal aggregates in dam concrete with different gradation. *J. Clean. Prod.* **2019**, *211*, 704–715. [CrossRef]
6. Zhao, J.; Ni, K.; Su, Y.; Shi, Y. An evaluation of iron ore tailings characteristics and iron ore tailings concrete properties. *Constr. Build. Mater.* **2021**, *286*, 122968. [CrossRef]
7. National Bureau of Statistics; Ministry of Environmental Protection. *China Statistical Yearbook on Environment*; China Statistics Press: Beijing, China, 2020.
8. Shettima, A.U.; Hussin, M.W.; Ahmad, Y.; Mirza, J. Evaluation of iron ore tailings as replacement for fine aggregate in concrete. *Constr. Build. Mater.* **2016**, *120*, 72–79. [CrossRef]
9. IFENG. Dam Break Accident in Xiangfen, Shanxi: Who is Responsible for the Disaster? 2008. Available online: https://news.ifeng.com/mainland/200809/0912_17_780544.shtml (accessed on 1 December 2022).
10. CHINANEWS. Shanxi Jiaokou Official Response 'A Tailings Dam Accident'. 2022. Available online: http://www.chinanews.com.cn/sh/2022/04-01/9717734.shtml (accessed on 1 December 2022).
11. Ahmari, S.; Zhang, L. Production of eco-friendly bricks from copper mine tailings through geopolymerization. *Constr. Build. Mater.* **2012**, *29*, 323–331. [CrossRef]
12. Kinnunen, P.; Ismailov, A.; Solismaa, S.; Sreenivasan, H.; Räisänen, M.L.; Levänen, E.; Illikainen, M. Recycling mine tailings in chemically bonded ceramics—A review. *J. Clean. Prod.* **2018**, *174*, 634–649. [CrossRef]
13. Zhang, N.; Tang, B.; Liu, X. Cementitious activity of iron ore tailing and its utilization in cementitious materials, bricks and concrete. *Constr. Build. Mater.* **2021**, *288*, 123022. [CrossRef]
14. Zareei, S.A.; Ameri, F.; Bahrami, N.; Shoaei, P.; Moosaei, H.R.; Salemi, N. Performance of sustainable high strength concrete with basic oxygen steel-making (BOS) slag and nano-silica. *J. Build. Eng.* **2019**, *25*, 100791. [CrossRef]
15. Kusuma, G.H.; Budidarmawan, J.; Susilowati, A. Impact of concrete quality on sustainability. *Procedia Eng.* **2015**, *125*, 754–759. [CrossRef]
16. National Bureau of Statistics of China. National Data: Output of Industrial Products in 2021. 2021. Available online: https://data.stats.gov.cn/english/easyquery.htm?cn=C01 (accessed on 1 December 2022).
17. Thomas, B.S.; Damare, A.; Gupta, R.C. Strength and durability characteristics of copper tailing concrete. *Constr. Build. Mater.* **2013**, *48*, 894–900. [CrossRef]
18. Sant'Ana Filho, J.N.; Da Silva, S.N.; Silva, G.C.; Mendes, J.C.; Fiorotti Peixoto, R.A. Technical and environmental feasibility of interlocking concrete pavers with iron ore tailings from tailings dams. *J. Mater. Civil. Eng.* **2017**, *29*, 04017104. [CrossRef]
19. Liu, J.; Zhou, Y.; Wu, A.; Wang, H. Reconstruction of broken Si-O-Si bonds in iron ore tailings (IOTs) in concrete. *Int. J. Min. Met. Mater.* **2019**, *26*, 1329–1336. [CrossRef]
20. Tang, C.; Li, K.; Ni, W.; Fan, D. Recovering Iron from iron ore tailings and preparing concrete composite admixtures. *Minerals* **2019**, *9*, 232. [CrossRef]
21. Carvalho Eugênio, T.M.; Francisco Fagundes, J.; Santos Viana, Q.; Pereira Vilela, A.; Farinassi Mendes, R. Study on the feasibility of using iron ore tailing (IoT) on technological properties of concrete roof tiles. *Constr. Build. Mater.* **2021**, *279*, 122484. [CrossRef]
22. Xu, F.; Wang, S.; Li, T.; Liu, B.; Li, B.; Zhou, Y. Mechanical properties and pore structure of recycled aggregate concrete made with iron ore tailings and polypropylene fibers. *J. Build. Eng.* **2021**, *33*, 101572. [CrossRef]
23. Ullah, S.; Yang, C.; Cao, L.; Wang, P.; Chai, Q.; Li, Y.; Wang, L.; Dong, Z.; Lushinga, N.; Zhang, B. Material design and performance improvement of conductive asphalt concrete incorporating carbon fiber and iron tailings. *Constr. Build. Mater.* **2021**, *303*, 124446. [CrossRef]
24. Lv, Z.; Jiang, A.; Liang, B. Development of eco-efficiency concrete containing diatomite and iron ore tailings: Mechanical properties and strength prediction using deep learning. *Constr. Build. Mater.* **2022**, *327*, 126930. [CrossRef]
25. Medjigbodo, S.; Bendimerad, A.Z.; Rozière, E.; Loukili, A. How do recycled concrete aggregates modify the shrinkage and self-healing properties? *Cem. Concr. Compos.* **2018**, *86*, 72–86. [CrossRef]
26. Carrasco, E.V.M.; Magalhaes, M.D.C.; Santos, W.J.D.; Alves, R.C.; Mantilla, J.N.R. Characterization of mortars with iron ore tailings using destructive and nondestructive tests. *Constr. Build. Mater.* **2017**, *131*, 31–38. [CrossRef]

27. Yao, G.; Wang, Q.; Wang, Z.; Wang, J.; Lyu, X. Activation of hydration properties of iron ore tailings and their application as supplementary cementitious materials in cement. *Powder Technol.* **2020**, *360*, 863–871. [CrossRef]
28. Tan, Y.; Zhu, Y.; Xiao, H. Evaluation of the hydraulic, physical, and mechanical properties of pervious concrete using iron tailings as coarse aggregates. *Appl. Sci.* **2020**, *10*, 2691. [CrossRef]
29. Feng, W.; Dong, Z.; Jin, Y.; Cui, H. Comparison on micromechanical properties of interfacial transition zone in concrete with iron ore tailings or crushed gravel as aggregate. *J. Clean. Prod.* **2021**, *319*, 128737. [CrossRef]
30. Zhao, S.; Fan, J.; Sun, W. Utilization of iron ore tailings as fine aggregate in ultra-high performance concrete. *Constr. Build. Mater.* **2014**, *50*, 540–548. [CrossRef]
31. Zhang, W.; Gu, X.; Qiu, J.; Liu, J.; Zhao, Y.; Li, X. Effects of iron ore tailings on the compressive strength and permeability of ultra-high performance concrete. *Constr. Build. Mater.* **2020**, *260*. [CrossRef]
32. Zhang, Z.; Zhang, Z.; Yin, S.; Yu, L. Utilization of iron tailings sand as an environmentally friendly alternative to natural river sand in high-strength concrete: Shrinkage characterization and mitigation strategies. *Materials* **2020**, *13*, 5614. [CrossRef]
33. Mendes Protasio, F.N.; de Avillez, R.R.; Letichevsky, S.; Silva, F.D.A. The use of iron ore tailings obtained from the Germano dam in the production of a sustainable concrete. *J. Clean. Prod.* **2021**, *278*, 123929. [CrossRef]
34. Zhu, Q.; Yuan, Y.; Chen, J.; Fan, L.; Yang, H. Research on the high-temperature resistance of recycled aggregate concrete with iron tailing sand. *Constr. Build. Mater.* **2022**, *327*, 126889. [CrossRef]
35. Chen, Z.; Chen, S.; Zhou, Y.; Zhang, C.; Meng, T.; Jiang, S.; Liu, L.; Hu, G. Effect of incorporation of rice husk ash and iron ore tailings on properties of concrete. *Constr. Build. Mater.* **2022**, *338*, 127584. [CrossRef]
36. Wang, C.; Ni, W.; Zhang, S.; Wang, S.; Gai, G.; Wang, W. Preparation and properties of autoclaved aerated concrete using coal gangue and iron ore tailings. *Constr. Build. Mater.* **2016**, *104*, 109–115. [CrossRef]
37. Ma, B.; Cai, L.; Li, X.; Jian, S. Utilization of iron tailings as substitute in autoclaved aerated concrete: Physico-mechanical and microstructure of hydration products. *J. Clean. Prod.* **2016**, *127*, 162–171. [CrossRef]
38. Cai, L.; Ma, B.; Li, X.; Lv, Y.; Liu, Z.; Jian, S. Mechanical and hydration characteristics of autoclaved aerated concrete (AAC) containing iron-tailings: Effect of content and fineness. *Constr. Build. Mater.* **2016**, *128*, 361–372. [CrossRef]
39. Cheng, Y.; Huang, F.; Li, W.; Liu, R.; Li, G.; Wei, J. Test research on the effects of mechanochemically activated iron tailings on the compressive strength of concrete. *Constr. Build. Mater.* **2016**, *118*, 164–170. [CrossRef]
40. Han, F.; Song, S.; Liu, J.; Huang, S. Properties of steam-cured precast concrete containing iron tailing powder. *Powder Technol.* **2019**, *345*, 292–299. [CrossRef]
41. Han, F.; Luo, A.; Liu, J.; Zhang, Z. Properties of high-volume iron tailing powder concrete under different curing conditions. *Constr. Build. Mater.* **2020**, *241*, 118108. [CrossRef]
42. Yang, M.; Sun, J.; Dun, C.; Duan, Y.; Meng, Z. Cementitious activity optimization studies of iron tailings powder as a concrete admixture. *Constr. Build. Mater.* **2020**, *265*, 120760. [CrossRef]
43. Gu, X.; Zhang, W.; Zhang, X.; Li, X.; Qiu, J. Hydration characteristics investigation of iron tailings blended ultra high performance concrete: The effects of mechanical activation and iron tailings content. *J. Build. Eng.* **2022**, *45*, 103459. [CrossRef]
44. Han, F.; Zhang, H.; Liu, J.; Song, S. Influence of iron tailing powder on properties of concrete with fly ash. *Powder Technol.* **2022**, *398*, 117132. [CrossRef]
45. Cheng, Y.; Huang, F.; Qi, S.; Li, W.; Liu, R.; Li, G. Durability of concrete incorporated with siliceous iron tailings. *Constr. Build. Mater.* **2020**, *242*, 118147. [CrossRef]
46. Cheng, Y.; Yang, S.; Zhang, J.; Sun, X. Test research on hydration process of cement-iron tailings powder composite cementitious materials. *Powder Technol.* **2022**, *399*, 117215. [CrossRef]
47. Zhang, R. Experimental Studies of Influence for Iron Tailings on Carbonation and Frost Resistance of Concrete. Master's Thesis, Hebei Agricultural University, Baoding, China, 2022.
48. Yang, M.; Sun, J.; Xu, Y.; Wang, J. Analysis of influence from iron tailings powder on sulfate corrosion resistance of concrete and its mechanism. *Water Resour. Hydropower Eng.* **2022**, *53*, 177–185.
49. Chen, X. Experimental Study on the Mechanical Properties of the Iron Tailing Sand Green Concrete Members. Ph.D. Thesis, Wuhan University of Technology, Wuhan, China, 2017.
50. Zhang, K. Experimental Research on Axial Compression of Ferrous Mill Tailing Concrete Short Columns. Master's Thesis, North China University of Science and Technology, Tangshan, China, 2016.
51. Li, H. Experimental Study on Seismic Bearing Capacity of Iron Tailorite Concrete Column. Master's Thesis, North China University of Science and Technology, Tangshan, China, 2018.
52. Ma, X.; Sun, J.; Zhang, F.; Yuan, J.; Meng, Z. Experimental studies and analyses on axial compressive properties of full iron tailings concrete columns. *Case Stud. Constr. Mat.* **2023**, *18*, e1881. [CrossRef]
53. *JGJ55-2011*; Specification for Mix Proportion Design of Ordinary Concrete. Architecture and Building Press: Beijing, China, 2011.
54. *GB/T 50081-2019*; Standard for Test Methods of Concrete Physical and Mechanical Properties. China Architecture and Building Press: Beijing, China, 2019.
55. Yan, P. Experimental study on mechanical behaviour of high performance iron tailing concrete columns. Master's Thesis, Hebei University of Architecture, Zhangjiakou, China, 2017. [CrossRef]
56. Lloyd, N.A.; Rangan, B.V. Studies on high-strength concrete columns under eccentric compression. *Aci Struct. J.* **1996**, *93*, 631–638. [CrossRef]

57. Tang, J. *Seismic Resistance of Joints in Reinforced Concrete Frames*; Southeast University Press: Nanjing, China, 1989.
58. Wang, L.; Kai-Leung Su, R. Theoretical and experimental study of plate-strengthened concrete columns under eccentric compression loading. *J. Struct. Eng.* **2013**, *139*, 350–359. [CrossRef]
59. *ACI 318-14*; Building Code Requirements for Structural Concrete and Commentary. American Concrete Institute: Detroit, MI, USA, 2014.
60. Li, Y. Experimental and Theoretical Research on Mechanical Behavior of RC Columns with 600MPa Reinforcing Bars. Ph.D. Thesis, Southeast University, Nanjing, China, 2019. [CrossRef]
61. *GB 50010-2010*; Code for Design of Concrete Structures. China Architecture and Building Press: Beijing, China, 2010.
62. Ge, W.; Chen, K.; Guan, Z.; Ashour, A.; Lu, W.; Cao, D. Eccentric compression behaviour of concrete columns reinforced with steel-FRP composite bars. *Eng. Struct.* **2021**, *238*, 112240. [CrossRef]
63. *SL191-2008*; Design Code for Hydraulic Concrete Structures. China Water and Power Press: Beijing, China, 2008.

Disclaimer/Publisher's Note: The statements, opinions and data contained in all publications are solely those of the individual author(s) and contributor(s) and not of MDPI and/or the editor(s). MDPI and/or the editor(s) disclaim responsibility for any injury to people or property resulting from any ideas, methods, instructions or products referred to in the content.

Article

Effect of Superfine Cement Modification on Properties of Coral Aggregate Concrete

Fei Wang [1], Jianmin Hua [1], Xuanyi Xue [1,*], Neng Wang [2], Feidong Yan [1] and Dou Feng [1]

[1] School of Civil Engineering, Chongqing University, Chongqing 400045, China
[2] School of Management Science and Real Estate, Chongqing University, Chongqing 400045, China
* Correspondence: xuexuanyi@cqu.edu.cn

Abstract: In marine engineering, using corals as aggregates to prepare concrete can reduce both the exploitation of stones and the transportation cost of building materials. However, coral aggregates have low strength and high porosity, which may affect the workability and mechanical properties of concrete. Hence, superfine cement is used innovatively in this study to modify coral aggregates; additionally, the effects of the water–cement ratio and curing time on the water absorption and strength of modified coral aggregates are investigated. Modified coral aggregate concrete is prepared, and the effect of using modified superfine cement on its workability and strength is investigated. Experimental results show that when the water-cement ratio exceeds 1.25, the slurry does not form a shell on the surface of the coral aggregates and the water absorption of the coral aggregates increases significantly. The strength of the modified coral aggregates cured for a short duration is slightly lower than that of unmodified coral aggregates, whereas that cured for 28 days is approximately 20% higher than that of unmodified coral aggregates. Using superfine cement to modify coral aggregate concrete can improve its workability, but not its compressive properties.

Keywords: coral aggregate concrete; superfine cement; aggregate modification; physical property; mechanical property; concrete workability

1. Introduction

Owing to the development and increased utilisation of the ocean, a significant amount of marine infrastructure is being constructed, which necessitates more building materials. Corrosion-resistant structural steels have been developed, such as fibre-reinforced polymer bars [1–3], stainless-clad bimetallic steel [4–8], and titanium-clad bimetallic steel [9–11]. The application of reinforced concrete structures in marine engineering is becoming increasingly extensive. Concrete, as one of the main raw materials for reinforced concrete structures, is in high demand. Aggregates constitute 70% to 80% of concrete volume and contribute significantly to concrete [12,13]. For construction on islands and reefs distant from the mainland, the transportation of sand aggregates significantly increases the construction cost [14]. Therefore, alternative materials that satisfy engineering requirements must be identified. Sea sand has been used as a fine aggregate to prepare concrete [15–17]. Coral is a relatively easy building material for marine engineering. In the 1830s, corals were first used as aggregates in concrete. After investigating coral aggregate concrete buildings in Guam, Howdyshell [18] concluded that coral aggregates can be used in concrete structures. In recent years, scholars have extensively investigated coral aggregate concrete. Wu et al. [19] investigated the physical and mechanical properties of different coral aggregates, revealed the pore morphology and distribution of coral aggregates. The results indicated that the mechanical properties of coral aggregate are poor due to its high porosity. Ma et al. [20] prepared high-strength coral aggregate seawater concrete and investigated its impact resistance, which illustrated that the high porosity of coral aggregate led to a large dynamic increase factor of concrete. Chen et al. [21] revealed

the internal damage and fracture mechanism of coral aggregate concrete via numerical simulations and discovered that microcracks initiated from concrete pores expanded into aggregates. Cai et al. [22] performed a comparison between coral aggregate concrete and ordinary Portland cement concrete, which revealed the low strength of coral aggregate under impact load. Furthermore, the energy required for coral aggregate concrete to initiate the main crack is much lower than that of ordinary Portland cement concrete. Zhou et al. [23] discovered that coral aggregates had a lower strength but higher porosity than sand gravel, which resulted in the strength of coral aggregate concrete being lower than that of ordinary concrete. The above research indicates that although coral aggregate can be used to prepare concrete, the workability and mechanical property of coral aggregate concrete are relatively poor.

To effectively improve the workability and mechanical properties of coral aggregate concrete, the coral aggregates to be used must be pretreated. Coral aggregates are characterised by low density, high crushing index, high porosity, and high water absorption, which are typical of lightweight aggregates; therefore, coral aggregates can be regarded as a lightweight aggregate. To modify lightweight aggregate concrete, researchers typically use inorganic cementitious materials to treat the aggregates, reduce the aggregate porosity, and improve the aggregate strength and durability. He et al. [24] formed a slurry comprising 80% cement, 10% fly ash, and 10% silica fume to modify recycled aggregates of broken brick concrete. The results showed that the slurry increased the apparent density and strength of recycled brick concrete aggregates. Yang et al. [25] investigated the strengthening effect of cement paste on the mechanical properties of broken brick aggregate concrete. It was found that the broken brick aggregate concrete demonstrated better mechanical properties and higher resistance to chloride ion migration after being treated with a cement-coal fly ash slurry via soaking for 4 h. Compared with before modification, the compressive strength of the brick aggregate concrete increased by 22–34% after modification. Hossein et al. [26] improved the durability of recycled concrete aggregate concrete by soaking it in silica fume paste. The results showed that the water absorption of recycled aggregates decreased by 14–22%, and the resistivity increased significantly after modification with silica fume paste. The above research shows that the inorganic cementitious materials modification method can improve the mechanical properties and durability of lightweight aggregate concrete. However, the above modification methods still cannot satisfy the requirements of coral aggregate concrete, due to the high porosity and low strength of coral aggregate. At present, the research on modification methods of coral aggregate is relatively limited, and the influence of modification on aggregate properties and concrete performance is still unclear.

Considering the high porosity of coral aggregates and the typical methods used to modify porous lightweight aggregates, superfine cement is used innovatively in this study to modify coral aggregates. Changes in the water absorption and strength of coral aggregates before and after modification are investigated. Modified and unmodified coral aggregates are used to prepare coral aggregate concrete. Subsequently, the effects of using modified superfine cement on the workability, strength, and failure mode of coral aggregate concrete are revealed. The aim of this study is to provide a new method for modifying coral aggregates, which can promote the use of coral aggregate in the construction of islands and reefs. The experimental results are conducive to the design and construction of modified coral aggregate concrete structures.

2. Materials and Methods

2.1. Materials

2.1.1. Coral

The coral aggregates used in this study were obtained from the South China Sea. As shown in Figure 1, corals have different shapes, which may be elongated, forked or irregular. In addition, there are many holes in the coral, showing a honeycomb-like surface. The aggregate particle size distribution indicates the proportion of aggregates with different particle size ranges relative to the total aggregate quality. The particle size distribution affects the slump, expansion, consistency, bleeding rate, and other performances of concrete mixtures [27,28], as well as the mechanical properties [29] and durability [30] of concrete. In accordance with GB/T 17431.1-2010 [31], the coarse aggregates used in this experiment were screened and analysed to obtain the gradation of the coral aggregates, as shown in Figure 2. The experimental coral aggregates satisfied the requirements of GB/T 17431.1-2010 [31] in terms of the upper and lower limits of gradation for lightweight aggregates measuring 5–20 mm. Table 1 lists the physical properties of the coral aggregates used in the experiment.

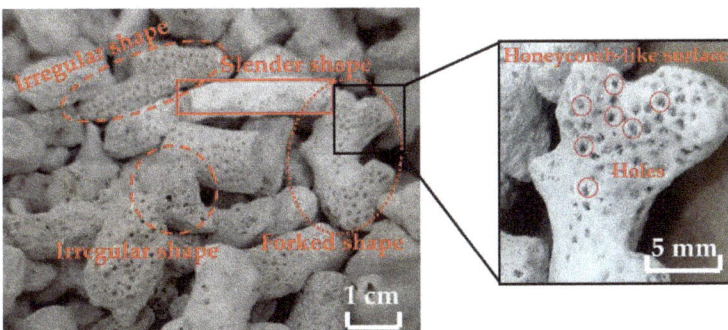

Figure 1. Coral.

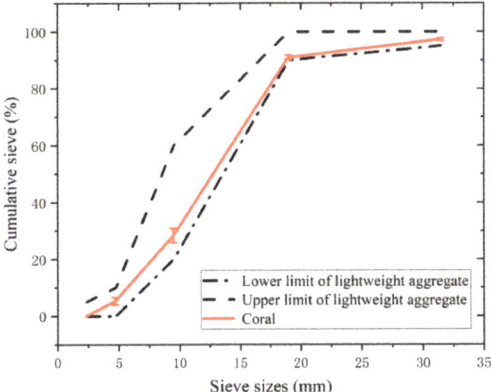

Figure 2. Coral aggregate gradation.

Table 1. Bulk density, apparent density, and porosity of coral aggregate samples.

Physical Properties	Coral Aggregate				
	Sample 1	Sample 2	Sample 3	Average	Standard Deviation
Bulk density (kg/m^3)	998	1065	1007	1023	36
Apparent density (kg/m^3)	2437	2509	2481	2475	36
Porosity (%)	59.0	57.6	59.4	58.7	1.0
Crush index (%)	31.9	31.3	32.0	31.7	0.4
Water absorption (%)	12.96	12.88	13.09	12.98	0.11

2.1.2. Cement and Superfine Cement

Owing to the high porosity of coral aggregates, the coral aggregates were modified by soaking them in a high-activity micro-inorganic cementitious material. The internal pores and microcracks of the coral aggregates were filled with slurry to improve the internal structure of the aggregates as well as the workability and mechanical properties of the concrete. Ordinary Portland cement, which is typically used for modification, features a large particle size; thus, the particles do not permeate easily into microcracks. Generally, particles of ordinary Portland cement can only penetrate into pores larger than 0.2 mm; therefore, the modification effect on porous aggregates is unsatisfactory, and improvements to fine pores require cement with a smaller particle size. Superfine cement is a high-performance cement-based grouting material with a much higher fineness than ordinary cement [32]. It has higher strength and better groutability. Because of the superfine grain size and larger surface area of cements, superfine cement slurry has good fluidity and particle filling [33]. Therefore, to fill the fine pores of coral aggregates and reduce the water absorption and crushing index of aggregates, superfine cement was adopted to strengthen and modify the coral aggregates. The superfine cement used in the current experiment was 1250 mesh superfine cement produced by China Resources, and its performance index is shown in Table 2. The cement used to prepare concrete was P.O 42.5 ordinary Portland cement produced by Sichuan Esheng Cement Plant, which satisfies the requirements of GB175-2020 [34], and its performance index is shown in Table 3.

Table 2. Performance index of superfine cement.

Specific Surface Area ($m^2 \cdot kg^{-1}$)	Particle Size Distribution (μm)			Compressive Strength (MPa)		Flexural Strength (MPa)	
	D50	D90	Average	3 d	28 d	3 d	28 d
800	≤3.5	≤10	3.5	≥40	≥70	≥8	≥11

Table 3. Performance index of cement.

Specific Surface Area ($m^2 \cdot kg^{-1}$)	Standard Consistency (%)	Setting Time (min)		Compressive Strength (MPa)		Flexural Strength (MPa)	
		Initial	Final	3 d	28 d	3 d	28 d
360	28.00	225	295	33.6	55.7	6.3	8.6

2.1.3. Seawater

Water is a basic component of concrete that also significantly affects its durability, strength, and workability [35–37]. Using seawater to prepare concrete can solve the problem of freshwater resource shortage on islands. However, seawater contains numerous ions, which may affect the performance of concrete. Previous studies have shown that Cl^- in seawater affects the strengthening of concrete [38] and that SO_4^{2-} may result in the expansion and cracking of concrete by affecting the formation time of gypsum and ettringite [39]. In the current experiment, artificial seawater was used to mix concrete and prepare a slurry of modified coral aggregates. The artificial seawater was prepared based on ASTM D1141-2013 [40]. The composition of the seawater is shown in Table 4. Similar configuration methods have been used in several studies [41–43].

Table 4. Chemical composition of artificial seawater used in current study.

Chemical	NaCl	$MgCl_2$	Na_2SO_4	$CaCl_2$
Concentration (g/L)	24.53	5.2	4.09	1.16

2.1.4. Sea Sand

The mineral composition of sea sand is similar to that of river sand; however, the content of shells and organic matter in sea sand deteriorates the strength and durability of concrete [44]. In addition, sea sand contains more ions than river sand. In this study, undiluted sea sand was obtained from the South China Sea. Based on the classification method presented in GB/T 14684-2011 [45], the grain size distribution of sea sand belongs to Zone II, and the fineness modulus is between 2.3 and 2.6. The particle size distributions are listed in Table 5. The mud, mica, and shell contents of sea sand were all less than 1.0%, 1.0%, and 5.0%, respectively.

Table 5. Sea sand particle gradation.

Sieve Diameter	4.75 mm	2.36 mm	1.18 mm	600 μm	300 μm
Cumulative percentages of sieve residue (%)	4.0	14.0	29.4	49.0	68.6

2.2. Coral Aggregate Modification

The method of preparing aggregates reinforced by modified slurry is to soak aggregates in the slurry for a period of time, remove the aggregates before the initial setting of the slurry, and perform natural curing after the surface slurry has solidified. This process ensures a tight slurry shell on the aggregate surface. However, it is disadvantageous to the aggregates in two aspects: the smooth shell on the aggregate surface reduces the friction between the aggregates and cement gels as well as reduces the bite force between the aggregates, which improves the mechanical properties of the concrete; additionally, the slurry shell forms additional pores, which increases the water absorption of the aggregates, thereby improving the water absorption of the modified aggregates to an extent greater than that before modification. The appropriate slurry water-cement ratio (W/C) and modification process must be identified. When the W/C of the slurry is relatively large, the slurry will be very thin, making it impossible to fill the aggregate pores when the slurry is solidified. On the contrary, the slurry will be very thick when the W/C is relatively small. The thick slurry will form a thick slurry shell outside the coral aggregate. The thick shell may affect the strength of the aggregate and lead to unexpected water absorption. Based on the results of pre-test, the values of W/C of slurry was determined as 1.0, 1.25, 1.5 and 2.0.

The superfine cement was placed in a stirrer, and artificial seawater was added to the stirrer based on the W/C values specified. Both materials were mixed well. After the slurry was prepared, coral aggregates were soaked in it and stirred to ensure that the aggregates were in full contact with the slurry. When the aggregates were soaked, they were stirred once every hour to avoid bonding between the aggregates and ensure the uniformity of the modification. Before the initial set was formed, the aggregates were removed from the slurry. The aggregates were continuously sieved using a dense mesh to filter out the excess modified slurry and avoid the formation of an excessively thick slurry shell on the aggregate surface. Subsequently, the aggregates were arranged flat on a net and dried at room temperature for 24 h, during which the aggregates were flipped to prevent them from bonding with each other. In order to reduce the impact of the environment [46,47], coral aggregates were placed in the standard environment (20 ± 2 °C and 95% relative humidity) for curing for 3 d, 7 d, 14 d and 28 d, respectively.

2.3. Physical and Mechanical Property Test

To investigate the effect of superfine cement modification on the coral aggregate concrete, the properties of coral aggregate concrete and coral aggregates before and after the modification were evaluated. When mixing concrete, the dry aggregate will absorb the moisture of the concrete mixture, resulting in a change in the effective W/C of concrete, thus affecting the construction process and mechanical properties of concrete. Therefore, the water absorption of aggregate must be considered when designing the concrete mix proportion. Furthermore, coral aggregate has a high porosity, which may cause its water absorption to be different from that of ordinary aggregate. Therefore, it is necessary to clarify its water absorption. In accordance with the GB/T 17431.2-2010 standard [48], dry coral aggregates were sieved and soaked; subsequently, their masses before and after water absorption were weighed. The water absorption rate of the coral aggregates was calculated using Equation (1) [48], where m_0 is the mass of the aggregates in the saturated surface condition, and m_1 is the mass of the dry aggregates. Concrete is composed of water, cement and aggregate, constituting 70% to 80% of concrete volume. Aggregate strength greatly affects concrete strength. Therefore, it is necessary to clarify the strength of the aggregate. The strength of the coral aggregates was evaluated using the crushing index (C_r). The mass of the coral aggregates (G_0) was weighed, and then loaded using a pressure tester. Subsequently, the assembly was pressurised to 200 kN at a rate of 1 kN/s and then unloaded. A 2.5 mm sieve was used to remove crushed fine particles, and the mass remaining on the sieve (G_1) was weighed. The C_r of the coral aggregates was calculated using Equation (2) [49].

The concrete shall meet the strength requirements to ensure the safety of the structure. In addition, it shall have good workability to facilitate transportation and construction. The physical and mechanical properties of the coral aggregates differed significantly from those of natural aggregates. Therefore, the workability and mechanical properties of seawater and sea sand concrete with coral aggregates differed significantly from those of seawater and sea sand concrete with natural aggregates. To investigate the effect of superfine cement modification on the coral aggregate concrete, modified coral aggregate concrete was prepared by replacing natural coarse aggregates with coral aggregates at replacement rates of 0%, 25%, 50%, 75%, and 100% by volume. There were three specimens in each aggregate replacement rate. The effects of the modification and replacement rate on the workability of coral aggregate seawater sea sand concrete were analysed. The mix ratio of the concrete is shown in Table 6. To reflect the workability of the concrete, the slump, expansion, and bleeding rate of the newly mixed modified coral aggregate concrete were tested based on GB/T 50080-2016 [50]. Axial compression experiments were performed to evaluate the mechanical properties of the concrete. Specimens measuring 100 mm × 100 mm × 300 mm were prepared by mixing coral aggregate concrete. The specimens were subjected to standard maintenance for 28 d, after which their compressive performance was evaluated using a pressure-testing machine. The loading rate was maintained at 0.5 MPa/s until the specimen was destroyed. The deformation of the specimen during compression was recorded using a displacement meter. The experimental procedure satisfied the requirements of GB/T 50081-2019 [51].

$$\rho = \frac{m_0 - m_1}{m_1} \times 100\% \tag{1}$$

$$C_r = \frac{G_0 - G_1}{G_0} \times 100\% \tag{2}$$

Table 6. Mix design of concrete (kg/m³).

Number	Replacement Rate of Aggregate	Water	Cement	Sea Sand	Aggregates			Water Reducer
					Gravel	Coral	Modified Coral	
P0	0%	200	500	830	1074	0	0	
P25W	25%	200	500	830	805.5	175.3	0	
P25G		200	500	830	805.5	0	175.3	
P50W	50%	200	500	830	537	350.6	0	
P50G		200	500	830	537	0	350.6	0.05%
P75W	75%	200	500	830	268.5	525.8	0	
P75G		200	500	830	268.5	0	525.8	

Table 6. Cont.

Number	Replacement Rate of Aggregate	Water	Cement	Sea Sand	Aggregates			Water Reducer
					Gravel	Coral	Modified Coral	
P100W	100%	200	500	830	0	701.1	0	
P100G		200	500	830	0	0	701.1	

Note: "W" and "G" refer to the coral aggregates before and after modification, respectively.

3. Results and Discussion

3.1. Water Absorption of Coral Aggregate

When preparing concrete, the water content of the aggregate affects the water and aggregate consumption of concrete. To ensure the workability and strength of concrete, the water absorption of the aggregate should be considered when designing the mix proportion. The surface of the modified coral aggregates (Figure 3) formed a layer of slurry shell, which appeared greyish-white. After modification, the aggregate became smoother and the number of pores reduced significantly, which considerably affected the water absorption of the coral aggregates. The water absorption of the coral aggregate was measured at 0.15, 0.5, 1, 6, and 24 h of immersion to understand its variation with immersion time, as shown in Figure 4. Compared with unmodified coral aggregates, the coral aggregates modified by superfine cement paste exhibited a significantly reduced level of water absorption after immersion. When W/C was 1.0, the water absorption of the modified coral aggregate was only 60% of that of the unmodified coral aggregate. Within 1 h of immersion, the water absorption capacity of the aggregates increased significantly. The water absorption of the aggregate reached 97% to 99% of the 24 h water absorption after soaking for 1 h. Therefore, the coral aggregates can be assumed to have reached the saturated-surface-dry state after 1 h of immersion. Thus, the coral aggregates should be soaked for more than 1 h in advance when used for fabricating coral aggregate concrete. As shown in Figure 4, the effect of the curing age of the modified coral aggregates on the water absorption is negligible. This is because the final setting time of the aggregate surface slurry is less than 3 d, and the slurry shell has already solidified during the curing time test on the third day, which effectively inhibited the water absorption and discharge effect of the aggregate. Therefore, the water absorption rate of the aggregate under different curing times was considerably lower than that before modification; however, the former did not differ significantly. Meanwhile, the W/C significantly affected the water absorption. Figure 5 shows the water absorption of the coral aggregates modified by cement slurry of different W/C values after 28 d of curing. The water absorption rate of the aggregate increased gradually with the W/C of the slurry. When the W/C increased from 1.0 to 1.25, the change range of aggregate water absorption was less than 2%. After the W/C exceeded 1.25, the water absorption of the aggregate increased significantly. When the W/C was 1.5, the water absorption increased by 11.13%. When the W/C was 2.0, the water absorption is 40.71% higher than when W/C = 1.0. This is because when the W/C was less than 1.25, although a denser slurry shell was formed on the surface of the aggregate, an overly thick shell formed additional pores, thus resulting in an insignificant decrease or a slight increase in the water absorption of the aggregate. When the W/C exceeded 1.25, the slurry was too thin to form a slurry shell on the surface of the coral aggregates, which resulted in an unsatisfactory filling and wrapping effect on aggregate pores and a significant increase in water absorption. Numerical fitting can show the evolution law of data and is often performed for data analysis [52,53]. Equation (3) was proposed to quantify the relationship between the W/C and the water absorption of modified coral aggregates via numerical fitting, where ρ represents the water absorption rate of the modified coral aggregates and ω represents the W/C of the slurry. The water absorption of different types of aggregates was compared (Figure 6). It can be seen that the water absorption of modified coral aggregate was bigger than that of recycled concrete aggregate, but less than that of recycled clay bricks aggregate.

$$\rho = 2.44 \times \omega^2 - 4.37 \times \omega + 8.96 \tag{3}$$

Figure 3. Appearance of coral aggregates: (**a**) unmodified; (**b**) modified.

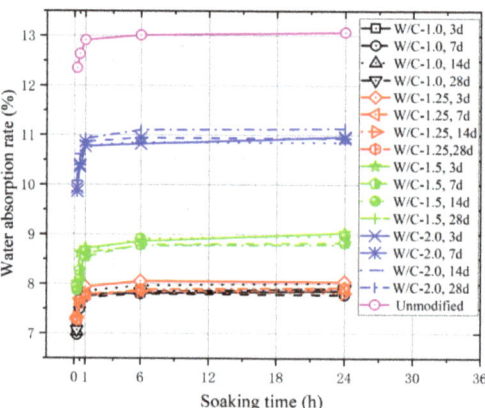

Figure 4. Water absorption of coral aggregates after different soak times.

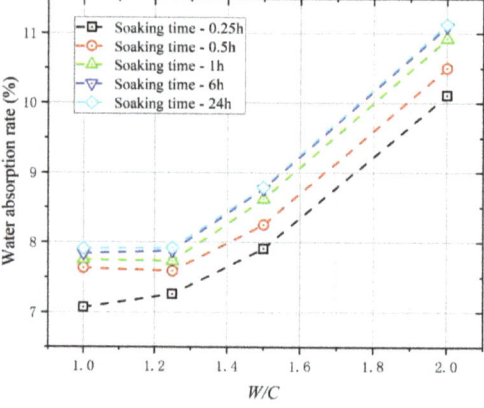

Figure 5. Effect of W/C of cement on water absorption of coral aggregates.

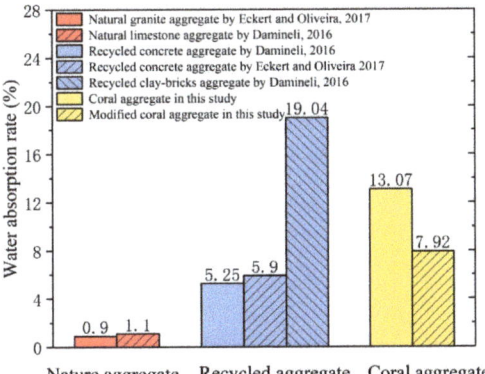

Figure 6. Comparison of water absorption of different types of aggregates [54,55].

3.2. Strength of Coral Aggregates

Tests were conducted to determine the effect of strength modification on the coral aggregates. Based on Figure 7, the crush value (C_r) of the aggregates at each W/C was greater than that before modification for curing times of 3 and 7 d. This is owing to the insufficient strength enhancement of the cement slurry encapsulating the aggregates in a short duration. The cement slurry was susceptible to breakage under the pressure of the testing machine and fine particles were filtered using a 2.5 mm sieve, which resulted in an increase in C_r. Figure 8 shows the C_r of the modified coral aggregates with different curing times. When the W/C of the slurry was 1.0 and 1.25, a dense slurry protective layer was formed on the surface of the aggregate; thus, the aggregate strength increased significantly after 14 and 28 d of curing. After 28 days of curing, the C_r of the aggregates with W/C of 1.0 was decreased by 21.98%. However, when the W/C was 1.5 and 2.0, the C_r of the coral aggregates decreased slightly with the increase of curing time. the C_r variation range of aggregate with W/C of 2.0 was within 4.43%. This may be due to the extremely high W/C of the superfine cement slurry and the inadequate slurry attached to and absorbed into the aggregate. In addition, a slurry with a high W/C cannot solidify for a long time, thus causing the slurry to be lost easily during the curing process; consequently, the formation of a compact slurry protective layer on the surface of the aggregate is hindered. Owing to these reasons, the C_r of the aggregates did not differ significantly from that before modification even at curing age of 14 and 28 d. The experimental results suggest that the slurry W/C should not be less than 1.25. Additionally, the modified aggregate must be maintained for 28 d, and until the strength of the slurry on the aggregate surface develops to a certain extent, it cannot be used for pouring concrete. The C_r of different types of aggregates was compared (Figure 9). It can be seen that the C_r of modified coral aggregate was between recycled concrete aggregate and recycled concrete and clay brick aggregate.

3.3. Workability of Concrete

Based on the discussion presented in Sections 3.1 and 3.2, concrete was fabricated using coral aggregates modified by a slurry with a W/C of 1.25 and cured for 28 d. The natural coarse aggregates were replaced by volume at 0%, 25%, 50%, 75%, and 100% replacement rates to analyse the effects of modification and replacement rate on the workability of seawater and sea sand concrete. The slump and expansion of concrete, which reflect the fluidity of concrete as well as the cohesion and water retention properties of concrete, are important indicators for determining the workability of concrete. Figures 10 and 11 illustrate the effects of the modification and replacement rate on the slump and expansion of seawater sea sand concrete. As the coral aggregate replacement rate increased, the slump and expansion increased; meanwhile, the slump and expansion of the modified coral concrete at the same replacement rate were significantly lower than those before modification. When the replacement rate was 100%, the slump and expansion of the modified concrete reduced by 16.14% and 14.97%, respectively. With the same concrete mix proportion, the slump of modified coral aggregate concrete was 2.97 times that of gravel aggregate concrete. This was primarily attributed to the significant water absorption and discharge effects of the coral aggregates. During mixing, the pre-wetted aggregate released water into the concrete mixture, thus increasing the concrete slump and expansion. After aggregate modification, the water absorption and discharge effects of the concrete improved; consequently, the slump and expansion of the concrete decreased and the cohesiveness and water retention properties of the concrete improved.

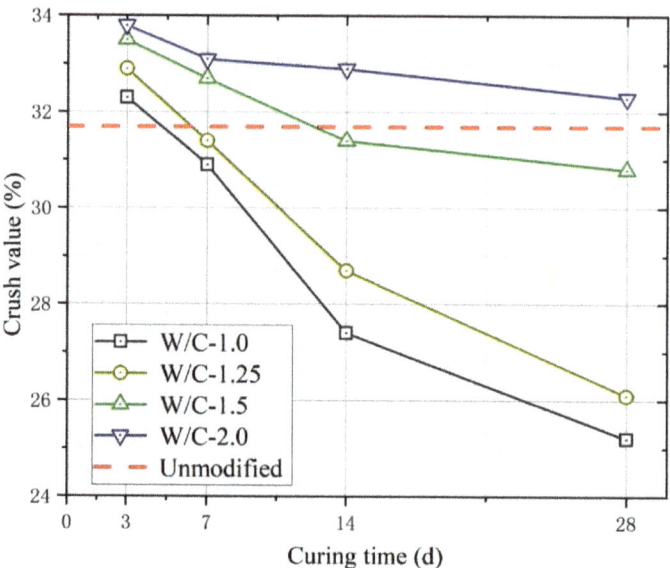

Figure 7. C_r of coral aggregates modified by slurry of different W/C values.

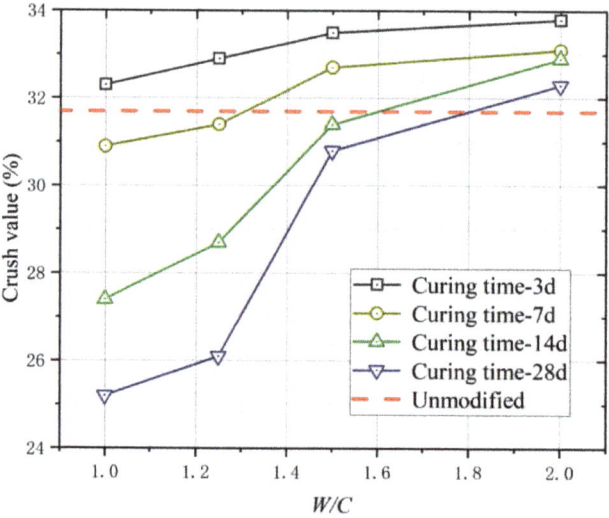

Figure 8. C_r of modified coral aggregates with different curing times.

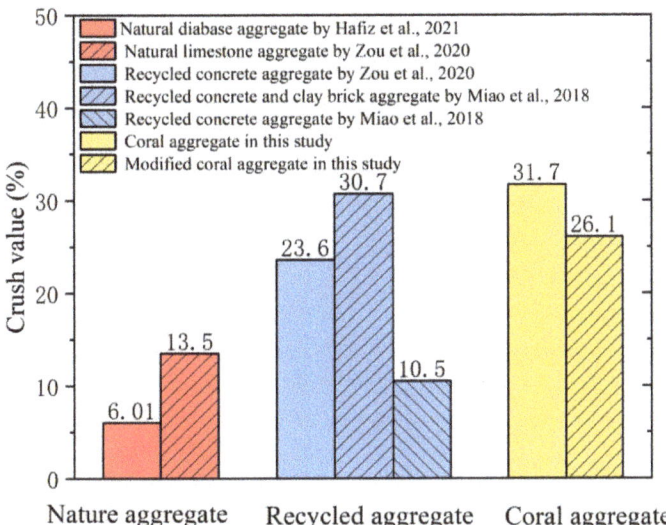

Figure 9. Comparison of C_r of different types of aggregates [56–58].

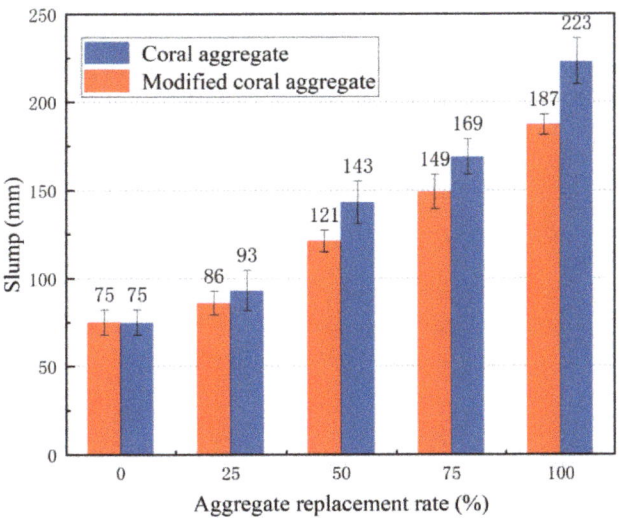

Figure 10. Relationship between aggregate replacement rate and slump.

The phenomenon where coarse aggregates sink due to gravity and water arises when the concrete mixture is placed for a long duration is referred to as water bleeding. Large cement particles, unreasonable aggregate gradation, excessive moisture release from aggregates, additives, etc. contribute to water bleeding in concrete. Water bleeding affects the workability and mechanical properties of concrete. Severe water bleeding results in severely deteriorated mechanical and durability performances of concrete as well as concrete deformation and the excessive development of incipient cracks. The water bleeding rates of coral aggregate concrete before and after modification were obtained experimentally, as shown in Figure 12. The water bleeding rate of the concrete increased significantly, which was consistent with the coral aggregate replacement rate. At the same replacement rate, the bleeding rate of the modified coral aggregate concrete was lower than that before the modification. At 0% replacement rate, water was completely used for aggregate pore filling and cement hydration, and no water bleeding occurred. When the replacement rate was 100%, the bleeding rate of the modified coral aggregate concrete decreased by 47.4% compared with that of coral aggregate concrete. This was primarily because the modified aggregate surface formed a dense slurry shell, thus significantly reducing the water absorption and discharge effects of the coral aggregates, and hence a significant improvement in the bleeding rate of the coral concrete. In addition, after the pre-wet treatment of the coral aggregates, a dense cement gel was formed at the aggregate-cement interface during the mixing process, which prevented the internal water escape of the coral aggregates and improved the water bleeding of the coral concrete.

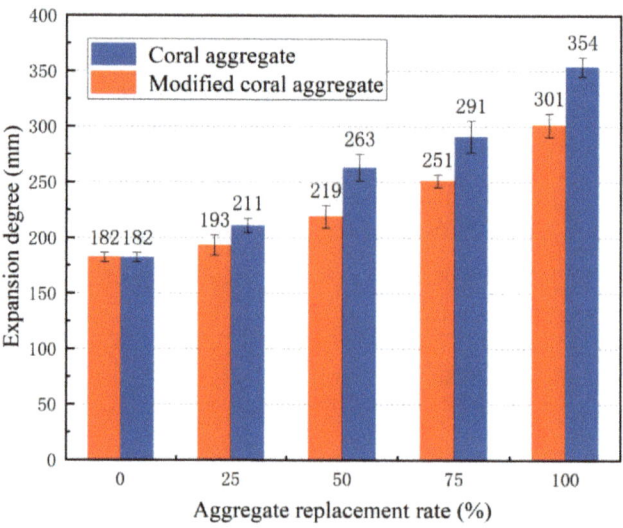

Figure 11. Relationship between aggregate replacement rate and expansion degree.

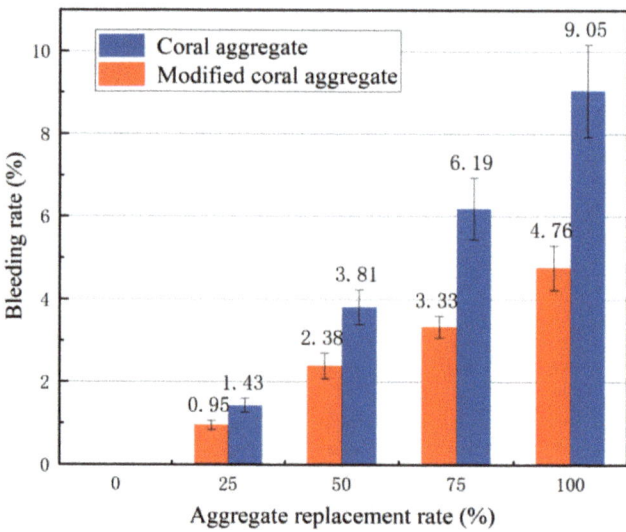

Figure 12. Relationship between aggregate replacement rate and bleeding rate.

3.4. Axial Compressive Strength of Concrete

Before the compressive strength test, the density of coral aggregate concrete and modified coral aggregate concrete was measured. As shown in Table 7, it can be seen that the density of modified coral aggregate concrete was slightly higher than that of coral aggregate concrete, and the density of both was bigger than that of ceramsite lightweight aggregate concrete. Axial compression tests were performed on gravel aggregate concrete, coral aggregate concrete, and modified coral aggregate concrete test specimens. During the loading process, numerous tiny cracks first appeared near the corners of the specimens. As the load continued to increase, the cracks continued to extend toward the opposite corner, whereas the number, length, and width of the cracks increased. Subsequently, the cracks continued to lengthen and widen, eventually penetrating through the entire specimen; additionally, cracks appeared obliquely along the specimen, as shown in Figure 13. The gravel aggregate concrete, coral aggregate concrete, and modified coral aggregate concrete test specimens indicated damage along the diagonal direction of the specimens. This was due to the frictional force between the upper and lower bottom surfaces of the specimen and the pressure on the specimen caused by the instrument during loading. Friction provides redundant constraints such that the maximum stress of the test specimens is near its diagonal. Despite the diagonal damage pattern of the specimens, more vertical cracks appeared in the coral aggregate concrete specimens. Due to the low strength of coral aggregate, cracks first formed near the coral aggregates. Then the miniscule cracks lengthened and widened with the load increased, thus resulting in more vertical cracks. For the gravel aggregate concrete, cracks appeared in the cement firstly. With the load increased, micro-cracks gradually converged and finally formed a continuous main crack. The axial compressive strength of the specimen was obtained via an axial compression test, as shown in Figure 14. The axial compressive strength of the coral aggregate concrete was much lower than that of the gravel aggregate concrete. The axial compressive strength of the gravel aggregate concrete was 46.92 MPa, whereas that of the coral aggregate concrete was only 28.99 MPa (coral aggregates) and 31.13 MPa (modified coral aggregates). The fracture surface of modified coral aggregate concrete passed through the aggregate. However, the gravel aggregate concrete broke along the interface between aggregate and cement, with the same loading conditions and mix proportion of concrete. Due to the low strength of the modified coral aggregate, the crack was less hindered during expansion. Therefore, the modified coral aggregate concrete was more likely to fail than gravel aggregate. The water absorption and release of the coral aggregates may affect the mechanical properties of the concrete in terms of two aspects. First, the water release of the aggregate will increase the W/C of concrete and degrade its mechanical properties. Second, during concrete hardening, water released from the aggregate is vital to internal curing and thus the concrete strength. Owing to these reasons, the modification of superfine cement does not significantly improve the compressive properties of coral aggregate concrete. To avoid the adverse effects of unstable water absorption and discharge, a precise proportion of concrete, which is difficult to achieve in practical engineering, must be ensured. Therefore, although superfine cement modification cannot significantly improve the strength of concrete, it can considerably diminish the water absorption and discharge of aggregates, thus avoiding the adverse effects of unstable water absorption and desorption.

Table 7. Density of different kinds of concrete.

	Coral Aggregate Concrete	Modified Coral Aggregate Concrete	Gravel Aggregate Concrete [59]	Ceramsite Lightweight Aggregate Concrete [60]
Density (kg/m^3)	2169	2233	2400–2600	≤1950

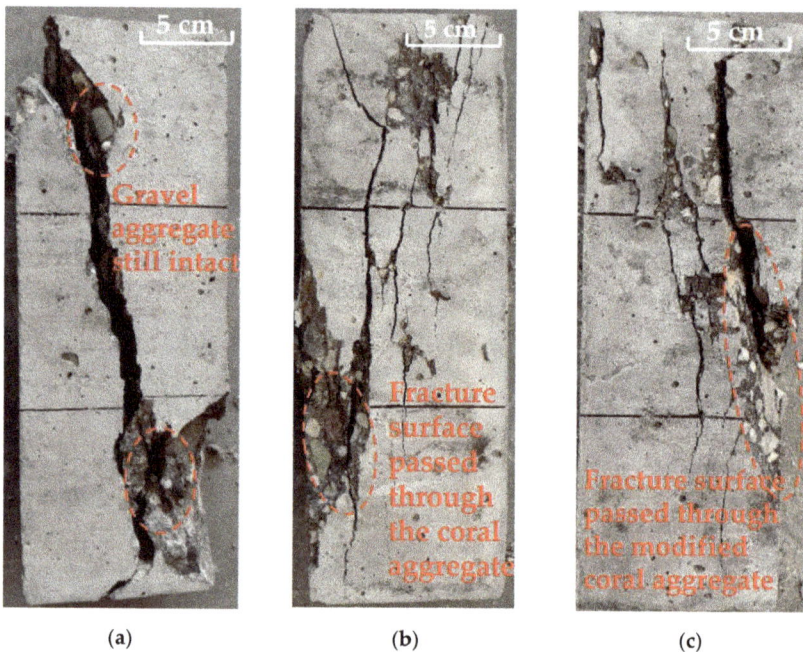

Figure 13. Failure mode of specimens: (**a**) gravel aggregate concrete specimen; (**b**) coral aggregate concrete specimen; (**c**) modified coral aggregate concrete specimen.

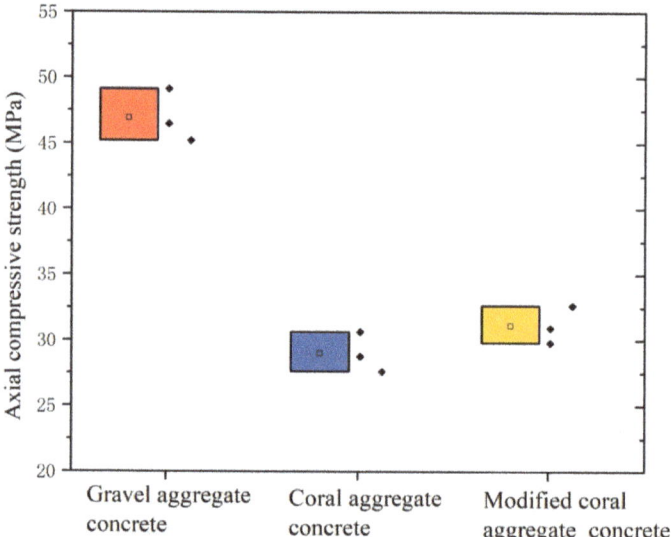

Figure 14. Axial compressive strength of specimens.

4. Conclusions

In this study, superfine cement was used to modify coral aggregates, and the physical and mechanical properties of modified coral aggregates were experimentally investigated. Subsequently, the effect of superfine cement modification on the workability and strength of coral aggregate concrete was investigated. The main results obtained were as follows:

(1) The water absorption of the modified coral aggregates increased with the slurry W/C. When the W/C was less than 1.25, the variation in the water absorption rate was not apparent; when the W/C exceeded 1.25, the slurry was overly thin, which prevented the formation of a slurry shell on the surface of the coral aggregate; consequently, the water absorption increased significantly with the W/C. After the coral aggregates were modified by superfine cement slurry, the water absorption of the aggregate reduced significantly, and the coral aggregate reached the saturated surface dry state after soaking for 1 h.

(2) The Cr of the modified coral aggregate was higher than that of the unmodified coral aggregate when the curing time was shorter. When the W/C of the cement slurry exceeded 1.25, the Cr of the modified coral aggregate started to be lower than that of the unmodified coral aggregate after 7 d of curing and then decreased significantly as the curing time increased. A slurry with a W/C greater than 1.25 could not be solidified for a long time; hence, a compact slurry protective layer could not be formed on the aggregate surface. The Cr of the aggregates did not change significantly as curing progressed.

(3) As the coral aggregate replacement rate increased, the slump and expansion of the concrete increased; meanwhile, at the same replacement rate, the slump and expansion of the modified coral concrete were significantly lower than those before modification. The slump and expansion of the modified concrete decreased by approximately 16.14% and 14.97% at 100% replacement rate, respectively. The dense slurry shell formed on the surface of the modified aggregate significantly reduced the water absorption and discharge effects of the coral aggregates, and the water secretion rate of the modified concrete was 47.4% lower compared with that before the modification.

(4) After the axial compression failure, the gravel aggregate concrete, coral aggregate concrete, and modified coral aggregate concrete exhibited damage along the diagonal direction of the specimens. Compared with the gravel aggregate concrete, the coral aggregate concrete showed more vertical cracks. The axial compressive strength of the coral aggregate concrete was approximately 60% that of the gravel aggregate concrete. The effect of superfine cement modification on improving the compressive properties of the coral aggregate concrete was insignificant.

(5) Based on the test results, it is suggested that the W/C of superfine cement shall be 1.25. The coral aggregate shall be cured for not less than 7 days after modification, and the soaking time of the modified coral aggregate shall not be less than 1 h.

Author Contributions: Investigation, F.W., F.Y. and D.F.; methodology, F.W., X.X. and F.Y.; supervision, J.H.; writing—original draft, F.W., X.X. and N.W.; writing—review and editing, J.H. and X.X. All authors have read and agreed to the published version of the manuscript.

Funding: This research was funded by the National Natural Science Foundation of China, grant number [52108115] and the APC was funded by Jianmin Hua.

Institutional Review Board Statement: Not applicable.

Informed Consent Statement: Not applicable.

Data Availability Statement: Data available on request.

Conflicts of Interest: The authors declare no conflict of interest.

References

1. Dong, Z.; Wu, G.; Zhao, X.-L.; Zhu, H.; Lian, J.-L. Durability test on the flexural performance of seawater sea-sand concrete beams completely reinforced with FRP bars. *Constr. Build. Mater.* **2018**, *192*, 671–682. [CrossRef]
2. Wang, Z.; Zhao, X.-L.; Xian, G.; Wu, G.; Raman, R.S.; Al-Saadi, S. Durability study on interlaminar shear behaviour of basalt-, glass- and carbon-fibre reinforced polymer (B/G/CFRP) bars in seawater sea sand concrete environment. *Constr. Build. Mater.* **2017**, *156*, 985–1004. [CrossRef]
3. El-Hassan, H.; El-Maaddawy, T.; Al-Sallamin, A.; Al-Saidy, A. Durability of glass fiber-reinforced polymer bars conditioned in moist seawater-contaminated concrete under sustained load. *Constr. Build. Mater.* **2018**, *175*, 1–13. [CrossRef]

4. Hua, J.; Wang, F.; Xue, X.; Ding, Z.; Chen, Z. Residual monotonic mechanical properties of bimetallic steel bar with fatigue damage. *J. Build. Eng.* **2022**, *55*, 104703. [CrossRef]
5. Shi, Y.; Wang, J.; Zhou, X.; Xue, X. Post–fire properties of stainless–clad bimetallic steel produced by explosive welding process. *J. Constr. Steel Res.* **2023**, *201*, 107690. [CrossRef]
6. Hua, J.; Wang, F.; Xue, X.; Fan, H.; Yan, W. Fatigue properties of bimetallic steel bar: An experimental and numerical study. *Eng. Fail. Anal.* **2022**, *136*, 106212. [CrossRef]
7. Hua, J.; Wang, F.; Xue, X. Study on fatigue properties of post-fire bimetallic steel bar with different cooling methods. *Structures* **2022**, *40*, 633–645. [CrossRef]
8. Shi, Y.; Luo, Z.; Zhou, X.; Xue, X.; Xiang, Y. Post-fire performance of bonding interface in explosion-welded stainless-clad bimetallic steel. *J. Constr. Steel Res.* **2022**, *193*, 107285. [CrossRef]
9. Luo, Z.; Shi, Y.; Zhou, X.; Xue, X.; Yao, X. Nonlinear patch resistance performance of hybrid titanium-clad bimetallic steel plate girder with web opening. *J. Build. Eng.* **2023**, *65*, 105703. [CrossRef]
10. Shi, Y.; Luo, Z.; Zhou, X.; Xue, X.; Li, J. Post-fire mechanical properties of titanium–clad bimetallic steel in different cooling approaches. *J. Constr. Steel Res.* **2022**, *191*, 107169. [CrossRef]
11. Luo, Z.; Shi, Y.; Xue, X.; Xu, L.; Zhang, H. Design recommendations on longitudinally stiffened titanium-clad bimetallic steel plate girder. *J. Constr. Steel Res.* **2023**, *201*, 107748. [CrossRef]
12. Chalangaran, N.; Farzampour, A.; Paslar, N.; Fatemi, H. Experimental investigation of sound transmission loss in concrete containing recycled rubber crumbs. *Adv. Concr. Constr.* **2021**, *11*, 447–454. [CrossRef]
13. Chalangaran, N.; Farzampour, A.; Paslar, N. Nano Silica and Metakaolin Effects on the Behavior of Concrete Containing Rubber Crumbs. *Civileng* **2020**, *1*, 17. [CrossRef]
14. Liu, G.; Hua, J.; Wang, N.; Deng, W.; Xue, X. Material Alternatives for Concrete Structures on Remote Islands: Based on Life-Cycle-Cost Analysis. *Adv. Civ. Eng.* **2022**, *2022*, 7329408. [CrossRef]
15. Wang, F.; Hua, J.; Xue, X.; Wang, N.; Yao, Y. Effects of Polyoxymethylene Fiber on Mechanical Properties of Seawater Sea-Sand Concrete with Different Ages. *Polymers* **2022**, *14*, 3472. [CrossRef]
16. Xue, X.; Wang, F.; Hua, J.; Wang, N.; Huang, L.; Chen, Z.; Yao, Y. Effects of Polyoxymethylene Fiber on Fresh and Hardened Properties of Seawater Sea-Sand Concrete. *Polymers* **2022**, *14*, 4969. [CrossRef]
17. Hua, J.; Yang, Z.; Xue, X.; Huang, L.; Wang, N.; Chen, Z. Bond properties of bimetallic steel bar in seawater sea-sand concrete at different ages. *Constr. Build. Mater.* **2022**, *323*, 126539. [CrossRef]
18. Howdyshell, P.A. *The Use of Coral as an Aggregate for Portland Cement Concrete Structures*; U.S. Army Construction Engineering Research Laboratory: Champaign, IL, USA, 1974.
19. Wu, Z.; Yu, H.; Ma, H.; Zhang, J.; Da, B. Physical and mechanical properties of coral aggregates in the South China Sea. *J. Build. Eng.* **2022**, *63*, 105478. [CrossRef]
20. Ma, H.; Yue, C.; Yu, H.; Mei, Q.; Chen, L.; Zhang, J.; Zhang, Y.; Jiang, X. Experimental study and numerical simulation of impact compression mechanical properties of high strength coral aggregate seawater concrete. *Int. J. Impact Eng.* **2020**, *137*, 103466. [CrossRef]
21. Chen, B.; Yu, H.; Zhang, J.; Ma, H. Evolution law of crack propagation and crack mode in coral aggregate concrete under compression: Experimental study and 3D mesoscopic analysis. *Theor. Appl. Fract. Mech.* **2022**, *122*, 103663. [CrossRef]
22. Cai, Y.; Ren, H.-Q.; Long, Z.-L.; Guo, R.-Q.; Du, K.-M.; Chen, S.-S.; Zheng, Z.-H. Comparison study on the impact compression mechanical properties of coral aggregate concrete and ordinary Portland concrete. *Structures* **2022**, *44*, 1403–1415. [CrossRef]
23. Zhou, W.; Feng, P.; Lin, H. Constitutive relations of coral aggregate concrete under uniaxial and triaxial compression. *Constr. Build. Mater.* **2020**, *251*, 118957. [CrossRef]
24. He, Z.; Shen, A.; Wang, W.; Zuo, X.; Wu, J. Evaluation and optimization of various treatment methods for enhancing the properties of brick-concrete recycled coarse aggregate. *J. Adhes. Sci. Technol.* **2022**, *36*, 1060–1080. [CrossRef]
25. Yang, J.; Shaban, W.M.; Elbaz, K.; Thomas, B.S.; Xie, J.; Li, L. Properties of concrete containing strengthened crushed brick aggregate by pozzolan slurry. *Constr. Build. Mater.* **2020**, *247*, 118612. [CrossRef]
26. Sasanipour, H.; Aslani, F.; Taherinezhad, J. Chloride ion permeability improvement of recycled aggregate concrete using pretreated recycled aggregates by silica fume slurry. *Constr. Build. Mater.* **2021**, *270*, 121498. [CrossRef]
27. Ashraf, W.; Noor, M. Performance-Evaluation of Concrete Properties for Different Combined Aggregate Gradation Approaches. *Procedia Eng.* **2011**, *14*, 2627–2634. [CrossRef]
28. Sokhansefat, G.; Ley, M.T.; Cook, M.D.; Alturki, R.; Moradian, M. Investigation of concrete workability through characterization of aggregate gradation in hardened concrete using X-ray computed tomography. *Cem. Concr. Compos.* **2019**, *98*, 150–161. [CrossRef]
29. Huang, Y.; Wang, J.; Ying, M.; Ni, J.; Li, M. Effect of particle-size gradation on cyclic shear properties of recycled concrete aggregate. *Constr. Build. Mater.* **2021**, *301*, 124143. [CrossRef]
30. Wang, J.; Sun, H.; Yu, L.; Liu, S.; Geng, D.; Yuan, L.; Zhou, Z.; Cheng, X.; Du, P. Improvement of intrinsic self-healing ability of concrete by adjusting aggregate gradation and sand ratio. *Constr. Build. Mater.* **2021**, *309*, 124909. [CrossRef]
31. *GB/T 17431.1-2010; Lightweight Aggregates and Its Test Methods. Part 1: Lightweight Aggregates.* China Standard Press: Beijing, China, 2010.
32. Chen, J.; Kwan, A. Superfine cement for improving packing density, rheology and strength of cement paste. *Cem. Concr. Compos.* **2012**, *34*, 1–10. [CrossRef]

33. Li, W.; Shaikh, F.U.; Wang, L.; Lu, Y.; Wang, B.; Jiang, C.; Su, Y. Experimental study on shear property and rheological characteristic of superfine cement grouts with nano-SiO2 addition. *Constr. Build. Mater.* **2019**, *228*, 117046. [CrossRef]
34. *GB 175-2007*; Common Portland Cement. China Standard Press: Beijing, China, 2007.
35. Ahmed, A.; Guo, S.; Zhang, Z.; Shi, C.; Zhu, D. A review on durability of fiber reinforced polymer (FRP) bars reinforced seawater sea sand concrete. *Constr. Build. Mater.* **2020**, *256*, 119484. [CrossRef]
36. Ding, Y.; Azevedo, C.; Aguiar, J.; Jalali, S. Study on residual behaviour and flexural toughness of fibre cocktail reinforced self compacting high performance concrete after exposure to high temperature. *Constr. Build. Mater.* **2012**, *26*, 21–31. [CrossRef]
37. Nath, P.; Sarker, P.K. Effect of GGBFS on setting, workability and early strength properties of fly ash geopolymer concrete cured in ambient condition. *Constr. Build. Mater.* **2014**, *66*, 163–171. [CrossRef]
38. Kucche, K.J.; Jamkar, S.S.; Sadgir, P.A. Quality of water for making concrete: A review of literature. *Int. J. Sci. Res. Publ.* **2015**, *5*, 1–10.
39. Wegian, F.M. Effect of seawater for mixing and curing on structural concrete. *IES J. Part A: Civ. Struct. Eng.* **2010**, *3*, 235–243. [CrossRef]
40. *ASTM D1141-1998(2013)*; Standard Practice for the Preparation of Substitute Ocean Water. ASTM International: West Conshohocken, PA, USA, 2013.
41. Xiao, J.; Qiang, C.; Nanni, A.; Zhang, K. Use of sea-sand and seawater in concrete construction: Current status and future opportunities. *Constr. Build. Mater.* **2017**, *155*, 1101–1111. [CrossRef]
42. Su, R.; Li, X.; Xu, S.-Y. Axial behavior of circular CFST encased seawater sea-sand concrete filled PVC/GFRP tube columns. *Constr. Build. Mater.* **2022**, *353*, 129159. [CrossRef]
43. Kazemi, H.; Yekrangnia, M.; Shakiba, M.; Bazli, M.; Oskouei, A.V. Bond-slip behaviour between GFRP/steel bars and seawater concrete after exposure to environmental conditions. *Eng. Struct.* **2022**, *268*, 114796. [CrossRef]
44. Zhao, Y.; Hu, X.; Shi, C.; Zhang, Z.; Zhu, D. A review on seawater sea-sand concrete: Mixture proportion, hydration, microstructure and properties. *Constr. Build. Mater.* **2021**, *295*, 123602. [CrossRef]
45. *GB/T 14684-2011*; Sand for Construction. China Standard Press: Beijing, China, 2011.
46. Farzampour, A. Temperature and humidity effects on behavior of grouts. *Adv. Concr. Constr.* **2017**, *5*, 659–669. [CrossRef]
47. Farzampour, A. Compressive behavior of concrete under environmental effects. In *Compressive Strength of Concrete*; IntechOpen: London, UK, 2019. [CrossRef]
48. *GB/T 17431.2-2010*; Lightweight Aggregates and Its Test Methods Part 2: Test Methods for Lightweight Aggregates. China Standard Press: Beijing, China, 2010.
49. *DL/T 5141-2014*; Code for Testing Aggregates of Hydraulic Concrete. China Electric Power Press: Beijing, China, 2014.
50. *GB/T 50080-2016*; Standard for Test Method of Performance on Ordinary Fresh Concrete. China Standard Press: Beijing, China, 2016.
51. *GB/T 50081-2019*; Standard for Test Methods of Concrete Physical and Mechanical Properties. Ministry of Construction of the PR China: Beijing, China, 2019.
52. Xue, X.; Shi, Y.; Zhou, X.; Wang, J.; Xu, Y. Experimental study on the properties of Q960 ultra–high–strength steel after fire exposure. *Structures* **2023**, *47*, 2081–2098. [CrossRef]
53. Hua, J.; Wang, F.; Xue, X.; Sun, Y.; Gao, Y. Post-fire ultra-low cycle fatigue properties of high-strength steel via different cooling methods. *Thin-Walled Struct.* **2023**, *183*, 110406. [CrossRef]
54. Damineli, B.L.; Quattrone, M.; Angulo, S.C.; Taqueda, M.E.S.; John, V.M. Rapid method for measuring the water absorption of recycled aggregates. *Mater. Struct.* **2016**, *49*, 4069–4084. [CrossRef]
55. Eckert, M.; Oliveira, M. Mitigation of the negative effects of recycled aggregate water absorption in concrete technology. *Constr. Build. Mater.* **2017**, *133*, 416–424. [CrossRef]
56. Hassan, H.M.Z.; Wu, K.; Huang, W.; Chen, S.; Zhang, Q.; Xie, J.; Cai, X. Study on the influence of aggregate strength and shape on the performance of asphalt mixture. *Constr. Build. Mater.* **2021**, *294*, 123599. [CrossRef]
57. Zou, G.; Zhang, J.; Liu, X.; Lin, Y.; Yu, H. Design and performance of emulsified asphalt mixtures containing construction and demolition waste. *Constr. Build. Mater.* **2020**, *239*, 117846. [CrossRef]
58. Miao, Y.; Yu, W.; Hou, Y.; Liu, C.; Wang, L. Influences of Clay Brick Particles on the Performance of Cement Stabilized Recycled Aggregate as Pavement Base. *Sustainability* **2018**, *10*, 3505. [CrossRef]
59. Joshi, T.; Dave, U. Construction of pervious concrete pavement stretch, Ahmedabad, India—Case study. *Case Stud. Constr. Mater.* **2022**, *16*, e00622. [CrossRef]
60. Zhang, S.; Yuan, K.; Zhang, J.; Guo, J. Experimental Study on Performance Influencing Factors and Reasonable Mixture Ratio of Desert Sand Ceramsite Lightweight Aggregate Concrete. *Adv. Civ. Eng.* **2020**, *2020*, 8613932. [CrossRef]

Disclaimer/Publisher's Note: The statements, opinions and data contained in all publications are solely those of the individual author(s) and contributor(s) and not of MDPI and/or the editor(s). MDPI and/or the editor(s) disclaim responsibility for any injury to people or property resulting from any ideas, methods, instructions or products referred to in the content.

Article

Determining Dynamic Mechanical Properties for Elastic Concrete Material Based on the Inversion of Spherical Wave

Huawei Lai [1,2,*], Zhanjiang Wang [3], Liming Yang [1], Lili Wang [1] and Fenghua Zhou [1,*]

1 MOE Key Lab. of Impact and Safety Engineering, Ningbo University, Ministry of Education, Ningbo 315211, China
2 College of Geomatics and Municipal Engineering, Zhejiang University of Water Resources and Electric Power, Hangzhou 310018, China
3 Northwest Institute of Nuclear Technology, Xi'an 710024, China
* Correspondence: laihw@zjweu.edu.cn (H.L.); zhoufenghua@nbu.edu.cn (F.Z.); Tel.: +86-574-87609958 (F.Z.)

Abstract: The paper presents a new method to study the dynamic mechanical properties of concrete under low pressure and a high strain rate via the inversion of spherical wave propagation. The dynamic parameters of rate-dependent constitutive relation of elastic concrete are determined by measured velocity histories of spherical waves. Firstly, the particle velocity time history profiles in the low stress elastic region at the radii of 100.6 mm, 120.6 mm, 140.6 mm, 160 mm, and 180.6 mm are measured in the semi-infinite space of concrete by using the mini-explosive ball and electromagnetic velocity measurement technology. Then, based on the universal spherical wave conservation equation and the fact that the accommodation relationship in state equation satisfies linear elastic law, the inverse problem analysis of spherical waves in concrete (called "NV + T0/SW") is proposed, which can obtain the dynamic numerical constitutive behavior of concrete in three-dimensional stress by measuring the velocity histories. The numerical constitutive relation is expressed in the form of distortion, and it is found that the distortion law has an obvious rate effect. Finally, the rate-dependent dynamic parameters in concrete are determined by the standard linear solid model. The results show that the strain rate effect of concrete cannot be ignored with the strain rate range of 10^2 1/s. This study can provide a feasible method to determine the dynamic parameters of rate-dependent constitutive relation of concretes. This method has good applicability, especially in the study of the dynamic behavior of multicomponent composite materials with large-size particle filler.

Keywords: spherical waves; wave propagation method; particle velocity histories; linear constitutive relation of concrete; rate-dependent

1. Introduction

Concrete is widely used in engineering, and these concrete engineering facilities are often subjected to various effects, such as earthquakes, weapon strike explosions, and engineering blasting. There are usually spherical wave problems such as point explosion and point impact. Then, it is necessary to deal with the propagation of spherical waves in concrete [1–3]. The dynamic response or spherical wave propagation in concrete under spherical impact completely depends on the dynamic properties of the concrete. Therefore, it is important to study the dynamic properties of concrete under a high strain rate, which has attracted the attention of many researchers [4–6]. Bischoff et al. [4] review experimental techniques commonly used for high strain rate testing of concrete in compression and characteristics of the dynamic compressive strength and deformation behavior. Malvar and Ross [5] undertake a literature review to characterize the effects of strain rate on the tensile strength of concrete. Cusatis [6] presents a previously developed meso-scale model for concrete, including the effect of loading rate, and the rate dependence of concrete behavior is assumed to be caused by two different physical mechanisms. Some studies [7–10] indicated that the different strain-rate sensitivity is determined in concrete under different strain rates. Al-Salloum et al. [7] studied the dynamic behavior of concrete experimentally by testing annular and solid concrete specimens using a split Hopkinson pressure bar (SHPB). Wang et al. [8] designed a large-diameter SHPB with a diameter of 100 mm used to carry out impact tests at different speeds. The results show that the increase in the strain rate has a hindering effect on the increase in damage variables and the increase rate (impact speeds of 5 m/s, 10 m/s, and 15 m/s). Wang et al. [9] provided guidance for selecting pulse shapers for concrete SHPB experiments. Grote et al. [10] applied SHPB and plate impact to achieve a range of loading rates and hydrostatic pressures.

Meanwhile, researchers have carried out many studies on rate-dependent materials of spherical waves [11–16]. Luk et al. [11] developed models for the dynamic expansion of spherical cavities from zero initial radii for elastic–plastic rate-independent materials with power-law strain hardening. Wegner et al. [12] presented a new formulation of the governing equations of spherical waves, in which the resulting system of five equations is treated as a strictly hyperbolic system of first-order hyperbolic partial differential equations, and the method of characteristics is adapted to obtain numerical solutions. Forrestal et al. [13] developed a spherical cavity-expansion penetration model for concrete targets, and predictions from the compressible penetration model are in good agreement with depth of penetration data. Lai et al. [14,15] used the ZWT linear and nonlinear visco-elastic constitutive model to set up the governing equations for linear and nonlinear visco-elastic spherical waves, and published numerical results using the characteristics method. Lu et al. [16] established the linear visco-elastic ZWT constitutive equation under a three-dimensional stress state by ignoring the relaxation effect of the low-frequency Maxwell element and the nonlinear spring element. The absorption and dispersion phenomena of the spherical wave propagation in the visco-elastic solid were analyzed. At present, with the development of experimental technology, researchers are interested in wave propagation technology (WPT) [17–20]. Zhu et al. [17] set up the error in the determination of dynamic stress–strain curve of rate-dependent brittle materials with the traditional SHPB techniques with either a three-wave method or a two-wave method, which is not accepted. Wang et al. [18] developed an experimental apparatus for spherical divergent wave propagation in solids. Liu et al. and Sollier et al. [19,20] completed a series of experiments to measure the shock initiation behavior using eleven embedded electromagnetic particle velocity gauges. The dynamic performance experiment of concrete is different from the quasi-static test. The behavior of materials under spherical impact cannot be separated from the analysis of spherical wave propagation (wave propagation effect). The core problem in carrying out this research is that the effects of wave propagation and strain rate are often coupled. When studying the dynamic constitutive relation of materials with high strain rates, the wave propagation effects in the experimental process, especially in the specimen, should not be ignored.

In order to solve the above-mentioned difficulties and deal with the coupling problem, people have developed WPT to study the dynamic properties of materials subjected to dynamic loads [21]. In various wave propagation analysis techniques, Lagrangian analysis has attracted the attention of many researchers [22–26], because there are no other pre-assumptions about the constitutive relation of the materials under study. In the case of spherical waves, the constitutive equation of spherical waves consists of two parts: the volumetric part and the distortional part [27]. The traditional Lagrange analysis of wave propagation is based on the conservation equations without any pre-assumption of material constitutive relation. However, when the radial particle velocity profiles are measured by velocity gauges at the Lagrangian coordinates r_i ($i = 1, 2, \ldots$), it is still difficult to solve the other two unknowns from the two constitutive equations with unknown dynamic parameters (Equations (1a), (1b), and (2)), which is different from the rate-independent elastic problem for parameters of constitutive equations, which are constant. In the work outlined in this paper, a series of particle velocity wave profiles of concrete in the far-field or low-pressure region under spherical impact loading is measured. Then, based on the universal spherical wave conservation equation and the fact that the volumetric part of constitutive relation satisfies linear elastic law, the Lagrangian inverse analysis of spherical wave problems and particle velocity history measurements (the inverse analysis) are carried out to obtain the numerical constitutive relation, expressed in the form of distortion. Furthermore, it is found that the rate-dependent characteristics of spherical wave distortion is different from the rate-independent case and therefore an appropriate rate-dependent constitutive model is chosen to describe this problem. Finally, the dynamic parameters in constitutive relation of concrete with high strain rates are obtained by the standard linear solid model.

2. Materials and Methods

2.1. Theoretical Concepts of Spherical Waves in Concrete

Many materials have significant rate correlation characteristics under the loading of short-duration explosion and impact [28–30]. Concrete materials also have relevant characteristics under short-history loading [31–33]. The fracture strain of concrete under a high strain rate is as low as a magnitude of 10^{-3}, and the behavior of concrete under one-dimensional and multidimensional stress under static load also shows great differences. Therefore, the concrete can be regarded as a linear viscose-elastic material, not just a linear elastic material.

First, the description system of spherical wave propagation is established in the spherical coordinate system (Figure 1a). The governing equation system of a linear viscose-elasticity (Figure 1b) spherical wave is composed of two parts: the conservation Equations (1a) and (1b) and the constitutive Equations (2a) and (2b) (the volumetric part 2a and the distortional part 2b), representing the physical properties [34]. The linear viscose-elasticity is reflected in the distortion relation of the constitutive Equation (2b):

$$\frac{\partial \varepsilon_r}{\partial t} = \frac{\partial v}{\partial r}, \tag{1a}$$

$$\frac{\partial \varepsilon_\theta}{\partial t} = \frac{v}{r}, \tag{1b}$$

$$\frac{\partial \sigma_r}{\partial r} + \frac{2(\sigma_r - \sigma_\theta)}{r} = \rho_0 \frac{\partial v}{\partial t}, \tag{1c}$$

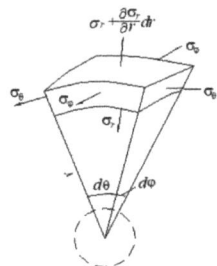
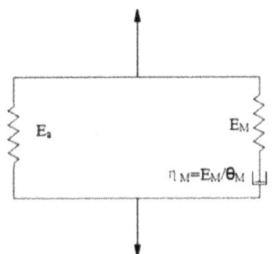

E_a static modulus of elasticity
E_M dynamic modulus of elasticity
θ_M relaxation time
η_M coefficient of viscosity

σ_r radial stress
$\sigma_\theta(\sigma_\varphi)$ circumferential stress
ε_r radial strain
$\varepsilon_\theta(\varepsilon_\varphi)$ circumferential strain

(a)   (b)

Figure 1. Schemes of governing equations: (a) micro-element in spherical coordinate system; (b) the standard linear solid constitutive model.

The linear viscose-elastic constitutive equation in differential form based on the standard linear solid model can be effectively used to describe the dynamic constitutive properties of concrete (3a) [35], and Figure 1b shows how the model works.

$$\frac{\partial \sigma_r}{\partial t} + 2\frac{\partial \sigma_\theta}{\partial t} - 3K\left(\frac{\partial \varepsilon_r}{\partial t} + 2\frac{\partial \varepsilon_\theta}{\partial t}\right) = 0, \tag{2a}$$

$$\frac{\partial \varepsilon_r}{\partial t} - \frac{\partial \varepsilon_\theta}{\partial t} = \frac{1}{2G}\left(\frac{\partial \sigma_r}{\partial t} - \frac{\partial \sigma_\theta}{\partial t}\right) + \frac{(\sigma_r - \sigma_\theta) - 2G_a(\varepsilon_r - \varepsilon_\theta)}{2G\theta_M}, \tag{2b}$$

The relevant material parameters are characterized as a linear elastic response (3b), volume deformation (3c), linear bulk modulus (3d), linear Young's modulus (3e), and linear shear modulus (3f). According to conventional considerations, it is assumed that Poisson's ratio u is constant, and the elastic stage is independent of other strains and strain rates.

$$\frac{\partial \sigma}{\partial t} + \frac{\sigma}{\theta_M} = (E_a + E_M)\frac{\partial \varepsilon}{\partial t} + \frac{E_a \varepsilon}{\theta_M}, \tag{3a}$$

$$\sigma = E_a \varepsilon \tag{3b}$$

$$\Delta = \varepsilon_r + 2\varepsilon_\theta \tag{3c}$$

$$K = \frac{E}{3(1 - 2\nu)} \tag{3d}$$

$$E = E_a + E_M \tag{3e}$$

$$G = \frac{E}{2(1 + \nu)}, \tag{3f}$$

In this way, in order to describe the linear viscose-elastic spherical wave propagation problem, based on the standard linear solid constitutive relation, the governing equation reflecting the linear and high strain rate effect of materials is established.

2.2. Experimental Method

In order to understand the propagation characteristics of spherical waves in concrete, an experimental method is developed to measure the particle velocity histories of spherical waves. The experiment adopts the electromagnetic method, and the sample is a cylinder with a diameter equal to the height. Because the arrangement of particle velocimeters have accurate representative characteristics, the method has strong advantages in studying the dynamic properties of multicomponent composites containing fillers, such as polymer–matrix composite materials, concrete, and rock in 3-D stress. In the spherical wave experiment, the characteristic size of the sample can be meters, which is more than ten times larger than the size of concrete coarse fillers, so that the information of wave histories can accurately reflect the wave propagation characteristics. A group of particle velocity waves v(r_i,t) at different radii distance r_i from the center of the sphere is measured by a series of embedded magneto-electric velocimeters.

In the experiment, a mini-charge is detonated in the center of a cylindrical concrete block with a diameter of 25 cm and a length of 25 cm, and a spherical impact is loaded by detonating an explosive with a weight of 0.1 g/0.8 g. The principle of the spherical particle velocity history device is shown in Figure 2 [36]. The experimental specimen consists of two equal-height cylinder parts. A series of concentric toroidal magneto-electric particle string gauges is arranged on the mating surface. Explosive charges are placed in the cavity at the center of the sample; the soft detonating cord for initiation is entered along the mini hole of the upper half of the sample, and the upper and lower parts are bonded with epoxy resin after the gauge and the explosive charge are placed. After initiation, the particle velocimeters move to cut the magnetic field to form voltage signals, and the particle velocity histories at a series of radii can be obtained from the calibration results.

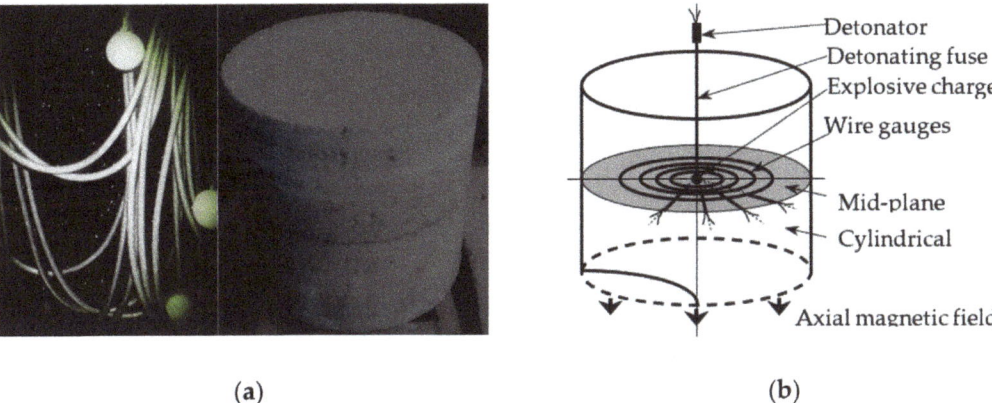

Figure 2. Scheme for velocity history test device in concrete with mini-charge: (**a**) mini-charge with soft detonating cord and long cylindrical block of concrete; (**b**) experimental concept for spherical wave experiments.

2.3. Inverse Method

The particle velocities in spherical wave propagation are easy to measure, but other physical quantities are difficult to measure directly at the same time. In order to obtain accurate information about other physical quantities during spherical wave propagation, and then obtain the constitutive relation of materials, Lagrangian inverse analysis is a good alternative, which is based on conservation equations and does not make any assumptions [37–40]. Next, the "second type inverse problem" in mathematics is dealt with to determine the dynamic constitutive properties of concrete. In the study of spherical waves, when the particle velocities at a series of different Lagrangian coordinates ri are obtained, it is difficult to calculate other unknown quantities from the former (2a, 2b). So we developed a new spherical wave analysis method "NV+T0/SW" to deal with this problem [14].

2.3.1. The Method Solving Strain (ε_r, ε_θ)

The differential relation of strain (ε_r, ε_θ) and the particle velocity is established by the conservation equation. Now, the initial condition t = 0, $v(r_i,t)$=0 is known, and $v(r_i,t)$ at different positions r_i (i = 1,2, ... ,) is also known. So, the time numerical integration operation can be performed to determine $\varepsilon_\theta(r_i, t)$. Then, the first derivative $\partial v(r_i,t)/\partial t$ can be obtained by numerical differential operation. Similarly, the strain $\varepsilon_r(r_i,t)$ can be determined by integrating time.

2.3.2. The Method Solving Strain (σ_r, σ_θ)

However, the stresses σ_r and σ_θ are still unknown. The system composed of volume and shape deformation is to be determined. The solving of σ_r and σ_θ in this way is not sufficient, and one of the equation relations must be known first. In the elastic range, it is accepted that the volume deformation satisfies the linear law of elasticity (2a) and is independent of the rate. Then, it is easy to determine this relationship under quasi-static conditions. The calculation process related to quantity ε_r and ε_θ, $\partial\sigma_\theta/\partial t$, and $\partial\sigma_r/\partial t$ can be expressed in Equations (4) and (5b).

In order to establish the magnitude relationship at each radius, the path-line processing method can be used to define the total derivative of a certain magnitude on the path-line (Grady, 1973), and the path-lines P1, P2, P3 ... Pi ... Pm can be established as shown in the figure. In this way, when the spherical particle velocity histories $v(r_i,t)$ at multiple

Lagrangian radii r=r$_i$ are provided, and their related other time and position differential components ∂v(r$_i$,t)/∂t can be easily determined.

$$\frac{\partial \sigma_\theta}{\partial t} = \frac{1}{2}\left(\left(3K\frac{\partial \varepsilon_r}{\partial t} + 2\frac{\partial \varepsilon_\theta}{\partial t}\right) - \frac{\partial \sigma_r}{\partial t}\right) \quad (4)$$

$$\left.\frac{d\sigma_r}{dr}\right|_p = \left.\frac{\partial \sigma_r}{\partial r}\right|_t + \left.\frac{\partial \sigma_r}{\partial t}\right|_r \frac{dt}{dr} = \left.\frac{\partial \sigma_r}{\partial r}\right|_t + \left.\frac{\partial \sigma_r}{\partial t}\right|_r \left.\frac{1}{r'}\right|_p \quad (5a)$$

substituting (4) into (5a), the calculation formula of partial derivative about stress ∂σ$_r$/∂t can be expressed as (5b).

$$\frac{\partial \sigma_r}{\partial t} = r'\left(\left.\frac{d\sigma_r}{dr}\right|_p - \rho_0 \frac{\partial v}{\partial t} + \frac{2(\sigma_r - \sigma_\theta)}{r}\right) \quad (5b)$$

The zero initial condition is known at different positions of wave propagation (σ$_r$ = 0 along path-line P1), and the stress σ$_r$ at different radius r=r$_i$ along the path-line P2 (Figure 3) is obtained through the integration of partial derivative ∂σ$_r$(r$_i$,t)/∂t by using the constructed path-lines (5). Then, ∂σ$_\theta$(r$_i$,t)/∂t is known from (4), and the circumferential stress at different positions r = r$_i$ on the path-line P1 σ$_\theta$(r$_i$,t) | $_{P=j}$ can be calculated by integrating ∂σ$_\theta$(r$_i$,t)/∂t. Similarly, the stress σ$_r$(r$_i$,t) | $_{P=j+1}$ and σ$_\theta$(r$_i$,t) | $_{P=j+1}$ on all path-lines can be determined by cycling in sequence. Note that this method can be used to load the whole process, which is called "NV + T0/SW" for short.

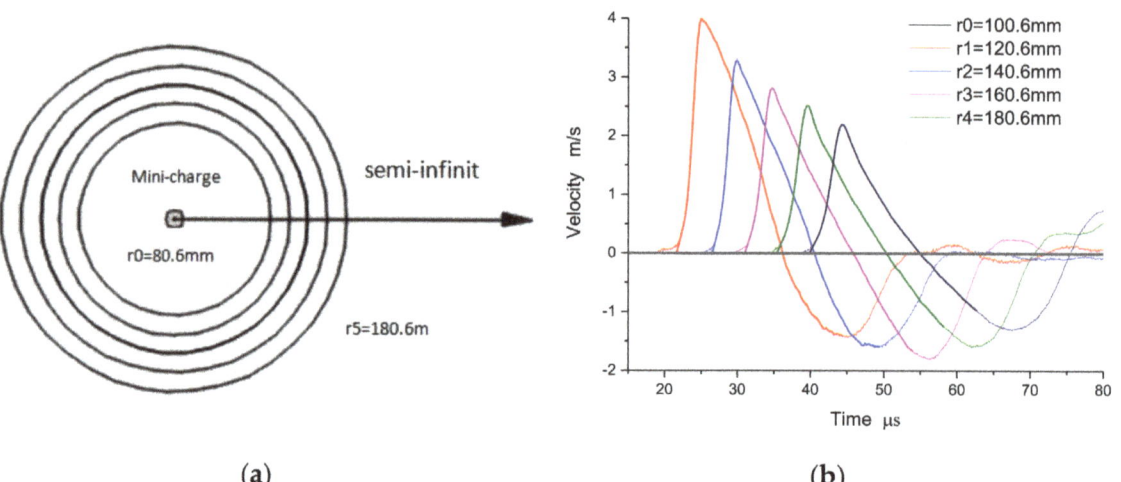

(a) (b)

Figure 3. Results of velocity histories in concrete: (a) schematic diagram of test location layout in Mid-plane; (b) the series of measured particle velocity profiles.

2.3.3. Solving for G and θ$_M$

An advantage of "NV + T0/SW" is that there are no assumptions of the constitutive equation of materials, directly giving the stress–strain numerical relation. However, now there is a next step to take when the dynamic properties of materials are known. So, its description with a known standard linear solid constitutive model is provided in future work, and the dynamic shear modulus G and one Maxwell element material parameters θ$_M$ can be determined by the following method (6).

$$G = \frac{\sigma_r - \sigma_\theta}{2(\varepsilon_r - \varepsilon_\theta)} \quad (6)$$

$$\theta_M = \frac{(\sigma_r - \sigma_\theta) - 2G_a(\varepsilon_r - \varepsilon_\theta)}{2G\left(\frac{\partial \varepsilon_r}{\partial t} - \frac{\partial \varepsilon_\theta}{\partial t}\right) - \left(\frac{\partial \sigma_r}{\partial t} - \frac{\partial \sigma_\theta}{\partial t}\right)} \quad (7)$$

3. Results

3.1. The Experimental Results

Based on the experimental method as described in the previous section, the particle velocity profiles (Figure 3) in the low stress elastic region at the radii of 100.6 mm, 120.6 mm, 140.6 mm, 160 mm, and 180.6 mm are measured accurately in the semi-infinite space of concrete by using the mini-explosive ball and electromagnetic velocity measurement technology. Here, the radius of the mini-explosive ball is 5 mm, with an explosive equivalent of 1.00 g TNT. As shown in Figure 3b, the maximum particle velocity is lower than 4 m/s, and the experimental model is a one-dimensional spherical symmetry problem. At same time, the static mechanical property parameters of concrete can be easily measured, as shown in Table 1.

Table 1. Concrete static parameters for 'NV+T_0/SW'.

Symbol	ρ	v	C_K	K	G_a	E_a
Units	kg/m^3	1	m/s	GPa	GPa	GPa
Value	2380	0.23	4347	19.26	12.68	31.20

3.2. The Inverse Numerical Results

In the series of measured particle velocity histories shown in the Figure 4, the path-line is constructed from the initial zero value line. The path-line is divided into regions by the peak value. The analysis value of the path-line is interpolated at equal time intervals in each region to serve as the basis of the inversion analysis framework. With these path-line values covering the particle velocity field, the physical quantities of the spherical wave can easily be solved by the aforementioned method, i.e., "NV + T_0/SW". Since the constitutive relation of materials is often described by volume deformation and shape deformation with multidimensional stress state, it is convenient to reflect the stress characteristics under 3-D stress. The results are expressed as spherical profiles of volumetric part and distortional part, such as stress histories $\sigma_r + 2\sigma_\theta$, strain histories $\varepsilon_r + 2\varepsilon_\theta$, stress histories $\sigma_r - \sigma_\theta$, and strain histories $\varepsilon_r - \varepsilon_\theta$. The numerical results are shown in Figures 5 and 6, and the numerical constitutive relation, expressed in the form of volume and distortion, is shown in Figure 7. The volumetric constitutive relation satisfies linear elastic law with linear bulk modulus K, but the distortional constitutive relation does not. It is not difficult to find that the latter relation has an obvious rate effect.

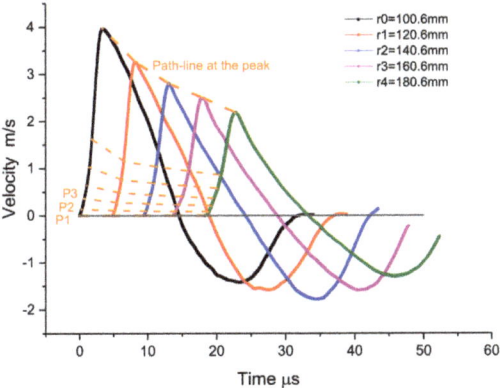

Figure 4. The schemes of inversion analysis with path-line.

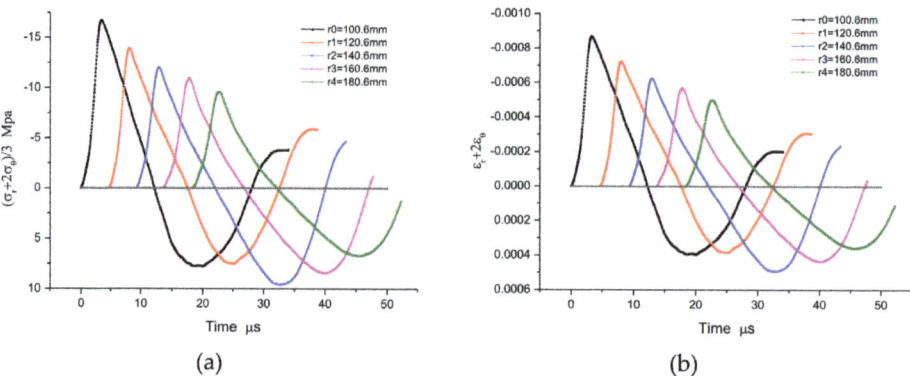

Figure 5. A comparison of positive and inverse results: (**a**) the volumetric part histories of stress $\sigma_r + 2\sigma_\theta$; (**b**) the volumetric part histories of strain $\varepsilon_r + 2\varepsilon_\theta$.

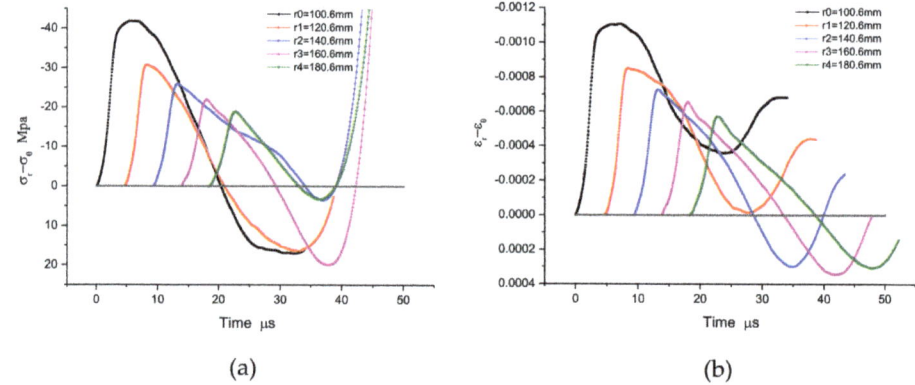

Figure 6. A comparison of positive and inverse results: (**a**) the distortional part histories of stress $\sigma_r - \sigma_\theta$; (**b**) the distortional part histories of strain $\varepsilon_r - \varepsilon_\theta$.

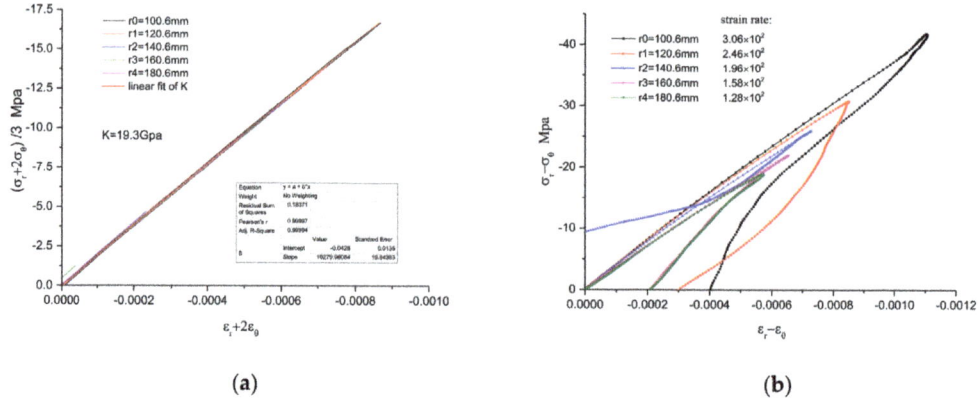

Figure 7. The numerical rate-dependent constitutive relation: (**a**) the volumetric relation of stress $(\sigma_r + 2\sigma_\theta)/3$ and strain $\varepsilon_r + 2\varepsilon_\theta$ stress $\sigma_r - \sigma_\theta$; (**b**) the distortional relation of stress $\sigma_r - \sigma_\theta$ and strain $\varepsilon_r - \varepsilon_\theta$.

3.3. The Determination of Dynamic Parameters G and θ_M

According to the above theory, the dynamic parameters G and θ_M can be determined from Equations (6) and (7), and the concrete static parameters and the numerical distortion relations of stress $\sigma_r - \sigma_\theta$ and strain $\varepsilon_r - \varepsilon_\theta$ are taken as the known conditions using the inverse method. The concrete static parameters used in the inverse analysis are the results of our experimental research on concrete under one-dimensional stress, and ρ and ν are measured from concrete samples, as shown in Table 1.

Note that Equation (6) is suitable for the series numerical distortion relations with different strain rates at each radius, so the average value of dynamic parameters G can be calculated easily with Equation (6), as shown in Table 2. Similarly, the dynamic parameters θ_M can be obtained through Equation (7), and the values of θ_M are also listed in Table 2. The results show that the dynamic shear modulus G is larger than the static modulus G_a and decreases with the reducing of strain rate (Figure 7b). At the same time, the dynamic relaxation time θ_M increases with a reducing strain rate and is in the magnitude range of 10^{-6} s.

Table 2. Dynamic parameters by 'NV+T_0/SW'.

Symbol	2G	G	E	θ_M
Units	GPa	GPa	GPa	µs
Value at r0	39.68	19.84	48.81	−1.14
Value at r1	39.51	19.75	48.60	−1.22
Value at r2	38.58	19.29	47.45	−1.41
Value at r3	35.90	17.95	44.16	−1.40
Value at r4	34.22	17.11	42.09	−1.42

4. Discussion

Firstly, a series of particle velocity histories of spherical waves in concrete is measured by magneto-electric velocimeters at each radius, which provides a basis for an experimental study on the dynamic properties of concrete in 3-D stress state under high strain rates. The particle velocimeter is a very thin ring coil, which is very suitable for measuring the physical quantities in spherical waves that change with the spherical radius. It is a good choice for measuring the signals of spherical waves for non-perspective materials, except for magnetic materials. Secondly, by analyzing the experimental data $v(r_i,t)$ of the spherical particle velocity wave of concrete, the Lagrangian "NV + T_0/ SW" inverse analysis is carried out using the path-line method, and the wave propagation information of each physical quantity of the spherical wave is obtained. The numerical constitutive relation is expressed in the form of distortion and has an obvious rate effect. The results shown in Figure 7 demonstrate the obvious different behaviors of concrete between dynamic loading and static loading normally, and the strain rate effect of concrete cannot be ignored with the strain rate range of 10^2 1/s. The numerical constitutive relation is deduced directly from the measurements and analyses of wave propagation signals, which should be more appropriate for the coupled effects between wave propagation and rate dependency, are considered. Next, the rate-dependent dynamic parameters in concrete are determined by the standard linear solid model, which is a typical and useful model for analyzing stress relaxation and creep behaviors of viscoelastic solids. The results of dynamic parameters show that the dynamic shear modulus G is larger than the static modulus G_a and decreases with the reducing of strain rate (Figure 7b). Furthermore, the dynamic relaxation time θ_M increases with reducing strain rate and is in the magnitude range of 10^{-6} s.

5. Conclusions

The goal of this research was to expand the knowledge about the possibilities of studying rate-dependent constitutive relation and the determination of dynamic parameters based on spherical waves in concrete. According to the former, the main conclusions drawn from the above results are as follows:

1. The series particle velocity of spherical waves in concrete specimens is measured by a magneto-electric velocimeter, which visually shows the propagation characteristics of spherical waves of particle velocity. It provides convenience for the interpretation of spherical wave information in concrete. At the same time, it also creates good support for the experimental study of constitutive relation of dynamic properties of strong impact in multidimensional stress state of concrete (inverse analysis).
2. The inverse problem is solved by the newly proposed "NV + T0/SW" Lagrangian analysis method, with the measured series velocity profiles as known conditions. The results provide a basis of a further study on how to determine accurately and effectively rate-dependent constitutive relation of concrete at high strain rate. When the numerical constitutive relation is expressed in the form of distortion, it is found that the distortion law has an obvious rate effect.
3. Based on a series of numerical constitutive relation with different strain rates at each radius, the rate-dependent dynamic parameters in concrete are determined by the standard linear solid constitutive model. The dynamic shear modulus G is larger than the static modulus and decreases with reducing strain rate. The dynamic relaxation time θ_M increases with the reducing strain rate and is in the magnitude range of 10^{-6} s.

It should be emphasized that, if more experimental data in the strain rate range and more continuous particle velocity profiles are measured through the improvement and development of experimental loading and data acquisition technology, the results obtained by this method will be enriched into a series. This method has good applicability, especially in the study of the dynamic behavior of multicomponent composite materials with large-size particle filler for the characteristic size of specimens in spherical wave experiments could be in the order of meters.

Author Contributions: Conceptualization, L.W. and F.Z.; Data curation, H.L.; Investigation, Z.W.; Methodology, H.L.; Project administration, Supervision L.Y.; Resources, Z.W. All authors have read and agreed to the published version of the manuscript.

Funding: This research was funded by the National Natural Science Foundation of China (NSFC 11390361, 11172244), by K. C. Wong Magna Fund in Ningbo University.

Institutional Review Board Statement: Not applicable.

Informed Consent Statement: Not applicable.

Data Availability Statement: Not applicable.

Conflicts of Interest: The authors declare no conflict of interest.

References

1. Wang, L.L. *Foundations of Stress Waves*, 1st ed.; National Defense Industry Press: Beijing, China, 2007; pp. 227–244.
2. Zukas, J.A. High Velocity Impact Dynamics. *Chem. Eng. Sci.* **1990**, *59*, 525–541.
3. Wang, L.L.; Hu, S.S.; Yang, L.M.; Dong, X.L. *Material Dynamics*, 1st ed.; China Science and Technology Press: Hefei, China, 2017; pp. 141–174.
4. Bischoff, P.H.; Perry, S.H. Compressive behaviour of concrete at high strain rates. *Mater. Struct.* **1991**, *24*, 425–450. [CrossRef]
5. Malvar, L.J.; Ross, C.A. Review of strain rate effects for concrete in tension. *ACI Mater. J.* **1998**, *95*, 735–739. [CrossRef]
6. Cusatis, G. Strain-rate effects on concrete behavior. *Int. J. Impact Eng.* **2011**, *38*, 162–170. [CrossRef]
7. Al-Salloum, Y.; Almusallam, T.; Ibrahim, S.M.; Abbas, H.; Alsayed, S. Rate dependent behavior and modeling of concrete based on SHPB experiments. *Cem. Concr. Compos.* **2015**, *55*, 34–44. [CrossRef]
8. Wang, W.; Zhang, Z.; Huo, Q.; Song, X.; Yang, J.; Wang, X.; Wang, J.; Wang, X. Dynamic Compressive Mechanical Properties of UR50 Ultra-Early-Strength Cement-Based Concrete Material under High Strain Rate on SHPB Test. *Materials* **2022**, *15*, 6154. [CrossRef] [PubMed]
9. Wang, J.; Li, W.; Xu, L.; Du, Z.; Gao, G.; Alves, M. Experimental study on pulse shaping techniques of large diameter SHPB apparatus for concrete. *Lat. Am. J. Solids Struct.* **2021**, *18*(1), e343. [CrossRef]
10. Grote, D.L.; Park, S.W.; Zhou, M. Dynamic behavior of concrete at high strain rates and pressures: I. Experimental characterization. *Int. J. Impact Eng.* **2001**, *25*, 869–886. [CrossRef]
11. Luk, V.K.; Forrestal, M.J.; Amos, D.E. Dynamic spherical cavity expansion of strain-hardening materials. *J. Appl. Mech. Trans. ASME* **1991**, *58*, 1–6. [CrossRef]

12. Wegner, J.L. Propagation of waves from a spherical cavity in an unbounded linear viscoelastic solid. *Int. J. Eng. Sci.* **1993**, *31*, 493–508. [CrossRef]
13. Forrestal, M.J.; Tzou, D.Y. A spherical cavity-expansion penetration model for concrete targets. *Int. J. Solids Struct.* **1997**, *34*, 4127–4146 . [CrossRef]
14. Lai, H.W.; Wang, Z.J.; Yang, L.M.; Wang, L.L. Characteristics analyses of linear visco-elastic spherical waves. *Explos. Shock Waves* **2013**, *33*, 1–10.
15. Wang, L.L.; Lai, H.W.; Wang, Z.J.; Yang, L.M. Studies on nonlinear visco-elastic spherical waves by characteristics analyses and its application. *Int. J. Impact Eng.* **2013**, *55*, 1–10. [CrossRef]
16. Lu, Q.; Wang, Z.J.; Wang, L.L.; Lai, H.W.; Yang, L.M. Analysis of linear visco-elastic spherical waves based on ZWT constitutive equation. *Baozha Yu Chongji/Explos. Shock Waves* **2013**, *33*, 463–470.
17. Zhu, J.; Hu, S.S.; Wang, L.L. Problems of SHPB technique used for rate-dependent concrete-sort materials. *Gongcheng Lixue/Eng. Mech.* **2007**, *24*, 78–087.
18. Wang, Z.; Li, X.; Zhang, R.; Zheng, X.; Zhu, Y.; Liu, X.; Hu, Z. Experimental apparatus for spherical wave propagation in solid. *Baozha Yu Chongji/Expolosion Shock Waves* **2000**, *20*, 103–109.
19. Liu, J.; Wang, Y.F.; Wang, G.J.; Zhang, R.; Zhong, B.; Zhao, F.; Zhang, X. Al-based electromagnetic particle velocity gauge technique of measuring the particle velocity of HMX-based PBX explosives. *Hanneng Cailiao/Chin. J. Energetic Mater.* **2016**, *24*, 300–305 . [CrossRef]
20. Sollier, A.; Hébert, P.; Letremy, R.; Mineau, V.; Boissy, X.; Pionneau, E.; Doucet, M.; Bouton, E. Experimental characterization of the shock to detonation transition in the Tx1 HMX-TATB based explosive using embedded electromagnetic particle velocity gauges. *AIP Conf. Proc.* **2020**, *2272*, 030028 . [CrossRef]
21. Wang, L.; Zhu, J.; Lai, H. A new method combining Lagrangian analysis with Hopkinson pressure bar technique. *Strain* **2011**, *47*, 173–182. [CrossRef]
22. Cowperthwaite, M.; Williams, R.F. Determination of constitutive relationships with multiple gauges in non-divergent waves. *J. Appl. Phys.* **1971**, *42*, 456. [CrossRef]
23. Grady, D.E. Experimental analysis of spherical wave propagation. *J. Geophys. Res.* **1973**, *78*, 1299–1307. [CrossRef]
24. Seaman, L. Lagrangian analysis for multiple stress or velocity gages in attenuating waves. *Journal of Applied Physics* **1974**, *45*(10), 4303–4314. [CrossRef]
25. Huawei, L.; Lili, W. Studies on dynamic behavior of Nylon through modified Lagrangian analysis based on particle velocity profiles measurements. *J. Exp. Mech.* **2011**, *26*, 221–226.
26. Wang, L.; Ding, Y.; Yang, L. Experimental investigation on dynamic constitutive behavior of aluminum foams by new inverse methods from wave propagation measurements. *Int. J. Impact Eng.* **2013**, *62*, 48–59. [CrossRef]
27. Wang, L.; Yang, L.; Dong, X.; Jiang, X. *Dynamics of Materials: Experiments 2019, Models and Applications*; Academic Press: London, UK, 2019. [CrossRef]
28. Limbach, R.; Rodrigues, B.P.; Wondraczek, L. Strain-rate sensitivity of glasses. *J. Non-Cryst. Solids* **2014**, *404*, 124–134 . [CrossRef]
29. Chen, T.H.; Tsai, C.K. The microstructural evolution and mechanical properties of Zr-based metallic glass under different strain rate compressions. *Materials* **2015**, *8*, 1831–1840 . [CrossRef]
30. Siviour, C.R.; Jordan, J.L. High Strain Rate Mechanics of Polymers: A Review. *J. Dyn. Behav. Mater.* **2016**, *2*, 15–32 . [CrossRef]
31. Shang, H.S.; Ji, G.J. Mechanical behaviour of different types of concrete under multiaxial compression. *Mag. Concr. Res.* **2014**, *66*, 870–876. [CrossRef]
32. Ye, Z.; Hao, Y.; Hao, H. Numerical study of the compressive behavior of concrete material at high strain rate with active confinement. *Adv. Struct. Eng.* **2019**, *22*, 2359–2372. [CrossRef]
33. Liu, P.; Liu, K.; Zhang, Q.B. Experimental characterisation of mechanical behaviour of concrete-like materials under multiaxial confinement and high strain rate. *Constr. Build. Mater.* **2020**, *258*, 119638 . [CrossRef]
34. Wang, L.L.; Zhu, J.; Lai, H.W. Understanding and Interpreting of the Measured Wave Signals in Impact Dynamics Studies. *Chin. J. High Press. Phys.* **2010**, *24*, 279–285.
35. Rossikhin, Y.A.; Shitikova, M.V.; Popov, I.I. Dynamic response of a viscoelastic beam impacted by a viscoelastic sphere. *Comput. Math. Appl.* **2017**, *73*, 970–984 . [CrossRef]
36. Zhanjiang, W.; Ruoqi, Z.; Xiaolan, L. Experimental Investigation on Tamped and Cavity Decoupled Explosion in Rock-Soil by Mili-Explosive Charge. Ph.D. Thesis, National University of Defense Science and Technology, Hunan, China, 2003.
37. Lin, Y.R.; Wang, Z.J.; Li, Y.L.; Men, C.J. Constitutive relation using particle velocity data of spherical waves. *Jiefangjun Ligong Daxue Xuebao/J. PLA Univ. Sci. Technol. (Nat. Sci. Ed.)* **2007**, *8*, 606–610.
38. Lai, H.W.; Wang, Z.J.; Yang, L.M.; Wang, L.L. Inversion of constitutive parameters for visco-elastic materials from radial velocity measurements of spherical wave experiments. *Gaoya Wuli Xuebao/Chin. J. High Press. Phys.* **2013**, *27*, 245–252. [CrossRef]
39. Lai, H.W.; Wang, Z.J.; Yang, L.M.; Wang, L.L. Analysis of nonlinear spherical wave propagation for concretes. In Proceedings of the Rock Dynamics: From Research to Engineering-2nd International Conference on Rock Dynamics and Applications, ROCDYN 2016, Suzhou, China, 18 May 2016. [CrossRef]
40. Lu, Q.; Wang, Z.J.; Ding, Y. Inversion for the complex elastic modulus of material from spherical wave propagation data in free field. *J. Sound Vib.* **2019**, *459*, 114851. [CrossRef]

Article

Blast Resistance of Reinforced Concrete Slabs Based on Residual Load-Bearing Capacity

Lijun Wang [1], Shuai Cheng [2], Zhen Liao [2], Wenjun Yin [2], Kai Liu [2], Long Ma [2], Tao Wang [1,*] and Dezhi Zhang [2,*]

[1] School of Nuclear Engineering, Rocket Force University of Engineering, Xi'an 710024, China
[2] Northwest Institute of Nuclear Technology, Xi'an 710024, China
* Correspondence: wtao009@163.com (T.W.); zhangdezhi@nint.ac.cn (D.Z.)

Abstract: In this paper, the blast-loading experiment and numerical simulation are carried out for RC slabs with two typical reinforcement ratios. The time history of reflected shockwave pressures and displacement responses at different positions on the impact surface of the specimens are obtained, and the influence of the reinforcement ratio on the dynamic responses and failure modes of the RC slabs is analyzed. Based on the experimental data, the simulation model of the RC slab is verified, and the results indicate good agreement between the two methods. On this basis, the residual load-bearing capacity of the damaged RC slabs is analyzed. The results show that the load distribution on the impact surface of the slab is extremely uneven under close-in blast loading. The resistance curve shape of the RC slabs varies markedly before and after blast loading, and its load bearing capacity and bending stiffness deteriorate irreversibly. Increasing the reinforcement ratio can impede crack extension, reduce the slab's residual displacement, and, at the same time, reduce the decrease of the damaged slab's load-bearing capacity. The findings of this study will provide insights into the anti-explosion design and damage evaluation of RC slabs.

Keywords: reinforced concrete slab; blast loading; residual load-bearing capacity; dynamic response; blast resistance

1. Introduction

The reinforced concrete (RC) slab is a typical load-bearing structural member in reinforced concrete buildings. Due to the influence of factors such as structural layer thickness, reinforcement ratio, and impact surface area, RC slabs tend to undergo larger deformation, more severe failure, and other forms of damage than RC beams/columns under the blast shockwave loading of the same intensity [1–3]. Figuring out the dynamic response and damage characteristics of RC slabs under shockwave loading induced by chemical explosions, and quantitatively and accurately evaluating the degradation of their load-bearing capacity, are of referential significance for the anti-explosion design of RC slabs and the evaluation of damage effects of weapons.

At present, considerable research efforts have been made by researchers at home and abroad concerning the failure modes and dynamic response of RC slabs under blast loading, yielding a series of results [4–6]. Lan et al. [7] conducted an experimental investigation of 74 groups of RC slabs under different charge quantities and stand-off distances and analyzed the differences of their damage and failure modes. Huff [8] also systematically studied the failure modes of a typical two-way RC slab under blast loading by experiments. Wang Wei et al. [9,10] carried out a series of experiments and numerical simulations to study the damage and failure of square RC slabs subjected to close-in explosions and obtained their damage–failure modes and criteria. Most experimental studies have also focused on determining new technologies to improve the deflection response and damage resistance of reinforced concrete slabs under explosive loads [11–14]. With respect to damage assessment of RC slabs, the P-I curve is often employed to assess the damage of structural members under blast loading [15–18]. As the equivalent single-degree-of-freedom (SDOF) analysis is simple and practical, it has been widely used in component damage assessment [17,19–21].

In fact, the dynamic response and failure modes of RC slabs are very complex because of the different blast-loading conditions (charge type, charge shape, scaled distance, etc.) and properties of the structure itself (concrete strength, reinforcement ratio, stirrup ratio, support boundary, etc.) [22–24]. Currently, the anti-explosion performance of RC slabs is evaluated mainly based on their dynamic response and the macro-damage modes of the damaged members. However, for the uneven load distribution in RC slabs subjected to close-in explosions, few analyses has been done using the performance index of the residual load-bearing capacity of damaged RC slabs to quantitatively evaluate the degradation of their bearing capacity.

In this paper, blast-loading experiments are conducted for RC slabs with two typical reinforcement ratios. The time history of the reflected shockwave pressures and displacement responses at different positions on the impact surface of the slabs are obtained, and the differences of the macro-damage modes of the two RC slabs at the same scaled distance are analyzed. The numerical simulation model of the slabs is constructed with finite element program LS-DYNA and corrected by the experimental data. On this basis, the residual load-bearing capacity of the damaged RC slabs with different reinforcement ratios is analyzed, the quantitative residual load-bearing capacity results are given, and the damage degrees of slab members with different reinforcement ratios are compared and analyzed.

2. Experimental Study

2.1. Test Specimen Design

According to the code for the design of concrete structures in China (2011 edition), RC slabs with two typical reinforcement ratios are designed and their geometric dimensions are both 1200 × 500 × 100 mm (length × width × thickness, respectively). The reinforcement drawings of the test specimens are shown in Figure 1 and both use double-layer two-way reinforcement. When the slab thickness is 100 mm~150 mm, the common diameter of the stressed reinforcement in the slab is 8~12 mm. Therefore, the longitudinal reinforcement bar diameters of the two RC slabs are 12 mm and 8 mm, and the corresponding reinforcement ratios are 1.41% and 0.63%. The stirrup diameter is 4 mm. The longitudinal bar is an HRB400 steel bar, the stirrup is an HPB235 steel bar, the concrete grade is C30, and the thickness of the protective layer is 50 mm.

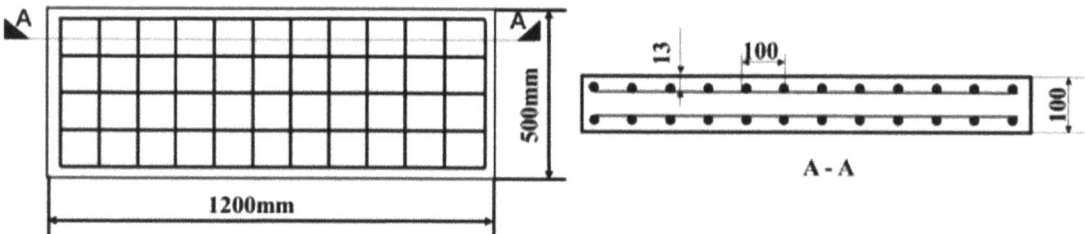

Figure 1. Schematic drawings of RC slab reinforcement.

During the experiment, the two RC slabs are placed vertically through independently designed supporting tools, as shown in Figure 2.

Slab A with the low reinforcement ratio, slab B with the high reinforcement ratio, and the pressure test tool are arranged in an equilateral triangle around the charge, with an interval of about 120° between them. The tool is fixed on the ground by multiple pegs and reinforced by welding the angle iron to the structure to ensure that the supporting tool does not move during the experiment. The four corners of the slab specimens are all clamped to give them full restraints. Steel plates are welded between the clamped corners to strengthen the restraint effect, so the boundary conditions can be regarded as fixed supports on four sides. The explosive used is a bare spherical TNT-RDX-Al explosive charge and its way of initiation is central detonation. The mass ratio of TNT to RDX was 4:6, which was composed of two nearly identical hemispheres. The TNT equivalent is 10 kg. The explosion origin or center is 1200 mm from both the centers of the RC slabs and the pressure tool and 700 mm above the ground. Explosives are positioned by the nylon rope hanging on the lifting bracket and fixed by the mesh pocket and white cloth belt to ensure that the ball center is at the center of the component's blast surface. In addition, a metal shield is also set behind the component, which is welded by several steel plates to prevent shock wave damage to the test sensors and cables behind the component.

Figure 2. Test site layout.

Figure 3 shows the layout of the pressure, acceleration and displacement sensors used in the experiment.

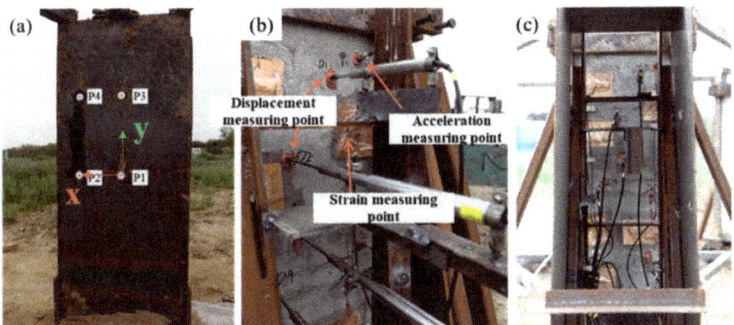

Figure 3. Layout of pressure and displacement measuring points, (**a**) the layout of the pressure sensors, (**b**) the layout of the acceleration and displacement sensors, and (**c**) overall survey point distribution.

Considering that the distribution of blast loads at different positions of the RC slabs subjected to close-in blast loading differs greatly, four PCB piezoelectric pressure sensors, numbered P1~P4, are arranged according to symmetry. Among them, the measuring point P1 is facing the explosion center, which is taken as the origin of the coordinate system, where the horizontal direction is the X-axis and the vertical direction is the Y-axis. The parameters of the pressure-measuring points are listed in Table 1.

Table 1. Coordinates of the pressure-measuring points.

Measuring Point	X/m	Y/m	Distance from Explosion Source/m	Scaled Distance/m/kg$^{1/3}$	Angle of Incidence/°
P1	0	0	1.2	0.557	90
P2	0.16	0	1.211	0.562	82.4
P3	0	0.3	1.237	0.574	75.96
P4	0.16	0.3	1.247	0.579	74.18

The displacement is measured by the contact pen-type displacement sensors, which are arranged on the back surface of the slabs. The embedded parts are fixed in the holes drilled beforehand using anchoring adhesive. The embedded parts and sensors are in threaded connection. Each slab is equipped with five displacement sensors, which are arranged at intervals along the length of the slab.

2.2. Experimental Results and Analysis

2.2.1. Blast-Load Measurement

The blast loading-time history of the measuring points P1~P3 are shown in Figure 4.

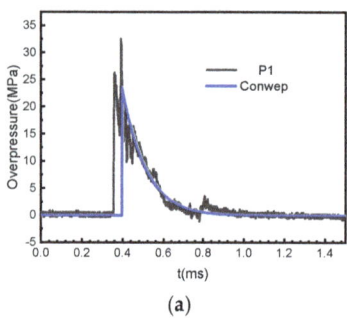
(a)

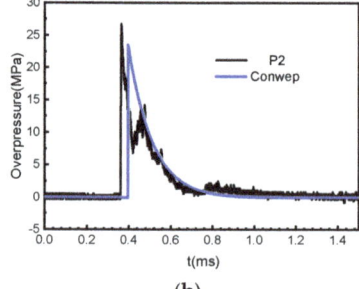
(b)

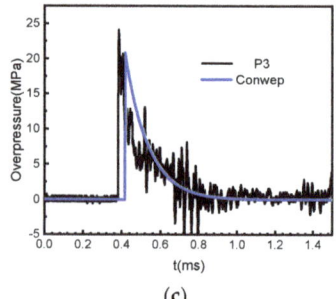
(c)

Figure 4. Comparison between measured overpressure curves of shock waves at different measuring points and ConWep calculation results: (a) the blast loading−time history of the measuring point P1, (b) the blast loading−time history of the measuring point P2, and (c) the blast loading−time history of the measuring point P3.

It can be seen from the figure that the trend of the reflected pressure waveforms at different measuring points are basically the same, exhibiting the typical law of exponential attenuation. Specifically, as P1 is closest to the explosion center, the reflected pressure here has a typical double-peak structure, and the peak overpressure of the next peak crest is remarkably higher than that of its previous peak crest. The analysis shows that the first wave crest is caused by the reflection of the incident wave on the wall, and the second one may be induced by the action of high-temperature detonation products on the sensitive surface of the sensor. Then the characteristic parameters of each measured curves are extracted as shown in Table 2.

Table 2. Characteristic parameters of pressure measuring points.

Measuring Point	Scaled Distance/m/kg$^{1/3}$	Peak Positive Pressure/MPa	Shock Wave Arrival Time/ms	Specific Impulse/MPa × ms	Angle of Incidence/°
P1	0.557	32.32	0.36	3.35	90
P2	0.562	26.47	0.363	3.02	82.4
P3	0.574	23.58	0.38	2.93	75.96

From the table, the peak overpressure and specific impulse of the shock waves decrease with the increasing scaled distance, the peak positive pressure and specific impulse of normal reflection are the largest, and the blast load distributed on the surface of the RC slabs is obviously uneven. Figure 4 gives the overpressure-time history obtained based on the Kingery–Bulmash [25] method to calculate the blast parameters from the air explosion. This empirical model has been embedded into a variety of calculation programs such as ConWep and LS-DYNA (i.e., the keyword *LOAD_BLAST) and has high reliability. In general, the peak overpressure and positive pressure action time of the measured overpressure-time history and the ConWep pressure-time history are relatively consistent, showing good agreement.

2.2.2. Displacement Response-Time History

Figure 5 presents the measured displacement-time history of slab A.

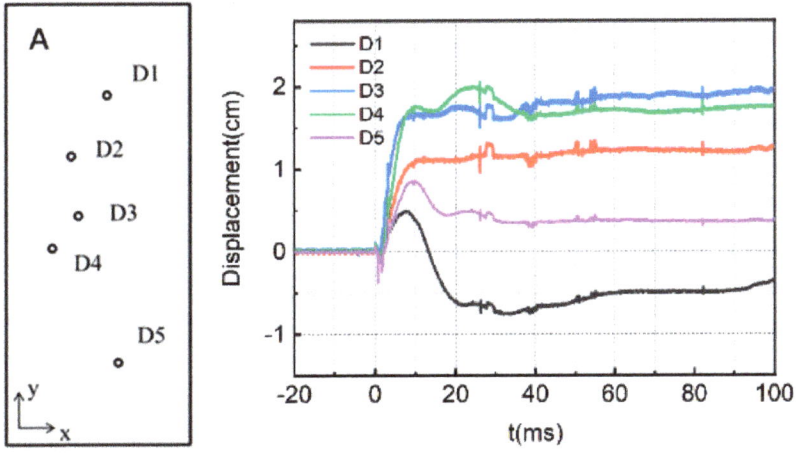

Figure 5. Measured displacement–time history of slab A.

The displacement data represent the displacement of the slab along the propagation direction of the shock wave, and the negative sign represents the displacement of the slab toward the detonation center. The peak displacement and residual displacement of each measured displacement curve of the slab A are extracted, as shown in Table 3.

Table 3. Parameters of slab A displacement measuring points.

Displacement Measuring Point	D1	D2	D3	D4	D5
Peak displacement/cm	0.51	1.18	1.97	1.98	0.89
Residual displacement/cm	−0.37	1.18	1.97	1.73	0.38

It can be seen from the table that when RC slab A with the low reinforcement ratio is subjected to explosion shock waves, its peak displacement at the center can reach 1.98 cm and the residual displacement is about 1.73 cm. The maximum displacement of the slab at the position farther away from the center gradually decreases. It can be seen from the diagram that with the increase of the distance from the center of the plate, the peak displacement of the plate gradually decreases, and with structural vibrations, the slab rebounds to some extent after reaching the peak displacement. The rebound is most obvious at D1 near the support, with a negative displacement of about 8 mm. It is important to note that both the distance from D5 and D1 to the support is basically the same, but as the ground has enhancement effect on shock waves, the blast loading strength below the slab is higher than that above the slab, resulting in some difference in the displacement response between the two points.

Similarly, Figure 6 shows the measured displacement curves of slab B with the high reinforcement ratio.

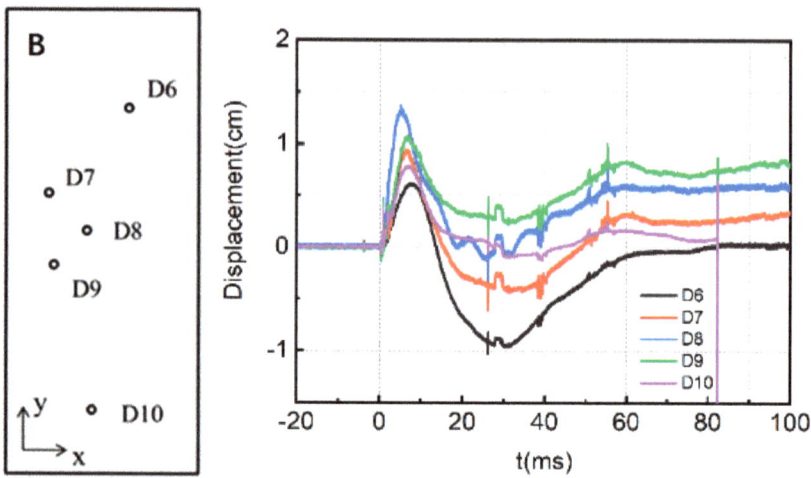

Figure 6. Measured displacement−time history of slab B.

The peak displacement and residual displacement of each measured displacement curve of slab B are extracted, as shown in Table 4.

Table 4. Parameters of slab B displacement measuring points.

Displacement Measuring Point	D6	D7	D8	D9	D10
Peak displacement/cm	0.67	0.92	1.41	1.09	0.79
Residual displacement/cm	0.03	0.31	0.58	0.80	0.08

It can be seen from the table that the peak displacement of the D8 measuring point in the center of the plate is 1.41 cm and the residual displacement is about 0.58 cm. The farther the measuring point is from the center of the plate, the smaller the positive displacement; the larger the rebound displacement is, and the longer the time to reach the maximum positive displacement peak, which conforms to the general deformation law of rectangular RC plate under an explosive shockwave.

The comparison of Figures 5 and 6 shows that the reinforcement ratio of the component will significantly affect the dynamic response of the RC plate under a shockwave. The peak displacement and residual displacement of the slab with high reinforcement ratio are smaller than that of the slab with a low reinforcement ratio, and the overall displacement response rebounds to a large extent, indicating that the degree of damage of the slab with a high reinforcement ratio is relatively small, and the deformation recovery ability is still strong after the shock wave.

2.2.3. Comparison of Damage Modes

Figure 7 gives a comparison of the damaged modes of the two RC slabs after the experiment.

Figure 7. Comparison of damage modes of the two RC slabs: (**a**) damage effect of slab A (low reinforcement ratio), and (**b**) damage effect of slab B (high reinforcement ratio).

The impact faces of the two plates are kept intact. As can be seen, there are multiple longitudinal and transverse cracks on the back surface of slab A, obvious oblique shear cracks on both ends of the slab, broken concrete in the middle near the edge, and exposed reinforcing bars (rebars), which means serious damage. There are only small longitudinal and transverse cracks appearing on the back surface of slab B, and the crack width is significantly smaller than that of slab A. There is no obvious concrete crushing phenomenon, which is mild damage.

3. Numerical Simulation

3.1. Finite Element Model

To further study the anti-explosion performance of the RC slabs under close-in blast loading, a finite element model (FEM) is constructed using the multi-material fluid-structure coupling algorithm, as shown in Figure 8.

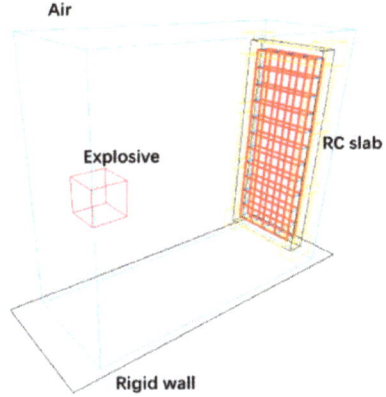

Figure 8. Finite element model.

The model is composed of rebars, concrete, clamps, and air. The separate modeling method is employed for the RC slabs, and the rebars and concrete are coupled through the keyword *CONSTRAINED_BEAM_IN_SOLID [26]. The rebars are modeled with beam element, and the rest with 3D solid element. The explosive and air are modeled using Eulerian grids, and the clamps and concrete slabs using Lagrangian grids. The keyword *RIGIDWALL_PLANAR [26] is adopted to define the influence of using a rigid wall to simulate the ground. The reflection-free boundary condition is used for the air domain boundary, and the model's unit system is m-kg-s.

3.2. Material Model

The explosive is described with the key word *MAT_HIGH_EXPLOSIVE_BURN [27] and the JWL equation of state (EOS), which is expressed as:

$$P = A\left(1 - \frac{\omega}{R_1 V}\right)e^{-R_1 V} + B\left(1 - \frac{\omega}{R_2 V}\right)e^{-R_2 V} + \frac{\omega E}{V} \quad (1)$$

where P is the pressure of detonation products, V is the relative volume, E is the initial internal energy density, and A, B, R_1, R_2, and ω are the EOS parameters. The JWL EOS parameters of the TNT explosive can be found in reference [25].

The air is defined by the combination of *MAT_NULL and *EOS_LINEAR_POLYNOMIAL equation of state [27], and the linear polynomial EOS is given by:

$$P = C_0 + C_1\mu + C_2\mu^2 + C_3\mu^3 + \left(C_4 + C_5\mu + C_6\mu^2\right)E_1 \quad (2)$$

where $\mu = \frac{\rho}{\rho_0} - 1$, ρ represents the current air density, ρ_0 is the initial air density, P is the air pressure, $C_0 \sim C_6$ are the coefficients of the polynomial equation, E_1 is the internal energy density, and V_0 is the initial relative volume. The parameters for air can be seen in reference [25].

The steel clamps, two-way HRB400 steel bars, and HPB235 stirrups are simulated with a type 3 bilinear elasto-plastic model (*MAT_PLASTIC_KINEMATIC) [27] in LS-DYNA, which is an isotropic-kinematic hardening or mixed isotropic-kinematic hardening model that can approximately simulate the elasto-plastic stage of rebars, and simplify the plastic and hardening stages into an oblique line [17]. The relevant material parameters are listed in Table 5.

Table 5. Material parameters of rebars and clamps.

Parameter	ρ/kg·m^{-3}	E/GPa	V_s	σ_y/MPa	E_t/GPa	C/s^{-1}	P_s	F_s
HRB400				400				
HPB235	7850	210	0.28	235	2.1	40	5	0.2
Clamp				300		0	0	

The concrete material is defined by the keyword *MAT_CONCRETE_DAMAGE_REL3 [27]. In the model, three strength failure surfaces (initial yield surface, ultimate strength surface, and softening strength surface) are employed to describe the plastic properties of concrete materials, which takes into account the characteristics of such materials such as elastic fracture energy, strain rate effect, and restraint effect. The input only requires three parameters of the concrete material, namely density, uniaxial ultimate compressive strength, and Poisson's ratio, and the remaining parameters are automatically generated by the system [27]. The material parameters include density $\rho_0 = 2300$ g/cm^{-3}, uniaxial compressive strength $f'_c = 30$ MPa, and Poisson's ratio $\nu = 0.2$. Under blast loading, the strain rate of the reinforced concrete is up to 100~10,000 s^{-1}. The strength of the materials under dynamic loading is fundamentally different from that under quasi-static loading. Therefore, the strain-rate effect of materials needs to be considered for blast-loading analysis. In the material model, the defined strain-rate curve can be called by LCRATE.

3.3. Finite Element Model Verification

Spherical explosive blasts propagate outward in the form of spherical shockwaves in free air, so the one-dimensional spherically symmetric model can be used to simulate the three-dimensional diffusion of blast shockwaves, which will greatly reduce the number of grids and improve the computational efficiency. Figure 9 shows the variation law of peak overpressure with scaled distance calculated by models with different grid sizes. From the figure, as the scaled distance increases, the sensitivity of peak overpressure to grid size gradually decreases.

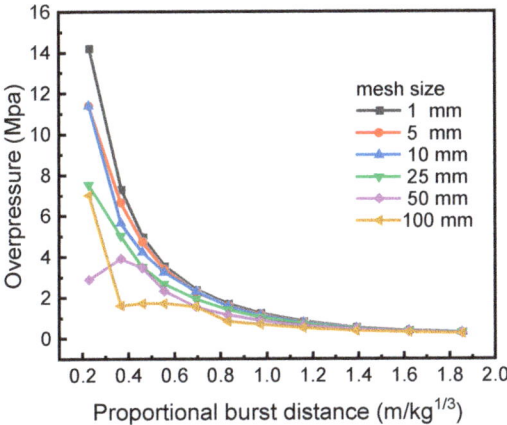

Figure 9. Comparison of calculation results under different grid sizes.

When the scaled distance is less than 0.8 kg/m^3, the peak overpressure is highly sensitive to grid size and has great discreteness, indicating that the grids need to be fine enough to ensure the computational accuracy. Considering the size of the three-dimensional model, and the computational accuracy and efficiency, the 10 mm grid size is selected for the subsequent finite element modeling and analysis.

To further verify the accuracy of the FEM, the time-history curve of shockwave overpressure obtained by finite element calculation is compared with the measured results, as shown in Figure 10.

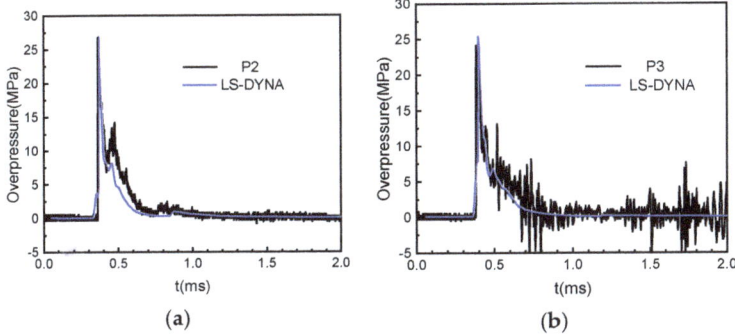

Figure 10. Comparison of pressure–time history, (a) the comparison of pressure–time history of the measuring point P2, (b) The comparison of pressure–time history of the measuring point P3.

It can be seen from the figure that the measured results are in good agreement with the calculated pressure curves. The calculated values and measured values of the peak overpressure at P2 and P3 have a deviation of 0.7% and 5%, respectively, and for the corresponding specific impulse at the two points, the deviation between the calculated and measured values is 27% and 13%, respectively.

Figure 11 gives the comparison between the measured displacement curves and the calculated results of D3 and D4.

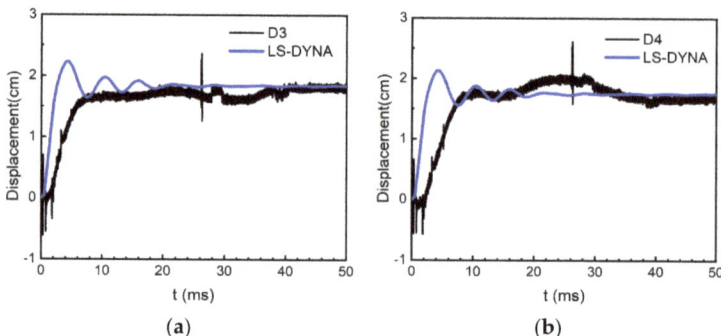

Figure 11. Comparison of displacement−time history: (**a**) the comparison of displacement−time history of the measuring point D3, and (**b**) the comparison of displacement−time history of the measuring point D4.

According to the figure, the peak displacement arrival time of the calculated curve is slightly earlier than the measured result, and the calculated peak displacement is larger than the measured value, but the residual displacement error of the two values is less than 1%, so the results have good agreement. Considering that the residual displacement of the RC slabs is the permanent displacement of the structure after the blast shockwaves, it can better characterize the degree of plastic damage of the slabs compared to the peak displacement. Therefore, it can be concluded that the calculation results can accurately reflect the dynamic mechanical response of the RC slabs. On the whole, the FEM proposed in this paper can accurately reflect the dynamic response characteristics of the RC slabs and the calculation results are reliable.

3.4. Analysis of Finite Element Calculation Results
3.4.1. Explosion Load Analysis

Based on the simulation results, the load distribution on the impact surface of the RC slabs at the scaled distance of 0.56 m/kg$^{1/3}$ is obtained, as shown in Figure 12. In the figure, the X-axis and Y-axis represent the directions of the short span and long span of the slab, respectively.

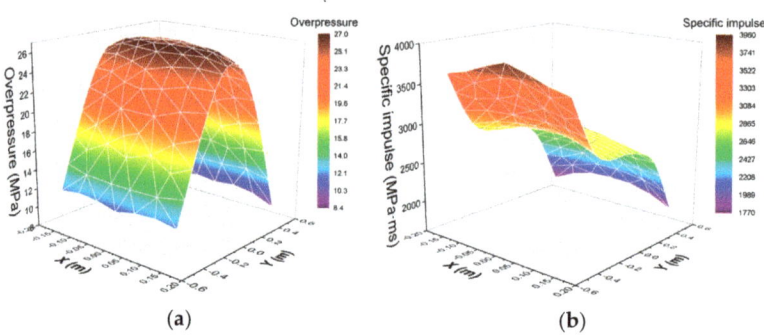

Figure 12. Contour plots of load distribution in RC slab: (**a**) peak overpressure, (**b**) specific impulse.

According to Figure 12a, the peak overpressure on the impact surface of the RC slab is large in the center and gradually decreases at both ends along the long-span direction. Due to the ground reflection effect, the peak overpressure at the end near the ground is slightly larger than that at the top, but along the short-span direction it is basically the same. Different from the peak overpressure distribution, the specific impulse of the impact surface of the RC slab in Figure 12b gradually reduces along the long-span direction, and the impulse near the ground is about two times that at the top. It shows that the ground makes the shockwave appear to be strengthening and creates a convergence effect at the bottom of the component. Therefore, the load distribution in the RC slab in the close-in explosion experiment is affected by many factors, such as the ground, the distance from the explosion center, and the angle of incidence. Although peak overpressure and impulse are unevenly distributed, there exists a distribution law.

3.4.2. Damage Analysis of RC Slab

The contour plots of the plastic damage of the two slabs are presented in Figure 13, where (a) front face and (b) rear face.

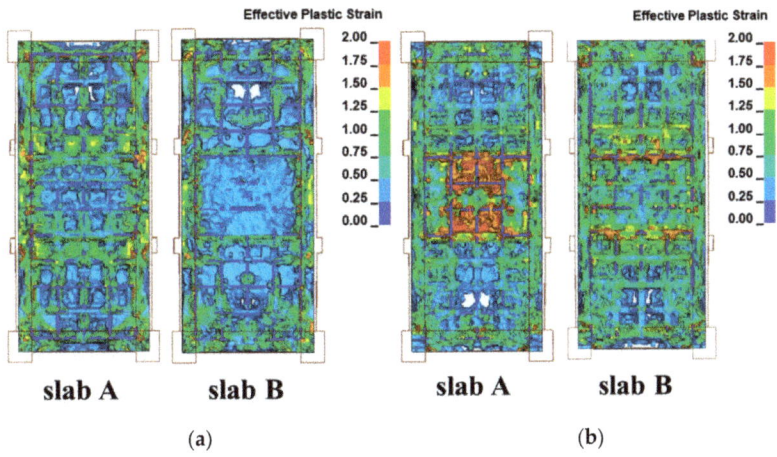

Figure 13. Comparison of damage contours of slab A and slab B: (**a**) Front face, (**b**) Rear face.

As can be seen from the figure, the damage distribution of the two RC slabs is similar. Taking the damage distribution of the blast face as an example, whether it is plate A or plate B, the damage is basically distributed along the reinforcement around the frame and the lateral constraint direction. The damage distribution of RC slabs is mainly along the transverse and longitudinal reinforcement and is obviously affected by frame constraints. The damage degree of RC plates with different reinforcement ratios is significantly different. The damage range and damage degree of the plates with small reinforcement ratios are significantly higher than those of the plates with large reinforcement ratios, regardless of the blast front or back surface. There is a large rectangular plastic strain zone in the center of the back blasting surface of the small reinforcement ratio plate, and several plastic strands appear along the short-span direction. The concrete in some areas is in almost complete failure, showing typical bending failure characteristics.

3.4.3. Residual Displacement Analysis

Figure 14 shows curves of the residual displacement distribution of the two slabs along the long-span direction.

From the figure, the curves of their residual displacement along the long-span direction under shockwave loading take the shape of a parabola, but the displacement of slab A is obviously smaller than that of slab B. At the symmetrical position along the slab center, the displacement at the end near the ground is slightly larger than that at the other end, mainly because of the ground reflection effect, which is consistent with the phenomenon observed in the experiment.

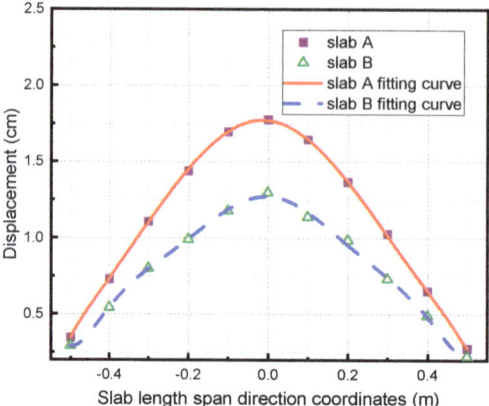

Figure 14. Comparison of residual displacement of the RC slabs along the long—span direction.

3.4.4. Residual Bearing Capacity Analysis

To further quantitatively evaluate the degradation of the bearing capacity of damaged RC slabs, the residual bearing capacity of damaged RC slabs was simulated by restarting in LS-DYNA, and the quasi-static loading was carried out by slowly applying the displacement perpendicular to the panel at each node of the component face. The method of applying a displacement load is shown in Figure 15.

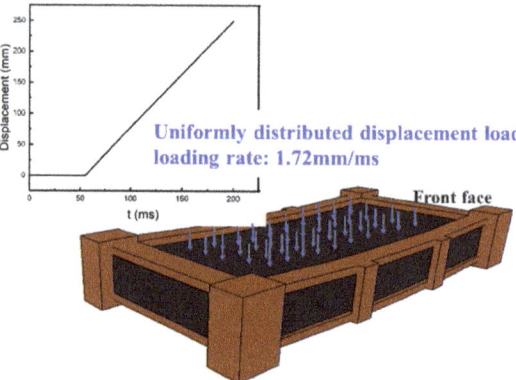

Figure 15. Schematic diagram of application method of displacement load.

Extract the reaction force at the support, and finally get the curve of the residual bearing capacity of the damaged member with displacement, as shown in Figure 16.

From Figure 16, the residual load-bearing capacity curves of the two damaged RC slabs under the same blast-loading conditions are obviously different. For slab A, when the mid-span displacement increases to 20 mm, its bearing capacity reaches the maximum, about 2000 kN; when the mid-span displacement increases to 50 mm, its bearing capacity almost decreases to 0, indicating that the slab has been completely damaged at this time. For slab B, when the mid-span displacement increases to 45 mm, its bearing capacity reaches the maximum, around 2250 kN, and when the displacement continues to increase to 105 mm, its bearing capacity is close to zero. The above data fully shows that increasing the reinforcement ratio can not only ensure that the RC slab has a high residual bearing capacity after explosive loading, but also ensures that the damaged members have better ductility and good energy-absorption effect.

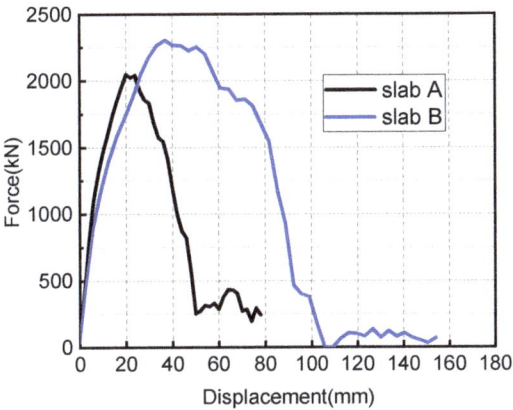

Figure 16. Residual load-bearing capacity curves of the two damaged RC slabs.

Figure 17 gives the comparison of the load-bearing capacity of the two RC slabs before and after the blast-loading test.

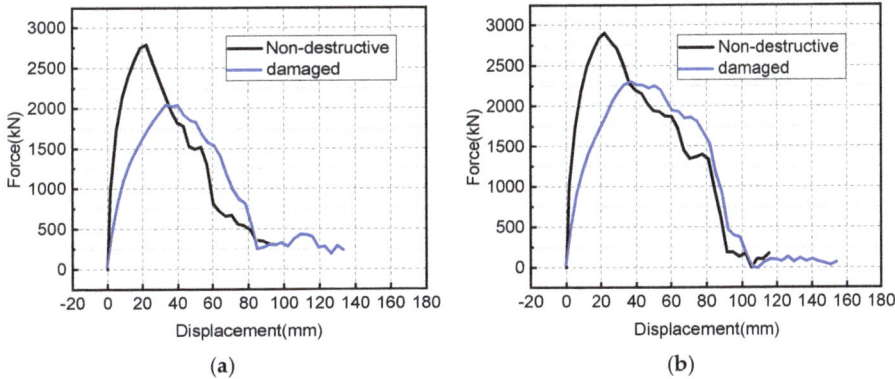

Figure 17. Comparison of load–bearing capacity between undamaged and damaged slabs: (**a**) slab A with low reinforcement ratio, and (**b**) slab B with high reinforcement ratio.

It can be seen from the figure that the shape of the bearing capacity-displacement curves of the RC slabs after blast loading have changed to some extent; specifically, the peak bearing capacity decreases to varying degrees, and the mid-span displacement corresponding to the peak bearing capacity increases, which is mainly due to the degradation of the bending stiffness after cracks appear in the damaged slabs. From the values, the bearing capacity of slab A decreases by 750 kN, or 26% compared to the undamaged slab; the residual load-bearing capacity of slab B decreases by 600 kN, or 20% compared to the undamaged slab, which is slightly smaller than that of slab A.

4. Conclusions

In this paper, the blast-loading experiment and numerical simulation were performed for reinforced concrete (RC) slabs with two typical reinforcement ratios. The environmental parameters of relevant loads and the damage data of displacement response of the specimens were obtained, and the calculation model was verified based on the measured data. Finally, the residual load-bearing capacity of the damaged RC slabs was analyzed. The main conclusions are as follows:

1. The load distribution in the RC slabs under close-in blast loading was extremely uneven. The peak overpressure on the impact surface of the slabs was large in the center along the long-span direction and gradually decreased at both ends. The specific impulse was gradually reduced along the long-span direction. The impulse near the ground was about two times that at the top.
2. The damage distribution of the two RC slabs with different reinforcement ratios was similar, but the degree of damage differed markedly. A large rectangular plastic strain zone appeared in the center of the back surface of the slab with a low reinforcement ratio, several plastic strands along the short-span direction could be observed, and the concrete in some areas almost completely failed, and the damage range and degree are significantly higher than those of the slab with a high reinforcement ratio.
3. Compared with the undamaged slabs, the shape of the resistance curves of the damaged RC slabs saw significant changes, and their load-bearing capacity and bending stiffness were irreversibly degraded. Increasing the reinforcement ratio can not only inhibit the crack extension and reduce the residual displacement of components, but also reduce the decrease of bearing capacity after damage.

Author Contributions: Writing—review and editing, L.W.; visualization, S.C.; supervision, Z.L.; project administration, K.L.; data curation, S.C. and W.Y.; resources, Z.L.; investigation, K.L.; funding acquisition, W.Y.; methodology, L.M.; validation, T.W. and D.Z. All authors have read and agreed to the published version of the manuscript.

Funding: This research received no external funding.

Institutional Review Board Statement: The study did not require ethical approval.

Informed Consent Statement: The study did not involve humans.

Data Availability Statement: The study did not report any data.

Conflicts of Interest: The authors declare no conflict of interest.

References

1. Li, J.; Hao, H. Numerical study of concrete spall damage to blast loads. *Int. J. Impact Eng.* **2014**, *68*, 41–55. [CrossRef]
2. Shi, Y.; Wang, J.; Cui, J. Experimental studies on fragments of reinforced concrete slabs under close-in explosions. *Int. J. Impact Eng.* **2020**, *144*, 103630. [CrossRef]
3. Wu, C.; Nurwidayati, R.; Oehlers, D.J. Fragmentation from spallation of RC slabs due to airblast loads. *Int. J. Impact Eng.* **2009**, *36*, 1371–1376. [CrossRef]
4. Kumar, V.; Kartik, K.V.; Iqbal, M.A. Experimental and numerical investigation of reinforced concrete slabs under blast loading. *Eng. Struct.* **2020**, *206*, 110125. [CrossRef]
5. Li, J.; Wu, C.; Hao, H. Investigation of ultra-high performance concrete slab and normal strength concrete slab under contact explosion. *Eng. Struct.* **2015**, *102*, 395–408. [CrossRef]
6. Yao, S.; Zhang, D.; Chen, X.; Lu, F.; Wang, W. Experimental and numerical study on the dynamic response of RC slabs under blast loading. *Eng. Fail. Anal.* **2016**, *66*, 120–129. [CrossRef]
7. Lan, S.; Lok, T.-S.; Heng, L. Composite structural panels subjected to explosive loading. *Constr. Build. Mater.* **2005**, *19*, 387–395. [CrossRef]
8. Huff, W.L. *Collapse Strength of a Two-Way-Reinforced Concrete Slab Contained within a Steel Frame Structure*; Final Report; US Army Engineer Waterways Experiment Station: Vicksburg, MI, USA, 1975.
9. Wang, W.; Zhang, D.; Lu, F.; Wang, S.-C.; Tang, F. Experimental study on scaling the explosion resistance of a one-way square reinforced concrete slab under a close-in blast loading. *Int. J. Impact Eng.* **2012**, *49*, 158–164. [CrossRef]
10. Wang, W.; Zhang, D.; Lu, F.; Wang, S.-C.; Tang, F. Experimental study and numerical simulation of the damage mode of a square reinforced concrete slab under close-in explosion. *Eng. Fail. Anal.* **2013**, *27*, 41–51. [CrossRef]
11. Lu, B. Improving the blast resistance capacity of RC slabs with innovative composite materials. *Compos. Part B Eng.* **2007**, *38*, 523–534.
12. Schenker, A.; Anteby, I.; Gal, E.; Kivity, Y.; Nizri, E.; Sadot, O.; Michaelis, R.; Levintant, O.; Ben-Dor, G. Full-scale field tests of concrete slabs subjected to blast loads. *Int. J. Impact Eng.* **2008**, *35*, 184–198. [CrossRef]
13. Thiagarajan, G.; Kadambi, A.V.; Robert, S.; Johnson, C.F. Experimental and finite element analysis of doubly reinforced concrete slabs subjected to blast loads. *Int. J. Impact Eng.* **2015**, *75*, 162–173. [CrossRef]
14. Wu, C.; Oehlers, D.J.; Rebentrost, M.; Leach, J.; Whittaker, A.S. Blast testing of ultra-high performance fibre and FRP-retrofitted concrete slabs. *Eng. Struct.* **2009**, *31*, 2060–2069. [CrossRef]

15. Abedini, M.; Mutalib, A.A.; Raman, S.N.; Alipour, R.; Akhlaghi, E. Pressure–Impulse (P–I) Diagrams for Reinforced Concrete (RC) Structures: A Review. *Arch. Comput. Methods Eng.* **2018**, *26*, 733–767. [CrossRef]
16. Krauthammer, T.; Astarlioglu, S.; Blasko, J.; Soh, T.B.; Ng, P.H. Pressure–Impulse diagrams for the behavior assessment of structural components. *Int. J. Impact Eng.* **2008**, *35*, 771–783. [CrossRef]
17. Li, Q.M.; Meng, H. Pulse loading shape effects on pressure–impulse diagram of an elastic–plastic, single-degree-of-freedom structural model. *Int. J. Mech. Sci.* **2002**, *44*, 1985–1998. [CrossRef]
18. Li, Q.M.; Meng, H. Pressure-Impulse Diagram for Blast Loads Based on Dimensional Analysis and Single-Degree-of-Freedom Model. *J. Eng. Mech.* **2002**, *128*, 87–92. [CrossRef]
19. Low, H.Y.; Hao, H. Reliability analysis of reinforced concrete slabs under explosive loading. *Struct. Saf.* **2001**, *23*, 157–178. [CrossRef]
20. Low, H.Y.; Hao, H. Reliability analysis of direct shear and flexural failure modes of RC slabs under explosive loading. *Eng. Struct.* **2002**, *24*, 189–198. [CrossRef]
21. Xu, J.; Wu, C.; Li, Z.-X. Analysis of direct shear failure mode for RC slabs under external explosive loading. *Int. J. Impact Eng.* **2014**, *69*, 136–148. [CrossRef]
22. Anas, S.M.; Alam, M.; Umair, M. Experimental and numerical investigations on performance of reinforced concrete slabs under explosive-induced air-blast loading: A state-of-the-art review. *Structures* **2021**, *31*, 428–461. [CrossRef]
23. Tai, Y.S.; Chu, T.L.; Hu, H.T.; Wu, J.Y. Dynamic response of a reinforced concrete slab subjected to air blast load. *Theor. Appl. Fract. Mech.* **2011**, *56*, 140–147. [CrossRef]
24. Zhao, C.; Wang, Q.; Lu, X.; Wang, J. Numerical study on dynamic behaviors of NRC slabs in containment dome subjected to close-in blast loading. *Thin Walled Struct.* **2019**, *135*, 269–284. [CrossRef]
25. Xin, C.; Wang, J.; Yu, D.; Li, T.; Song, J. Empirical Formula and Numerical Simulation of TNT Explosion Shock Wave in Free Air. *Missiles Space Veh.* **2018**, *3*, 98–102. [CrossRef]
26. Hallquist, J. *LS-DYNA Keyword User's Manual, Version: 970*; Livermore Software Technology Corporation (LSTC): Livermore, CA, USA, 2003.
27. Zhao, C.; Wang, Q.; Wang, J.; Zhang, Z. Blast Resistance of Containment Dome Reinforced Concrete Slab in NPP under Close-in Explosion. *Chin. J. High Press. Phys.* **2019**, *33*, 143–155. [CrossRef]

MDPI
St. Alban-Anlage 66
4052 Basel
Switzerland
www.mdpi.com

Materials Editorial Office
E-mail: materials@mdpi.com
www.mdpi.com/journal/materials

Disclaimer/Publisher's Note: The statements, opinions and data contained in all publications are solely those of the individual author(s) and contributor(s) and not of MDPI and/or the editor(s). MDPI and/or the editor(s) disclaim responsibility for any injury to people or property resulting from any ideas, methods, instructions or products referred to in the content.

www.ingramcontent.com/pod-product-compliance
Lightning Source LLC
LaVergne TN
LVHW070418100526
838202LV00014B/1480